暨南大学本科教材资助项目（普通教材资助项目）

高等学校“十三五”规划教材

无机化学实验

Inorganic Chemistry Experiment

（中英双语版）

杨 芳　郑文杰　编著

化学工业出版社

·北京·

本书为无机化学实验中英双语教材，每部分均按先中文后英文的模式安排内容，有利于学生进行中英对照。全书分基本操作训练、常数测定、化合物的制备与提纯、元素及其化合物的性质、离子的分离与鉴定和创新实验六部分，共 34 个实验。每个实验设有“实验提要”，利于学生对实验内容的基本理解和把握；“问题与讨论”帮助学生在实验前更好地理解实验原理，把握实验重点，抓住实验关键，在实验后分析实验现象和实验结果，深入思考并进一步扩展知识。本书可满足综合性大学多个学科专业学生的教学需要，不同学科专业，可根据需要选择所侧重的内容。此外，一些实验还设有选做部分，具有一定难度，可进一步培养学生解决问题的能力。

本书可作为高等学校化学类、化工类及相关专业的无机化学实验课程教材。

图书在版编目(CIP)数据

无机化学实验：汉英对照 / 杨芳，郑文杰编著. —2 版. —北京：化学工业出版社，2020.7（2024. 1 重印）

ISBN 978-7-122-36525-5

Ⅰ. ①无… Ⅱ. ①杨… ②郑… Ⅲ. ①无机化学-化学实验-高等学校-教材-汉、英 Ⅳ. ①O61-33

中国版本图书馆 CIP 数据核字（2020）第 050390 号

责任编辑：宋林青　　文字编辑：刘志茹

责任校对：宋　夏　　装帧设计：关　飞

出版发行：化学工业出版社（北京市东城区青年湖南街 13 号　邮政编码 100011）

印　　装：北京盛通数码印刷有限公司

787mm×1092mm　1/16　印张 16¾　字数 412 千字　　2024 年 1 月北京第 2 版第 4 次印刷

购书咨询：010-64518888　　售后服务：010-64518899

网　　址：http://www.cip.com.cn

凡购买本书，如有缺损质量问题，本社销售中心负责调换。

定　　价：42.00 元

前　言

本书第一版已出版六年，在此基础上，编著者修改后推出了第二版。本书可作为高等学校化学、化工类及相关专业的无机化学实验课教材和参考书。

在教学实践和教材修订过程中，我们致力于提高实验课的教学质量，引导学生进行创新实验。本书精选了 34 个实验，不仅体现了无机化学学科的基础性和系统性，也体现了探究性与创新性。本书系统介绍了实验室安全、化学实验常用仪器、实验基础操作、误差与数据处理等内容，这些基础内容涵盖了无机化学的主要操作，学生只有掌握了正确的基础实验操作，掌握实验中药品的安全性和化学废弃物的处理，掌握自我防护的方法，才能从容应对要求更高、难度更大的综合实验和创新实验。

本书的创新实验主要涉及纳米化合物的制备与表征，有助于学生了解化学前沿和现代实验技术，例如学习使用元素分析仪、红外光谱仪、电感耦合等离子光谱仪、透射电子显微镜等。

我们参考了许多专业的、系统的国外英文原版教材和国内的中英双语教材，保证了知识体系和文字描绘的准确性、系统性和科学性。有些专业词汇美式英语与英式英语的拼写不同，我们对每个单词进行推敲，进行了细致的修改，力求全书英文表达一致、准确。

编著者感谢暨南大学教务处和化学与材料学院的资助和支持，感谢历年来化学系老师提出的许多宝贵意见。

本书编写过程中，我们力求实验内容选取得当、英文翻译准确达意，但限于编者水平，疏漏和不当之处难免，请各位同行专家及读者批评指正!

编著者

2020 年 3 月

第一版前言

在经济全球化、社会信息化的大背景下，化学是一门实验科学，无机化学实验是理工类学生进入大学后的第一门实验课程，是学生在化学实验技能方面受到系统和严格训练的开端，并为学生今后独立进行科学实验和科学研究起着开路铺石的作用。

21 世纪对国际化人才需求日益迫切，对人才综合素质的要求不断提高。教育部要求高校进一步加大力度推广双语教学、全英教学，不断改革教育体制和教育模式。暨南大学化学系根据教育部的要求，自 2004 年起在相关教学改革研究中，做了大量的工作，特别是双语和全英的教学实践和探索。中英文双语版《无机化学实验》教材的编写出版，既是一种探索方式，也是一个阶段性成果。本书总结了暨南大学化学系近二十年来无机化学实验教学研究的相关成果，是无机化学教研室全体教师集体智慧的结晶。

全书分基本操作训练、常数测定、化合物的制备与提纯、元素及其化合物的性质、离子的分离与鉴定和创新实验六部分，可满足综合性大学多个学科专业学生的教学需要。不同学科专业，可根据需要选择所侧重的内容。此外，一些实验还设有选做部分，具有一定难度，可进一步培养学生解决问题的能力。

每个实验设有“实验提要”，目的在于加强学生对实验的内容基本理解和把握；“问题与讨论”帮助学生在实验前更好地理解实验原理，把握实验重点，抓住实验关键；在实验后分析实验现象和实验结果，深入思考并进一步扩展知识。

为了加强培养学生动手、分析的能力，进一步提高实验课的质量，本书增加了创新实验，如纳米硒、纳米硫的制备等，这些内容是编著者的一些最新研究成果，也有助于学生了解化学新知识，培养开拓创新能力。

本书的实验既各自独立又相互联系，通过适当安排可以形成系列实验。既可激发学生的兴趣，又可节省实验经费，减少环境污染，收到更好的教学效果。

本书编写过程中，借鉴了许多国内外的文献资料；暨南大学教务处和化学系给予了大力支持和热情帮助；本教研室老师提出了许多宝贵意见；美国的 K.Cox 先生审读了全书英文。在此一一致以衷心的感谢！

本书编写过程中，我们力求实验内容选取得当、英文翻译准确达意，但限于编者水平，不当之处在所难免，请各位同行专家及读者批评指正！

编　者

2013 年 9 月于暨南园

目　录

第1章　绪论……1
Chapter 1　Introduction……1
1.1　无机化学实验的目的……1
1.1　Objectives of *Inorganic Chemistry Experiment*……1
1.2　无机化学实验的学习方法……2
1.2　Methods to Study *Inorganic Chemistry Experiment*……4
1.3　学生实验守则……6
1.3　Laboratory Rules for Students……6
1.4　实验室安全守则……8
1.4　Laboratory Safety Rules……9
1.5　化学实验常用仪器介绍……12
1.5　Frequently Used Laboratory Equipment……16

第2章　基本操作训练……20
Chapter 2　Basic Laboratory Techniques……20
无机化学实验基本操作……20
Basic Laboratory Techniques for Inorganic Chemistry Experiment……20
2.1　常用玻璃仪器的洗涤和干燥……20
2.1　Cleaning and Drying Glassware……22
2.2　加热方法……25
2.2　Heating……28
2.3　容量仪器及其使用……31
2.3　Volumetric Glassware……34
2.4　化学试剂及其取用……39
2.4　Transferring Chemical Reagents……39
2.5　溶液的配制……41
2.5　Preparing Solutions……41
2.6　气体的发生、净化、干燥和收集……42
2.6　Preparing, Purifying, Drying and Collecting Gases……44
2.7　蒸发（浓缩）、结晶（重结晶）……47
2.7　Evaporation (Concentration) and Crystallization (Recrystallization)……47
2.8　结晶（沉淀）的分离和洗涤……49
2.8　Separating a Liquid or Solution From a Solid……52
2.9　试纸的使用……56
2.9　Using Test Paper……56

2.10 误差与数据处理……58
2.10 Errors and Treatment of Data……62
2.11 有效数字……68
2.11 Significant Figures……70
实验部分……74
Experiments……74
实验 1 滴定操作练习……74
Experiment 1 Titrimetric Analysis……76
实验 2 酸碱滴定……79
Experiment 2 Acid-Base Titration……80
实验 3 EDTA 标准溶液的配制与标定……83
Experiment 3 Preparation and Standardization of an EDTA Solution……84
实验 4 水中钙、镁含量的测定（配位滴定法）……86
Experiment 4 An Analysis of Concentrations of Calcium and Magnesium Ions in a Water Sample……87

第 3 章 常数测定……89
Chapter 3 Determination of the Constant……89
实验 5 化学反应速率和活化能……89
Experiment 5 Determination of a Rate Law and Activation Energy……92
实验 6 弱酸电离常数测定（pH 法）……95
Experiment 6 Determination of Ionization Constant of Acetic Acid……96
实验 7 硫酸钡溶度积的测定（电导法）……98
Experiment 7 Determination of the Solubility Product Constant of Barium Sulfate……100
实验 8 磺基水杨酸铁配合物的组成及稳定常数的测定……104
Experiment 8 Determination of the Composition and the Stability Constant of an Iron(Ⅲ)-Sulfosalicylate Complex……106
实验 9 氧化还原反应……110
Experiment 9 Oxidation-Reduction Reactions……111

第 4 章 化合物的制备与提纯……114
Chapter 4 Preparation and Purification of Compounds……114
实验 10 硫酸铜的提纯……114
Experiment 10 Purification of Copper(Ⅱ) Sulfate……116
实验 11 氯化钠的提纯……118
Experiment 11 Purification of Sodium Chloride……119
实验 12 氯化钾的提纯……122
Experiment 12 Purification of Potassium Chloride……123
实验 13 硝酸钾的制备与提纯……126
Experiment 13 Preparation and Purification of Potassium Nitrate……127

实验 14　明矾的制备 130
Experiment 14　Synthesis of Potassium Alum 131
实验 15　莫尔盐（硫酸亚铁铵）的制备 134
Experiment 15　Synthesis of Mohr's Salt (Ammonium Ferrous Sulfate) 135
实验 16　草酸配合物的合成 138
Experiment 16　Synthesis of Oxalate Complexes 139
实验 17　硫代硫酸钠的制备 141
Experiment 17　Preparation of Sodium Thiosulfate 143

第 5 章　元素及其化合物的性质 145
Chapter 5　Properties of Elements and Compounds 145
实验 18　s 区元素及其化合物 145
Experiment 18　Elements in s-block and Their Compounds 147
实验 19　卤素及其化合物 151
Experiment 19　Halogens and Their Compounds 155
实验 20　过氧化氢和硫的化合物 159
Experiment 20　Hydrogen Peroxide and Compounds of Sulfur 163
实验 21　氮和磷 167
Experiment 21　Nitrogen and Phosphorus 171
实验 22　ds 区元素的化合物 177
Experiment 22　Elements in ds-block and Their Compounds 180
实验 23　钛、钒、铬和锰的化合物 184
Experiment 23　Compounds of Ti, V, Cr and Mn 187
实验 24　铁、钴、镍的化合物 192
Experiment 24　Compounds of Fe, Co and Ni 196
实验 25　配合物 202
Experiment 25　Coordination Compounds 203

第 6 章　离子的分离与鉴定 206
Chapter 6　Separation and Identification of Ions 206
实验 26　混合溶液中阳离子的分析（Ⅰ） 206
Experiment 26　Qual Ⅰ. Al^{3+}, Zn^{2+}, Cr^{3+}, Mn^{2+}, Fe^{3+}, Co^{2+}, Ni^{2+} 207
实验 27　混合溶液中阳离子的分析（Ⅱ） 210
Experiment 27　Qual Ⅱ. Ag^{+}, Pb^{2+}, Cu^{2+}, Bi^{3+}, Cd^{2+} 213
实验 28　混合溶液中阴离子的分析 215
Experiment 28　Analysis of Anions 216

第7章　创新实验……219
Chapter 7　Innovative Experiments……219
实验 29　葡萄糖酸锌的制备……219
Experiment 29　Preparation of Zinc Gluconate……220
实验 30　从茶叶和紫菜中分离与鉴定某些元素……221
Experiment 30　Separation and Identification of Elements from Tea and Laver……222
实验 31　固相合成硒芳香杂环化合物……224
Experiment 31　Solid-Phase Synthesis of Heterocyclic Aromatic Selenium Compounds……225
实验 32　纳米硫的制备……227
Experiment 32　Preparation of Sulfur Nanoparticles……228
实验 33　纳米硒的制备……230
Experiment 33　Preparation of Selenium Nanoparticles……231
实验 34　立德粉（锌钡白）的制备……233
Experiment 34　Preparation of Lithopone……235

附录……238
附录 1　化学试剂的规格……238
附录 2　市售酸碱浓度……238
附录 3　水的饱和蒸气压……238
附录 4　常用的一些酸碱指示剂……239
附录 5　常见沉淀物的 pH……240
附录 6　常见无机化合物在水中的溶解度……241
附录 7　普通有机溶剂性质……242
附录 8　常用配合物的稳定常数表……243
附录 9　半微量定性分析的基本操作……245
附录 10　常见阳离子的分析……248
附录 11　常见阴离子的分析……256

参考文献……258

第1章 绪 论

Chapter 1 Introduction

1.1 无机化学实验的目的

化学是一门实验科学。任何化学规律的发现和化学理论的建立，都必须以严格的实验为基础，并受实验的检验，所以化学实验是研究化学的重要手段和方法。无机化学实验是学生进入大学后在化学实验技能方面受到系统和严格训练的开端，它不仅为学习无机化学的基础理论提供了依据，而且为后续课程的实验打好基础，也为以后独立进行科学实验和科学研究起着开路铺石的作用。

通过实验，可以更好地理解和掌握无机化学的基本理论和基础知识，掌握典型元素及其化合物的重要性质和反应性能；掌握测定常数以及制备、分离、提纯和鉴定物质等方法，从而巩固和扩大课堂所学到的知识，学会用实验的方法去认识和验证化学定律。

通过实验，可以正确地掌握无机化学实验的基本操作和技能技巧，学会正确观察和记录实验现象，分析、归纳、判断和评价实验结果，并学会处理实验数据和撰写实验报告。

通过实验，可培养实事求是的科学作风，严肃认真、一丝不苟的科学态度，知难而进、百折不挠的科学精神，善于观察、善于思考的科学习惯，有条不紊、周密准确的科学修养。

1.1 Objectives of *Inorganic Chemistry Experiment*

Chemistry is an experimental science. All chemical principles, tools, and techniques are developed in the laboratory. *Inorganic Chemistry Experiment* is a primary course of chemistry laboratory. The systematic and strict training of laboratory skills and techniques is helpful for students to learn chemistry and to do scientific research in the future.

At the successful completion of chemistry experiments, you will have a better understanding of the chemical principles and knowledge, the characterizations of typical elements and their compounds, and the methods of preparing, separating, purifying and identifying compounds. The laboratory gives students the advantage of a research experience in which questions stemming from the literature lead to the formulation of hypotheses. Answers are sought through experiments.

At the successful completion of chemistry experiments, you will develop proper laboratory skills and techniques for acquiring reliable and reproducible data in a safety-conscience laboratory environment. You will also be able to demonstrate an understanding of the key chemical principles through the observation, collection, and summarization of experimental data using the scientific method.

At the successful completion of chemistry experiments, you will develop the abilities in finding, analyzing and solving problems, and possess good habits of working carefully, methodically and systematically.

1.2 无机化学实验的学习方法

无机化学实验的学习方法，主要抓住下列三个环节。

1.2.1 预习

为了确保实验顺利并达到预期效果，实验前必须认真预习。并要求：

（1）阅读实验教材、教科书和参考文献的有关内容。

（2）明确实验目的，了解实验的内容、原理、操作方法、步骤以及安全注意事项。

（3）对于设计实验，应反复认真思考，预先拟定出合理的实验方案。

（4）在认真预习的基础上，写出实验预习报告（包括实验步骤、操作要领、注意事项及对问题的思考等），作为进行实验的依据。

1.2.2 实验

根据实验教材（或自己的设计方案）所拟定的方法、步骤和试剂用量进行操作。要求做到：

（1）认真操作，细心观察，及时地、如实地、详细地记录实验现象。

（2）若发现异常实验现象时应主动进行分析，找出原因解决问题，必要时重做实验。逐步培养自己独立思考和解决实际问题的能力。

（3）发现疑难问题，可提请教师帮助，共同研究、解决。

（4）实验过程中应保持肃静，严格遵守实验室的工作守则和安全守则。

1.2.3 实验报告

实验报告要求简明扼要，标题明显，文理通顺，字迹清楚，整齐洁净。实验报告要求包括以下的内容。

（1）实验名称。

（2）实验目的：扼要地简述实验的目的。

（3）实验原理：测定实验或制备实验应扼要叙述其原理。

（4）实验步骤：尽量用简图、表格、化学式或化学符号表示。

（5）实验现象和数据的记录：清晰地描述实验现象，如实记录每一实验数据，做到严谨、认真、实事求是。

（6）现象解释与数据处理：根据实验的现象进行分析、解释，并得出结论，写出有关的反应方程式。或根据记录的数据进行计算或作图。

（7）问题讨论：针对实验遇到的异常或特别的现象或疑难问题提出自己的见解；或总结实验中某方面的收获或教训；对定量实验应分析产生误差的原因；对教学方法、实验内容、实验方法都可提出自己的意见。

下面列举出不同类型的实验报告形式，以供参考。

【类型 1】 无机物制备或提纯实验

无机化学实验报告

实验名称____________________室温____________________

班组________________姓名____________日期____________

实验目的

简要原理

主要现象记录

产品纯度检验（根据实验要求进行）

产品外观

产　量

产　率

问题和讨论

【类型 2】 测定实验

无机化学实验报告

实验名称________________室温____________气压____________

班组________________姓名____________日期____________

实验目的

简要测定原理

数据记录和结果处理

问题和讨论

【类型 3】 性质实验

无机化学实验报告

实验名称____________________室温____________________

班组________________姓名____________日期____________

实验内容与步骤	实验现象	解释、反应方程式和结论

小结

问题和讨论

1.2 Methods to Study *Inorganic Chemistry Experiment*

1.2.1 Preview

In order to complete the experiments successfully, you need to preview carefully.

(1) Read the laboratory manual, textbook, and reference, etc.

(2) Understand the experimental objectives, procedures, and safety rules.

(3) If you need to design an experiment, think and write the protocol carefully.

(4) Finish the prelaboratory assignment.

1.2.2 Experiment

Do the experiment according to the experimental procedures or your own protocols.

(1) Perform and observe the experiment carefully. Record data honestly.

(2) If something goes wrong during the experiment, you should find the reason and solve the problem. If necessary, do the experiment again.

(3) Consult your instructor when necessary.

(4) Be sure that you follow laboratory safety and other laboratory rules.

1.2.3 Lab report

Your lab report should include the following sections with clear, distinct headings:

(1) the title of the experiment.

(2) the objectives of the experiment.

(3) a brief introduction or description.

(4) a brief, but clearly written experimental procedure that includes the appropriate balanced equations for the chemical reactions.

(5) a section for the data that is recorded. This data section must be planned and organized carefully and honestly.

(6) a section for data analysis that includes representative calculations, graphical analyses, and organized tables.

(7) a section for results and discussion.

1.3　学生实验守则

（1）实验前应认真预习，写好实验预习报告，上课时交指导教师检查和签字。

（2）遵守纪律，保持肃静，集中思想，认真操作，积极思考，仔细观察，如实记录。

（3）爱护各种设备和仪器，节约水电和药品。实验过程中有任何事故，均应报告老师。

（4）实验后，废纸、火柴梗和废液、废渣应倒入指定的回收容器内，严禁倒入水槽，以防水槽腐蚀、堵塞和造成环境污染。废玻璃应放入废玻璃箱内。

（5）使用试剂应注意下列几点。

① 按教材规定量使用，如无规定用量，应适量取用，注意节约。

② 公用试剂和试剂架上的试剂使用后，应立即盖上原来的瓶塞，并放回原处。确保洁净和放置有序。

③ 取用固体试剂时，注意避免洒落。

④ 试剂从瓶中取出后，不应倒回原瓶中。滴管未经洗净时，不准在试剂瓶中吸取溶液，以免带入杂质而使瓶中试剂变质。

⑤ 规定要回收的药品都应倒入指定的回收瓶内。

（6）使用精密仪器时，必须严格按操作规程操作，细心谨慎，避免粗心大意而损坏仪器。发现仪器有故障，应立即停止使用，报告指导教师，及时排除故障。

（7）注意安全操作，遵守实验室安全守则。

（8）实验后应将仪器洗净，放回原处，清理实验台面。

（9）值日生应按规定打扫整个实验室，清洗水槽，检查并关闭水源、电源、门窗。

1.3　Laboratory Rules for Students

(1) You should do the preview and finish the prelaboratory assignment before lab.

(2) Whenever you are in the laboratory, follow the direction of your instructor and the laboratory manual. Perform experiments carefully, and record data honestly.

(3) Report all accidents, no matter how minor, to the instructor.

(4) Use the waste container for used solution, etc. Use the recycling container for chemicals that need to be recycled. Do not pour solution in the sink due to corrosion, clog, or environmental pollution.

(5) When using chemicals, pay attention to the following:

① Use the proper amount of chemicals according to the instructions.

② Put the reagent bottle stoppers back on as soon as you finish using the bottle, and then put the bottle back to its original place. All chemicals in the laboratory must be clearly labeled.

③ Clean up any spill immediately. Report any unusual spill or breakage to your instructor.

④ Never put chemicals back into the original bottles. Never put an unclean pipet into a reagent bottle. It is better to waste a small amount of the chemical than to risk contaminating the entire chemical in the bottle.

⑤ Recycle the chemicals to a recycling container according to the instruction.

(6) When using precise instruments, perform strictly and carefully according to the operating instructions. Report to the instructor if anything goes wrong.

(7) Abide by the laboratory safety rules.

(8) Clean, rinse, and put back all the glassware.

(9) After the experiment, students on duty should clean the laboratory, close the windows, and make sure the faucets and electric switches are turned off.

1.4 实验室安全守则

进入化学实验室，每个人都务必十分重视安全问题，决不能麻痹大意。进入化学实验室的每一个人，都必须十分熟悉实验室的一般安全守则；熟悉易燃、易爆、具有腐蚀性的药物及毒物的使用规则；熟悉化学实验意外事故的处理及急救措施。在每一个实验前都应充分了解该实验的有关安全注意事项，在整个实验过程中，都应集中注意力，严格遵守操作规程和各项安全守则，避免事故的发生。

1.4.1 实验室的一般安全守则

（1）师生务必了解实验室内及周围环境各项灭火和救护设备（如沙箱、灭火器、急救箱等）及安放的位置，以及水管阀门、电闸的位置；熟悉各类灭火器的性能和使用方法。

（2）严禁在实验室内饮食、吸烟。

（3）使用电器时，要谨防触电。不要用湿手、湿物接触电器设备。

（4）加热试管时，试管口不要对着自己和别人，也不要俯视正在加热的液体，以免因液体溅出而受到伤害。

（5）不要直接用手触及毒物。实验完毕，洗净双手方可离开实验室。

（6）实验室内所有药品不得携带出室外。

（7）进入实验室应穿实验服，穿鞋，不穿凉鞋、高跟鞋等。应戴防护眼镜。隐形眼镜应换成普通眼镜。戴普通眼镜者也应使用防护眼镜。

1.4.2 易燃、易爆、具有腐蚀性的药物及毒物的使用规则

（1）涉及氢气的实验，操作时要远离明火，点燃氢气前，必须先检查氢气的纯度。

（2）银氨溶液久置后会变成氮化银而发生爆炸，用剩的银氨溶液必须酸化后回收。

（3）某些强氧化剂（如氯酸钾、过氧化钠、硝酸钾、高锰酸钾）或其混合物（如氯酸钾与红磷、碳、硫等的混合物）不能研磨，以防爆炸。

（4）钾、钠暴露在空气中或与水接触易燃烧，应保存在煤油中，并用镊子取用。

（5）白磷在空气中易自燃且有剧毒，能灼伤皮肤，切勿与人体接触，应保存在水中，在水下切割并用镊子取用。

（6）有机溶剂（乙醇、乙醚、苯、丙酮等）易燃，使用时要远离明火，用后立即盖紧瓶塞并放置阴凉处。

（7）浓酸、浓碱具有强腐蚀性，切勿使其溅在皮肤或衣服上，尤其要注意保护眼睛。稀释时（特别是浓硫酸），应将它们慢慢倒入水中而不能相反进行，以避免迸溅。

（8）能产生有毒、有刺激性恶臭气体的实验（如硫化氢、氯气、一氧化碳、二氧化碳、二氧化氮、二氧化硫、溴等），都要在通风橱或台面通风口下面进行操作。

（9）嗅闻气体时，用手轻拂气体，把少量气体扇向自己的鼻孔，绝不能将鼻子直接对着瓶口。

（10）可溶性汞盐、铬(Ⅵ)的化合物、氰化物、砷盐、锑盐、镉盐和钡盐都有毒，不得进入口内或接触伤口，其废液也不能倒入下水道，应集中统一处理。

（11）金属汞易挥发，它在人体内会蓄积引起慢性中毒。一旦汞洒落在桌上或地面，必须尽可能收集起来，并用硫黄粉盖在洒落的地方，使汞转变成不挥发的硫化汞。

1.4.3 意外事故的处理及救护措施

（1）割伤 在伤口上抹消炎药水或药粉，贴上止血贴。严重玻璃扎伤、伤口有玻璃碎片，需送往医院救治。

（2）烫伤 可用冷自来水冲洗，或冰敷。严重烫伤，送往医院救治。

（3）受酸腐蚀 先用大量水冲洗，再用饱和碳酸氢钠溶液或稀氨水洗，最后再用水洗。如果酸溅入眼内，用大量水或洗眼器冲洗。

（4）受碱腐蚀 先用大量水冲洗，再用醋酸（20g/L）洗，最后再用水冲洗。如果碱溅入眼中，用大量水或洗眼器冲洗。

（5）受溴腐蚀 用苯或甘油洗，再用水洗。

（6）受白磷灼伤 用1%（质量分数）硫酸银溶液、1%（质量分数）硫酸铜溶液，或浓高锰酸钾溶液洗后进行包扎。

（7）吸入刺激性气体 吸入硫化氢等气体而感到不适时，立即到室外呼吸新鲜空气。并立即报告老师，避免长时间暴露在有害气体中。

（8）毒物进入口内 严重中毒的救护最好在医院急救室进行。

（9）触电 首先切断电源，必要时进行人工呼吸并送医院治疗。

（10）起火 起火后，要立即一面灭火，一面防止火势扩大（如采取切断电源，停止加热、停止通风，移走易燃、易爆物品等措施）。灭火方法要根据起火原因采取不同的扑灭方法。

① 一般的小火可用湿布、石棉布或沙土覆盖在燃烧物上。

② 火势大时可用泡沫灭火器喷射起火处。

③ 由电器设备引起的火灾，不能用泡沫灭火器，以免触电，只能用四氯化碳气体或二氧化碳灭火器扑灭。

④ 实验人员衣服着火时，切勿惊慌乱跑，赶快脱下衣服。

⑤ 伤势重者，立即送医院。

⑥ 如果着火面积较大，及时报火警“119”。

1.4 Laboratory Safety Rules

Laboratory safety should be a constant concern to everyone in the laboratory. Every one should abide by the rules of laboratory safety. You should be aware of the potential dangers of chemicals, and be additionally careful in handling and storing chemicals which are flammable, explosive and corrosive. You should know how to deal with an accident or emergency. You should be fully prepared and aware of the safety rules before every experiment. Perform your experiment carefully and strictly to avoid an accident.

1.4.1 General laboratory rules

(1) Be sure that you know about the location of important safety equipment such as fire extinguishers and first-aid cases. Make sure that you know how to use the fire extinguishers. You should also know where the electric switches and water valves are.

(2) Eating, drinking, smoking are not permitted at any time.

(3) To prevent electric shock, do not touch electrical equipment with wet hands or wet things.

(4) Never point the test tube at anyone when heating a solution in a test tube over flame because the contents in the test tube may be ejected violently. Do not look down directly at a liquid which is heating over a flame.

(5) Do not touch chemicals directly. Wash your hands before leaving the laboratory.

(6) Never take any chemicals out of the laboratory.

(7) The wearing of laboratory coats is encouraged. Shoes must be worn. Sandals, canvas shoes, and high-heeled shoes are not permitted. Safety goggles must be worn at all times. Contact lenses should be replaced with glasses. People wearing glasses must also wear goggles.

1.4.2 Rules of handling flammable, explosive and corrosive chemicals

(1) Hydrogen should be kept away from flames. Be sure to check the purity of hydrogen before igniting.

(2) Diaminesilver solution can be explosive after storage for a long time. Recycle it after acidizing.

(3) Some strong oxidizing agents, such as potassium chlorate, sodium peroxide, potassium nitrate, and potassium permanganate, or the mixture of potassium chlorate with red phosphorus, carbon, or sulfur, can not be ground due to their explosive properties.

(4) Potassium and sodium should be stored in kerosene and handled with tweezers. Both of them are easy to ignite when exposed to the air or reacting with water.

(5) White phosphorus can cause spontaneous ignition in the air. Avoid any skin contact. White phosphorus should be stored and handled in water. Handle with tweezers.

(6) Organic solvents, such as ethanol, diethyl ether, benzene, and acetone, are flammable. Handle them away from flame. Put bottle stoppers on immediately and tightly after using.

(7) Concentrated acids and alkalies are dangerous and corrosive, and can produce painful burns on skin and eyes, which may not heal. These reagents must be dispensed with great care. Never add them directly to other chemicals unless you are sure that it is safe to do so. When diluting concentrated sulfuric acid, add sulfuric acid slowly to the water. Never add water to the concentrated sulfuric acid because the heat produced can cause the acid to spatter.

(8) Do experiments in the fume hood if poisonous or irritating gases are produced. These gases are H_2S, Cl_2, CO, CO_2, NO_2, SO_2, Br_2, etc.

(9) To note the odor of a substance, waft the fumes gently with your hand toward your nose. Never smell concentrated fumes.

(10) Do not touch poisonous chemicals or dump them into the sink directly. A few poisonous chemicals, commonly used in the laboratory, are as follows: mercury and its compounds, chromium (Ⅵ) compounds, cyanides, arsenic and its compounds, antimonic salt, cadmium salt, and barium salt.

(11) Mercury (Hg) is volatile and can cause chronic poisoning by accumulating in the human body. Mercury spills should be treated with sulfur powder to change the mercury into mercuric sulfide.

1.4.3 Accidents and first-aid

(1) Cuts. Put antibiotic medicine on the wound, and put on a bandage. Severe cuts,

especially with glass in the wound, seek medical attention.

(2) Burns. The affected area should be placed under running water, or rubbed with ice to remove the heat from the burned area. Severe burns require medical attention.

(3) Acid corrosion. Flush the affected area with tap water. Then wash with saturated sodium bicarbonate solution or diluted ammonia water followed by water again. If acid gets into the eyes, flush the eyes with copius amounts of water for at least ten minutes. If available use eye irrigation equipment.

(4) Alkali corrosion. Flush the affected area with tap water. Then wash with acetic acid (20g/L) followed by water again. If alkali gets into the eyes, flush the eyes with copius amounts of water for at least ten minutes. If available use eye irrigation equipment.

(5) Bromine corrosion. Wash with benzene or glycerin first, and then with water.

(6) White phosphorus burns. Wash with 1% silver sulfate solution, or 1% copper sulfate solution, or concentrated potassium permanganate solution. Then put on bandages.

(7) Inhalation of irritating gas. Get fresh air. Report to the instructor and avoid prolonged exposure to poisonous gas.

(8) Ingestion of chemicals. Most severe poisoning should be treated in the hospital emergency room.

(9) Electric shock. Cut off the main power switch to the lab first. Seek medical attention immediately and start resuscitation if necessary.

(10) Fire. Cut off the main power switch to the lab if possible and put the fire out with the proper method if possible.

① Small flames can be smothered with a wet towel or sand.

② Discharge a fire extinguisher at the base of the flames.

③ Use carbon dioxide extinguisher for fire caused by electrical equipment.

④ If your clothes catch on fire, immediately remove all clothing or, if possible, use water or fire extinguisher.

⑤ Severe injury, seek medical attention immediately.

⑥ If the fire is severe, pull the fire alarm, exit the building, and call 119.

1.5 化学实验常用仪器介绍

仪器	规格	作用	注意事项
普通试管　离心试管	玻璃质。分硬质试管，软质试管；普通试管，离心试管 无刻度的普通试管规格以管口外径(mm)×管长(mm)表示。离心试管规格以容量(mL)表示	普通试管用作少量试剂的反应容器，便于操作和观察。也可用于少量气体的收集 离心试管主要用于沉淀分离	普通试管可直接用火加热。硬质试管可加热至高温 加热时应用试管夹夹持 加热后不能骤冷 离心试管只能用水浴加热
烧杯	玻璃质。分普通型，高型；有刻度，无刻度 规格以容量(mL)表示	用作较大量反应物的反应容器，反应物易混合均匀。也用作配制溶液时的容器或简易水浴的盛水器	加热时应置于石棉网上，使其受热均匀。刚加热后不能直接置于桌面上，应垫以石棉网
锥形瓶	玻璃质 规格以容量(mL)表示	反应容器，振荡方便，适用于滴定操作	加热时应置于石棉网上，使其受热均匀。刚加热后不能直接置于桌面上，应垫以石棉网
蒸馏烧瓶	玻璃质 规格以容量(mL)表示	用于液体蒸馏，也可用作少量气体的发生装置	加热时应置于石棉网上
普通圆底烧瓶 磨口圆底烧瓶	玻璃质 分普通型和标准磨口型 规格以容量(mL)表示 磨口的还以磨口标号表示其口径大小，如10、14、19等	反应物较多，且需长时间加热时常用它作反应容器	加热时应放置在石棉网上

续表

仪 器	规 格	作 用	注意事项
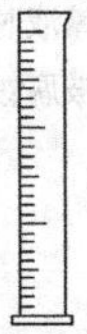 量筒	玻璃质 规格以刻度所能量度的最大容积(mL)表示 上口大、下部小的称作量杯	用于量度一定体积的液体	不能加热 不能量热的液体 不能用作反应容器
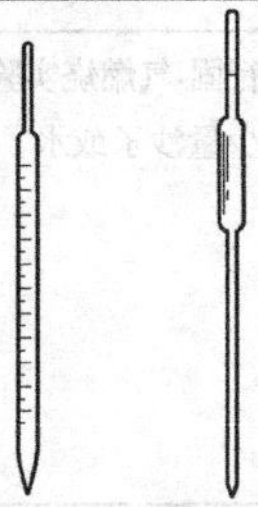 移液管 吸量管	玻璃质 移液管为单刻度，吸量管有分刻度 规格以刻度最大标度(mL)表示	用于精确移取一定体积的液体	不能加热 用后应洗净，置于吸管架（板）上，以免沾污
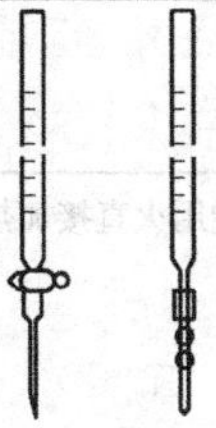 （酸式） （碱式） 滴定管	玻璃质 分酸式和碱式两种；管身颜色为棕色或无色 规格以刻度最大标度(mL)表示	用于滴定分析，或用于量取准确体积的液体	不能加热及量取热的液体。不能用毛刷洗涤内管壁 酸、碱管不能互换使用。酸管与碱管的玻璃活塞配套使用，不能互换
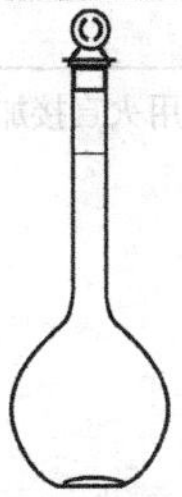 容量瓶	玻璃质 规格以刻度以下的容积(mL)表示 有的配以塑料瓶塞	配制准确浓度的溶液时用	不能加热。不能用毛刷洗刷 瓶的磨口瓶塞配套使用，不能互换
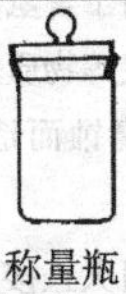 称量瓶	玻璃质 分高型和矮型 规格以外径(mm)×瓶高(mm)表示	需要准确称取一定量的固体样品时用	不能直接用火加热 盖与瓶配套，不能互换
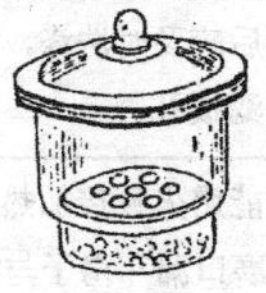 干燥器	玻璃质 分普通干燥器和真空干燥器 规格以上口内径(mm)表示	内放干燥剂，用作样品的干燥和保存	小心盖子滑动而打破 灼烧过的样品应稍冷后才能放入，并在冷却过程中要每隔一定时间开一下盖子，以调节器内压力
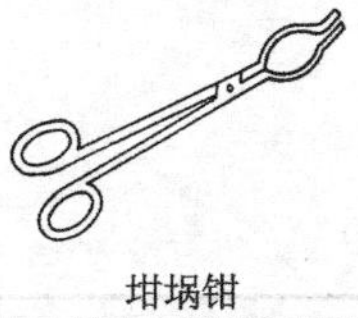 坩埚钳	金属（铁、铜）制品。有长短不一的各种规格。习惯上以长度（cm）表示	夹持坩埚加热，或往热源（煤气灯、电炉、马福炉）中取、放坩埚	使用前钳尖应预热；用后钳尖应向上放在桌面或石棉网上

续表

仪器	规格	作用	注意事项
滴瓶　细口瓶　广口瓶 试剂瓶	玻璃质 带磨口塞或滴管，有无色和棕色。规格以容量(mL)表示	滴瓶、细口瓶用于盛放液体药品。广口瓶用于盛放固体药品	不能直接加热。瓶塞不能互换。盛放碱液时要用橡皮塞，防止瓶塞被腐蚀而粘在一起
集气瓶	玻璃质。无塞，瓶口面磨砂并配毛玻璃盖片 规格以容量(mm)表示	用作气体收集或气体燃烧实验	进行固-气燃烧实验时，瓶底应放少量沙子或水
表面皿	玻璃质 规格以口径(mm)表示	盖在烧杯上，防止液体进溅或其他用途	不能用火直接加热
漏斗　长颈漏斗	玻璃质或搪瓷质。分长颈、短颈 以斗径(mm)表示	用于过滤操作以及倾注液体。长颈漏斗特别适用于定量分析中的过滤操作	不能用火直接加热
抽滤瓶和布氏漏斗	布氏漏斗为瓷质。规格以容量(mL)或斗径(cm)表示 抽滤瓶为玻璃质。规格以容量(mL)表示	两者配套，用于无机制备晶体或粗颗粒沉淀的减压过滤	不能用火直接加热
砂芯漏斗	又称烧结漏斗、细菌漏斗 漏斗为玻璃质。砂芯滤板为烧结陶瓷 其规格以砂芯板孔的平均孔径(μm)和漏斗的容积(mL)表示	用作细颗粒沉淀以及细菌的分离。也可用于气体洗涤和扩散实验	不能用于含氢氟酸、浓碱液及活性炭等物质体系的分离，避免因腐蚀而造成微孔堵塞或沾污 不能用火直接加热 用后应及时洗涤，以防滤渣堵塞滤板孔
分液漏斗	玻璃质。规格以容量(mL)和形状(球形、梨形、筒形、锥形)表示	用于互不相溶的液-液分离。也可用于少量气体发生器装置中加液	不能用火直接加热。玻璃活塞、磨口漏斗塞子与漏斗配套使用，不能互换

续表

仪器	规格	作用	注意事项
蒸发皿	瓷质，也有玻璃、石英或金属制成 规格以口径,(mm)或容量(mL)表示	蒸发浓缩液体用。随液体性质不同可选用不同质地的蒸发皿	能耐高温但不宜聚冷
坩埚	有瓷、石英、铁、镍、铂及玛瑙等材质 规格以容量(mL)表示	灼烧固体用。随固体性质不同而选用	可直接灼烧至高温 灼热的坩埚置于石棉网上
泥三角	用铁丝弯成，套以瓷管有大小之分	灼烧坩埚时放置坩埚用	铁丝已断裂的不能使用。灼热的泥三角不能直接置于桌面上
石棉网	由铁丝编成，中间涂有石棉 规格以铁网边长(cm)表示，如 16×16、23×23 等	加热时垫在受热仪器与热源之间，能使受热物体均匀受热	用前检查石棉网是否完好，石棉脱落的不能使用。不能与水接触或卷折
燃烧匙	铁或铜制品	检验物质可燃性，进行固气燃烧实验	用后应立即洗净、擦干
铁夹（烧瓶夹） 铁环 铁架（台）	铁制品。烧瓶夹也有铝或铜制成的	用于固定或放置反应容器 铁环还可代替漏斗架使用	使用前检查各旋钮是否可旋动 使用时仪器的重心应处于铁架台底盘中部
研钵	用瓷、玻璃、玛瑙或金属制成 规格以口径(mm)表示	用于研磨固体物质及固体物质的混合。按固体物质的性质和硬度选用	不能用火直接加热 研磨时，不能捣碎只能碾压 不能研磨易爆物质
点滴板	透明玻璃质、瓷质。分黑釉和白釉两种 按凹穴的多少分有四穴、六穴、十二穴等	用作同时进行多个不需分离的少量沉淀反应的容器，根据生成的沉淀以及反应溶液的颜色选用黑、白或透明点滴板	不能加热 不能用于含氢氟酸溶液和浓碱液的反应
碘量瓶	玻璃质。瓶塞、瓶颈部为磨砂玻璃 规格以容量(mL)表示	主要用作碘的定量反应的容器	瓶塞与瓶配套使用

1.5 Frequently Used Laboratory Equipment

Item	Specification	Application	Notes
test tube　centrifuge tube	Made of glass The test tube is divided into hard quality test tube and soft quality test tube; general test tube and centrifuge tube	Test tubes are used as reaction containers for small quantities of reagents. They are easy to operate and observe, and they can also be used to collect small amounts of gases Centrifuge tubes are mainly used to separate precipitates	General test tubes can be heated directly A test tube clamp should be used when heating the test tube The centrifuge tubes can only be heated in a water bath
beaker	Made of glass Some beakers have graduation of volume	Beakers are used as reaction containers for large amounts of reagents because of the ease in mixing the reagents evenly They can also be used to prepare solutions or as a water bath	In order to heat the beaker evenly, the wire/asbestos gauze should be used
Erlenmeyer flask/ conical flask	Made of glass	Erlenmeyer flasks are used as reaction containers. They are generally used in titration	To heat a flask evenly, the wire/asbestos gauze should be used
distilling flask	Made of glass	Distilling flasks are used for distillation, and as chemical reaction containers for large amounts of gases	To heat a flask evenly, wire/asbestos gauze should be used
round flask ground-in round flask	Made of glass There are general models and standard ground-in models with different volumes	Round flasks are used as chemical reaction containers for large amounts of reagents which need to be heated for a long period of time	To heat a flask evenly, wire/asbestos gauze should be used

continuous

Item	Specification	Application	Notes
graduated cylinder / measuring cylinder	Made of glass	Graduated cylinders measure quantitative volumes of solutions	Graduated cylinders can not be heated directly, or measure hot liquid, or be used as reaction containers
pipet	Made of glass Pipets are also called transferring pipets or measuring pipets. There are mono-graduated pipets with a big belly, and graduated tube pipets	Pipets accurately measure quantitative volumes of solutions	Pipets can not be heated
acid buret base buret	Made of glass The buret is divided into acid buret (Geiser buret) and base buret (Mohr buret), brown color buret and colorless buret	Burets are used for titrimetric analysis, or for accurately measuring quantitative volumes of solutions	Burets can not be heated directly, or measure hot liquid The use of the acid buret and the base buret should not be mixed
volumetric flask	Made of glass	Volumetric flasks prepare accurate concentrations of solutions	Volumetric flasks can not be heated Each volumetric flask is matched with its own ground mouth stopper
weighing bottle	Made of glass There are tall models and short models	Weighing bottles accurately measure the weight of solids	Weighing bottles can not be heated Each ground mouth weighing bottle is matched with its own stopper
desiccator	Made of glass There are general desiccators and vacuum desiccators	Desiccators dry and store chemicals with desiccants at the bottom	Be careful to slide the lid in order to open the desiccator Hot chemicals should be cooled before being put into the desiccator. Open the lid occasionally to adjust the pressure of the desiccator
crucible tongs	Made of iron or copper	Crucible tongs are used to clasp crucibles	Crucible tongs should be preheated before use

continuous

Item	Specification	Application	Notes
dropping bottle, narrow-mouth bottle, wide-mouth bottle reagent bottles	Made of glass There are ground mouth bottles with stoppers or droppers, colorless bottles and brown color bottles	Dropping bottles and narrow-mouth bottles store liquid reagents. Wide-mouth bottles store solid reagents	Reagent bottles can not be heated directly. Each bottle is matched with its stopper. Do not store an alkaline solution in a reagent bottle with a glass stopper because the bottle and the stopper could stick to each other
gas-jar	Made of glass The opening of the jar is ground, and there is a piece of ground glass covering the opening	Gas-jars are used to collect gases or do gas combustion experiments	To do gas combustion experiments, a thin layer of water or sand should be put at the bottom of the bottle to avoid breakage
watch glass	Made of glass	Watch glasses are used to cover beakers to avoid liquid splashing	Watch glasses can not be heated directly
funnel	Made of glass The funnel is divided into long neck funnel and short neck funnel	Funnels filter or transfer liquids. The long neck funnels are usually used for filtration in quantitative analysis	Funnels can not be heated directly
filter flask and Büchner funnel	The Büchner funnel is made of porcelain and the filter flask is made of glass	A Büchner funnel set into a filter flask and connected to a water aspirator is the apparatus used for vacuum filtration	Can not be heated directly
glass sand funnel	Also called sintered glass funnel and made of sintered glass	Glass sand funnels are used for separation of fine precipitates and bacteria, or gas filtration and dispersion	Glass sand funnels can not be used for hydrofluoric acid, concentrated alkali, or activated carbon due to blockage and contamination of the pores Can not be heated directly Wash the funnel after use to avoid blockage of the pores
separating funnel	Made of glass The separating funnel is divided into ball shape, pear shape and tube shape	Separating funnels are used to add liquids or separate solutions with multi-phases	The funnel can not be heated directly. Its upper mouth glass stopper and the lower valve stopper are both ground

continuous

Item	Specification	Application	Notes
evaporating dish	Made of porcelain, glass, quartz or metal	Evaporating dishes are used to evaporate and concentrate solutions	Do not cool the evaporating dish rapidly
crucible	Made of porcelain, quartz, iron, nickel, platinum or agate	Crucibles are used to heat substances vigorously	Crucibles can reach very high temperature
clay triangle	The clay triangle is made of iron wire with porcelain tubes	Clay triangles are used to support crucibles	Check if the clay triangle is broken before use. Do not put a hot clay triangle directly on the bench
asbestos gauze	The asbestos gauze is made with asbestos spreading in the center of a iron net	To heat a container evenly, put the asbestos gauze between the container and the flame	Check if the asbestos gauze is broken before use. Do not bend it or wet it
combustion spoon	Made of iron or copper	Combustion spoons are used to do combustion experiments	Wash and dry them thoroughly after use
flask clamp iron ring ring stand	Usually made of iron. Some clamps are made of copper or aluminum	They are used to clamp or place reaction containers. The iron ring can also be used as the funnel support	Pay attention to the balance to avoid tipping
mortar	Made of porcelain, glass, agate or metal	Mortars grind and mix solid substances	Mortars can not be heated directly. They can only be used for grinding, not pounding. Do not grind explosive substances
spot plate	Made of porcelain It is black glazed or white glazed. It is divided into 4 wells, 6 wells, or 12 wells	Spot plates are used as the container for preparing small amount of precipitates which don't need separation	Spot plates can not be heated directly, and can not be used for hydrofluoric acid and concentrated alkali
iodine flask	Made of glass. The stopper is ground	It is used as the container for quantitative reaction of iodine	Each flask is matched with its own stopper

第2章　基本操作训练

Chapter 2　Basic Laboratory Techniques

无机化学实验基本操作

Basic Laboratory Techniques for Inorganic Chemistry Experiment

2.1　常用玻璃仪器的洗涤和干燥

2.1.1　仪器的洗涤

化学实验必须在干净的反应容器中进行，才能得到正确可靠的结果。因此，在开始实验之前，必须把仪器洗涤干净。

洗涤仪器的方法很多，应根据实验的要求、污物的性质和沾污的程度来选用。一般说来，附着在仪器上的污物既有可溶性物质，也有尘土和其它不溶性物质，还有油污和有机物质。针对这种情况，可以分别采用适当的洗涤方法来洗涤。常见污物处理方法见表 2-1。

表 2-1　常见污物处理方法

污　物	处　理　方　法
可溶于水的污物、灰尘等	自来水清洗
不溶于水的污物	肥皂、合成洗涤剂
氧化性污物（如 MnO_2、铁锈等）	浓盐酸、草酸洗液
油污、有机物	碱性洗液（Na_2CO_3、NaOH 等）、有机溶剂、铬酸洗液、碱性高锰酸钾洗涤液
残留的 Na_2SO_4、$NaHSO_4$ 固体	用沸水使其溶解后趁热倒掉
高锰酸钾污垢	酸性草酸溶液
黏附的硫黄	用煮沸的石灰水处理
瓷研钵内的污迹	用少量食盐在研钵内研磨后倒掉，再用水洗
被有机物染色的比色皿	用体积比 1∶2 的盐酸-酒精混合液处理
银迹、铜迹	硝酸
碘迹	用 KI 溶液浸泡，温热的稀 NaOH 或用 $Na_2S_2O_3$ 溶液处理

常见玻璃仪器的洗涤方法有以下几种。

（1）用水刷洗

用毛刷和自来水刷洗，可洗去可溶性物质，使附着在仪器上的尘土和不溶性物质脱落下来，但往往洗不去油污和有机物质。

（2）用去污粉或合成洗涤剂洗

去污粉中含有碳酸钠，合成洗涤剂含有表面活性剂，它们都能除去仪器上的油污。去污粉中还含有白土和细沙，刷洗时起摩擦作用，使洗涤效果更好。刷洗后，再用自来水反复冲洗，以除去附着在仪器内外壁上的白土、细沙或洗涤剂。

（3）用铬酸洗液洗

对于口颈细小的仪器，很难用上述方法洗涤，可用铬酸洗液洗。铬酸洗液由浓硫酸和重铬酸钾溶液配成，具有很强的氧化性，对有机物和油污的去污能力特别强。

洗涤时，在仪器中加入少量洗液，倾斜容器，来回旋转，使器壁全部为洗液润湿，静置片刻，待洗液与污物充分作用，然后把洗液倒回原瓶，再用自来水把残留洗液冲洗干净。如果用洗液把仪器浸泡一段时间，或用热的洗液洗，则效果更好。

使用洗液时必须注意：

① 尽量把待洗容器内的积水去掉后，再注入洗液，以免水把洗液冲稀。

② 使用后的洗液应倒回原来瓶内，可以反复使用至失效为止。失效的洗液呈绿色（重铬酸钾还原为硫酸铬的颜色）。

③ 决不允许将毛刷放入洗液中刷洗。

④ 洗液具有很强的腐蚀性，会灼伤皮肤和破坏衣物。若不慎把洗液洒在皮肤、衣物或实验桌上，应立即用水冲洗。

⑤ Cr(Ⅵ) 有毒，残液排放到下水道，会污染环境、造成公害，要尽量避免使用。若使用时，清洗器壁的第一、二遍残液回收处理，不要直接排放到下水道 [处理方法：加入 $FeSO_4$ 使 Cr(Ⅵ) 还原成无毒的 Cr(Ⅲ) 再排放]。

使用市售厨用洗洁精代替铬酸洗液刷洗分析仪器，效果很好。厨用洗洁精是以一种非离子表面活性剂为主要成分的中性洗液，具有较强的去污、去油能力。实验室使用时配成 1%～2%的溶液，按常规方法刷洗。

（4）特殊物质的去除

应根据沾在器壁上的污物的性质，采取“对症下药”的办法进行处理。如 MnO_2 用 $NaHSO_3$ 或草酸溶液洗净。

用以上各种方法洗涤后的仪器，经自来水冲洗后，往往还残留有 Ca^{2+}、Mg^{2+}、Cl^-等离子，如果实验中不允许有这些杂质存在，则应该用蒸馏水或去离子水把它们洗去，一般以冲洗 3 次为宜。每次用量不必太多，采用“少量多次”的洗涤方法效果更佳，既洗得干净又不浪费。

已洗净的仪器，可以被水润湿，将水倒出后并把仪器倒置，可观察到仪器透明、器壁不挂水珠，否则仪器尚未洗净。

已经洗净的仪器不能用手指、布或纸擦拭内壁，以免重新沾污容器。

2.1.2　仪器的干燥

洗净的仪器如需要干燥可采用以下方法。

（1）加热烘干

洗干净的仪器可以放在电烘箱（控制在 105℃左右）内烘干（图 2-1），仪器放入之前应尽量把水倒净，然后小心放入，应注意仪器口朝下倒置，不稳的仪器应平放，并在烘箱下层放一个搪瓷盘，以承接从仪器上滴下的水珠。

一般常用的烧杯、蒸发皿等可置于石棉网上用小火烤干（外壁水珠应先揩干）。试管可以直接用小火烤干，如图 2-2 所示。操作时，试管口要略微向下倾斜，以免水珠倒流炸裂试管。火焰不要集中于一部位，可从底部开始缓慢向下移至管口。如此烘烤到不见水珠时，再使管口朝上，以便把水汽赶尽。

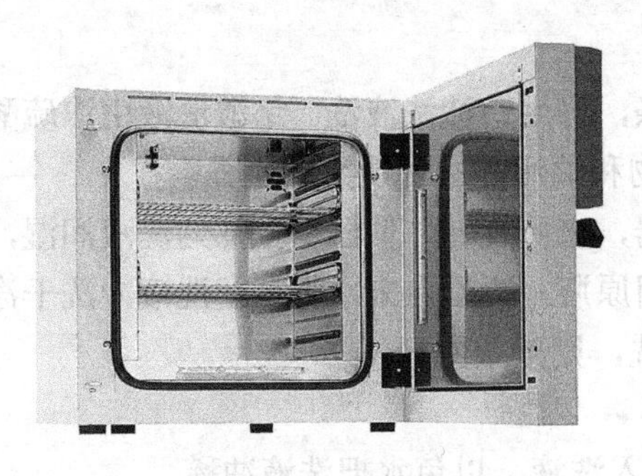

图 2-1　电烘箱

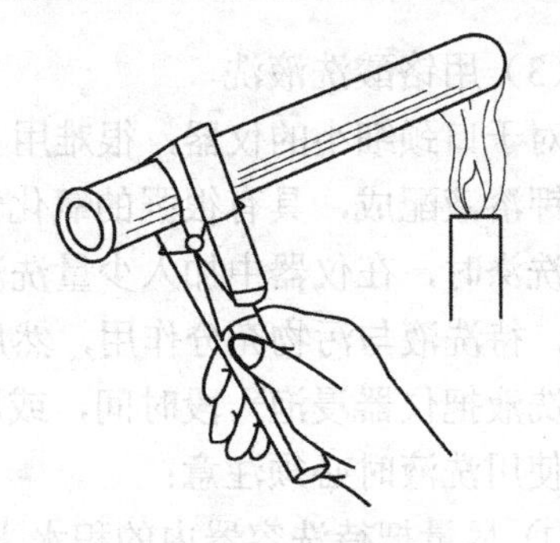

图 2-2　烤干试管

（2）晾干

不急用的、洗净的仪器可倒置放在实验柜内或仪器架上，任其自然晾干。

（3）吹干

用压缩空气或吹风机把仪器吹干。

（4）用有机溶剂干燥

带有刻度的计量仪器，不能用加热的方法进行干燥，因为加热会影响仪器的精密度。可以加一些易挥发的有机溶剂（最常用的是酒精或酒精与丙酮按体积比 1∶1 的混合物）到洗净的仪器中，倾斜并转动仪器，使器壁上的水与这些有机溶剂互相溶解混合，然后倾出它们，少量残留在仪器中的混合物很快就挥发了。若用吹风机往仪器中吹风，将干得更快。

2.1　Cleaning and Drying Glassware

2.1.1　Cleaning glassware

Cleanliness is extremely important in minimizing errors in the precision and accuracy of data. Therefore, glassware must be thoroughly cleaned before the experiment.

The proper method to clean glassware depends on what contaminant is present in the glassware. Table 2-1 lists different methods to clean glassware.

Table 2-1　Different methods to clean glassware

Contaminants	Methods
Water-soluble contaminants or dust	Tap water
Water-insoluble contaminants	Soap or detergent
MnO_2, rust	Concentrated hydrochloric acid, oxalate acid cleaning solution
Oil and organic contaminants	Alkaline solution such as Na_2CO_3 or NaOH, organic solvent, chromic acid cleaning solution, basic permanganate solution
Solid Na_2SO_4、$NaHSO_4$	Dissolve with hot water
Permanganate	Oxalate acid
Sulfur	Boiled lime water
Contaminants in porcelain mortar	Grind salt in mortar, and rinse with water
Cuvette contaminated by organics	HCl-ethanol (1∶2, vol/vol)
Silver or copper stain	Nitric acid
Iodine stain	Soak with KI solution, and wash with warm diluted NaOH or $Na_2S_2O_3$ solution

The common cleaning solutions are as follows.

(1) Water

Water and a brush are used to wash away soluble contaminants and dust from glassware. Oil and organic contaminants can not be washed away with water.

(2) Cleaning powders or detergents

Most cleaning powders or detergents contain sodium carbonate or surfactant, so the oil on the glassware can be washed off.

(3) Chromic acid cleaning solution

$K_2Cr_2O_7$ dissolved in concentrated sulfuric acid is traditionally used for difficult cleaning tasks.

Clean glassware thoroughly with a small amount of chromic acid solution. Roll each rinse around the entire inner surface of the glassware for a complete rinse. Then decant the solution back to its original reagent bottle. Rinse the glassware several times with water. Soaking the glassware in a chromic acid solution will obtain better results.

Pay attention to the following when using a chromic acid solution:

① Eliminate the water in the glassware before using the chromic acid solution. This prevents chromic acid dilution.

② The chromic acid solution can be used repeatedly until it turns green.

③ Never use a brush to the chromic acid.

④ The chromic acid solution is very corrosive. Wash immediately with water if the solution touches your skin.

⑤ Cr(Ⅵ) is poisonous. Don't discard it into the sink directly.

Dishwashing liquid used in the kitchen is a good cleaning solution for oil and contaminants. The chromic acid solution can be replaced by dishwashing liquid in the laboratory.

(4) Other cleaning solutions

Proper cleaning solutions should be used for different contaminants. For example, MnO_2 should be washed with $NaHSO_3$ or with an oxalate acid solution.

Once the glassware is thoroughly cleaned, first rinse several times with tap water and then once or twice with small amounts of deionized water.

The glassware is clean if no water droplets adhere to the cleaned area of the glassware after the final rinse.

2.1.2 Drying glassware

If you need to dry the glassware after cleaning, use the following methods.

(1) Heating

An electric drying oven (Figure 2-1) is commonly used in the laboratory to remove water or other solvents from chemical samples and to dry glassware. Eliminate water in the glassware before putting it into the oven. Heat the glassware to about 105℃.

Beakers and evaporating dishes can be dried with a cool flame on wire/asbestos gauze. Test tubes can be dried by heating directly with a cool flame (Figure 2-2).

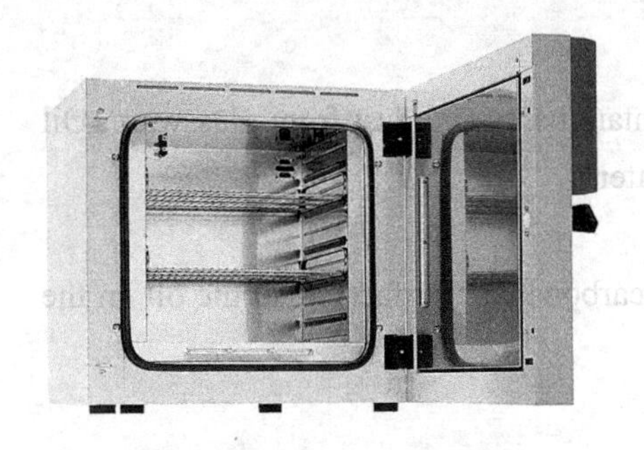

Figure 2-1 Electric drying oven

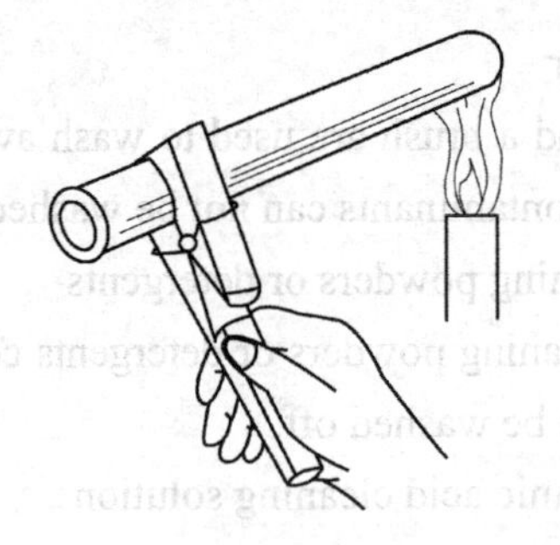

Figure 2-2 Position when drying a test tube with a cool flame

(2) Airing

Invert the glassware and allow it to dry in ambient air.

(3) Blow-drying

Blow-dry the glassware with compressed air or hair drier.

(4) Organic solvents

Glassware with graduations can not be dried by heating. Organic solvents, such as ethanol, or a mixture of ethanol and acetone (1∶1, vol/vol), can be used to rinse the wet glassware. Those organic solvents and water are completely miscible. Therefore, the water is removed from the glassware.

2.2 加 热 方 法

2.2.1 常用加热装置及其使用方法

（1）酒精灯

酒精灯的加热温度为 400～500℃，适用于温度不太高的实验。

酒精灯是由灯帽、灯芯和盛有酒精的灯壶所组成。灯的颈口与灯头（瓷套管）连接是活动的。使用酒精灯时应注意：

① 灯内酒精不可装得太满，一般不应超过酒精灯容积的 2/3，以免移动时洒出或点燃时受热膨胀而溢出。

② 点燃酒精灯之前，先将灯头提起，吹去灯内的酒精蒸气。

③ 点燃酒精灯时，要用火柴引燃（图 2-3），不能用燃着的酒精灯引燃，避免灯内的酒精洒在外面着火而引起事故。

④ 熄灭酒精灯时要用灯罩盖熄火焰，不能用嘴吹灭。待火焰熄灭片刻，还需再提起灯盖一次，通一通气再罩好，以免下次使用时揭不开盖子。

⑤ 添加酒精时，应把火焰熄灭，然后借助于漏斗把酒精加入灯内（图 2-4）。

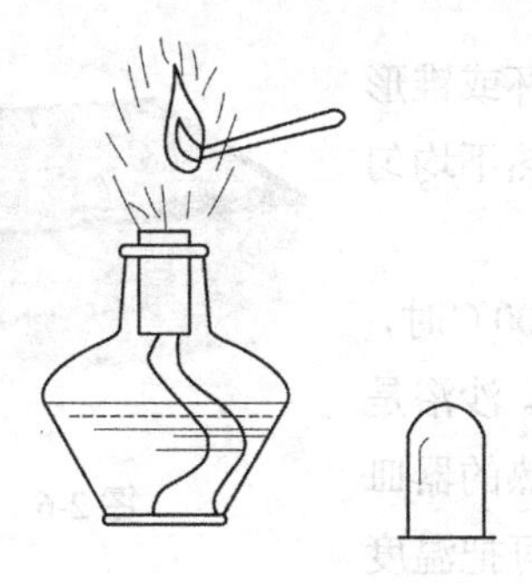

图 2-3　点燃酒精灯

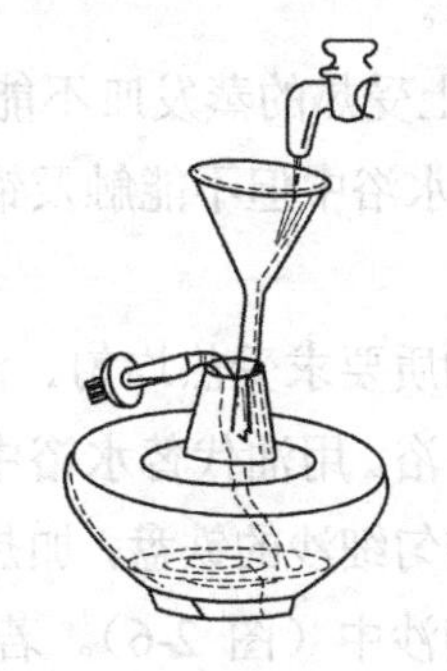

图 2-4　往酒精灯内添加酒精

（2）酒精喷灯

酒精喷灯是用酒精作燃料的加热器。国外一般用本生灯使用时，先将酒精汽化后与空气混合，点燃混合气体，故其火焰温度高，约 900℃。常用于需要温度高的实验。

酒精喷灯有挂式和座式两种，这里着重介绍挂式酒精喷灯（图 2-5）。其喷灯部分由金属制成，除灯座外还有预热盆和灯管。灯管处有一蒸气开关，预热盆下方有一支管为酒精入口，支管经橡皮管与酒精储罐相连。使用时，先将储罐悬挂在高处，打开储罐下的开关，在预热盆中注上酒精并点燃，以预热灯管。待盆内酒精将近燃完时，开启蒸气开关，由于灯管已被灼热，进入灯管的酒精立即汽化，酒精蒸气与气孔进来的空气混合，即可在管口点燃。调节灯管处的蒸气开关可控制火焰的大小。使

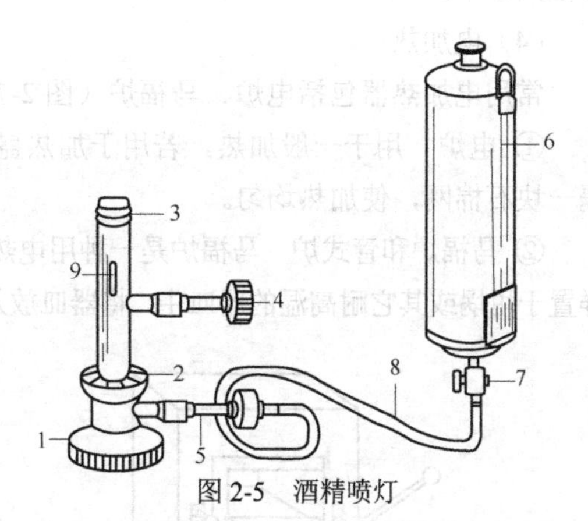

图 2-5　酒精喷灯

1—灯座；2—预热盆；3—灯管；4—蒸气开关；5—支管；6—酒精储罐；7—储罐开关；8—橡胶管；9—空气入口

用完毕，关上蒸气开关及储罐下的酒精开关，火焰即自行熄灭。

必须注意：

① 在点燃前，灯管必须充分预热，否则酒精在管内不能完全汽化，开启蒸气开关时，会有液态酒精从管口喷出，形成“火雨”，四处洒落酿成事故。这时应立即关闭蒸气开关，重新预热。

② 酒精蒸气喷出口，应经常用特制的金属针穿通，以防阻塞。

③ 不用时，必须将储罐口用盖子盖紧，关好储罐的酒精开关，以免酒精漏失造成危险。

（3）水浴、油浴和沙浴

当被加热物质要求受热均匀、温度在 100℃以下时，可用水浴加热。例如，蒸发浓缩溶液时，把溶液放在蒸发皿中，将蒸发皿置于水浴锅上，利用蒸汽加热。

实验室也常用大烧杯代替水浴锅。

使用水浴锅加热应注意以下几点：

① 水浴锅中的水量不要超过容量的 2 / 3，水量不足时用少量的热水补充，绝对不能把水烧干。

② 水浴锅上根据承受的不同器皿，选择不同的铜圈，但应注意增大器皿的受热面积，保持水浴的严密。

③ 在水浴上受热的蒸发皿不能浸入水里。烧杯或锥形瓶，可直接浸入水浴中但不能触及锅底，以防因受热不均匀而破裂。

当被加热物质要求受热均匀、温度又需高于 100℃时，可使用油浴或沙浴。用油代替水浴中的水即是油浴。沙浴是一个铺有一层均匀细沙的铁盘，加热铁盘，把被加热的器皿的下部埋置在细沙中（图 2-6）。若要测量温度，可把温度计插入沙中。

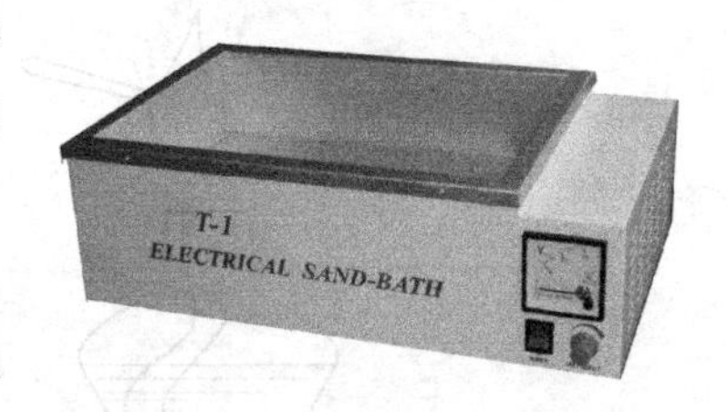

图 2-6　电沙浴

（4）电加热

常用电加热器包括电炉、马福炉（图 2-7）和管式炉（图 2-8）等。

① 电炉　用于一般加热。若用于加热盛于玻璃器皿中的液体，玻璃器皿与电炉之间要隔一块石棉网，使加热均匀。

② 马福炉和管式炉　马福炉是一种用电热丝或硅碳棒加热的炉子。它的炉膛是长方体，试样置于坩埚或其它耐高温的器皿中，将器皿放入炉膛加热。最高使用温度可达 950℃或 1300℃。

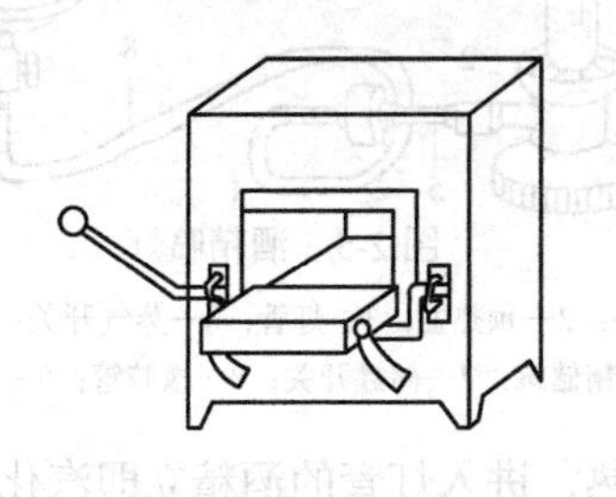
图 2-7　马福炉

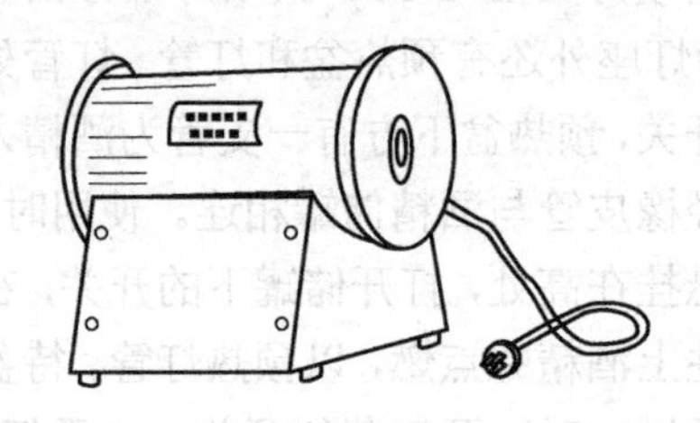
图 2-8　管式炉

管式炉有一管状炉膛，也是用电热丝或硅碳棒来加热，温度可调节，最高使用温度可达 950℃或 1300℃。炉膛中可插入一根耐高温的瓷管或石英管，加热试样放在瓷管或瓷盘中，并推入炉膛加热。试样可在空气气氛或其它气氛中受热。

马福炉和管式炉的温度测量，不是用一般玻璃质的温度计，而是用一种热电偶高温计。将两种不同的金属丝或不同成分的合金丝（例如一根是镍铬丝，另一根是镍铝丝）两端焊好，构成闭合回路。当这两个焊接端温度不同时，回路中就产生电动势，称为温差电动势（或称热电势），并有电流流过。这一金属或合金的组合体就称为热电偶。对确定的热电偶，其热电势只与两焊接端的温度差有关。如果将一焊接端置于冰水混合液中以保持 0℃，另一端置于待测的物体中，用毫伏计测出回路中的热电势，就可求出待测温度。

马福炉和管式炉带有温度控制器，可以把炉温控制在某一温度附近。只要把热电偶和温度控制器连接起来，待炉温升到所需温度时，控制器就会把电源自动切断，使炉子的电热丝断电停止工作，炉温就停止上升。由于炉子的散热，炉温稍低于所需温度时，控制器又把电源接通，使电热丝工作，则炉温上升，不断交替，就可把炉温控制在某一温度附近。

2.2.2 常用的加热操作

（1）加热烧杯、烧瓶中的液体

液体一般不超过容器容量的一半。仪器必须放在石棉网上，使之受热均匀。

（2）直接加热试管中的液体

在火焰上直接加热试管中的液体操作如图 2-9 所示。

① 用试管夹夹住试管的中上部，不要用手拿住试管加热，以免烫手。

② 试管应稍微倾斜，管口向上。加热过程中管口不能对着自己或别人，以免在溶液煮沸时迸溅而造成烫伤。

③ 液体量不能超过试管高度的 1 / 3。

④ 先加热液体的中上部再慢慢往下移动，然后不时地上下移动，使溶液各部分受热均匀。

（3）直接加热试管中的固体

在火焰中直接加热试管中的固体，操作如图 2-10 所示。试管口略向下倾，防止释放出来而冷凝的水珠流到试管的灼热处，而使试管破裂。

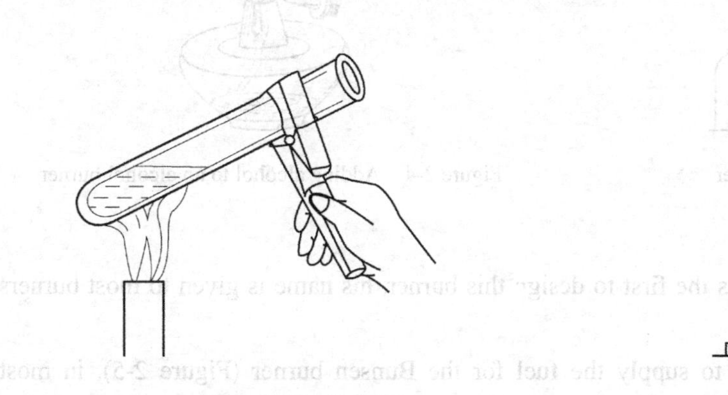

图 2-9　加热试管中的液体

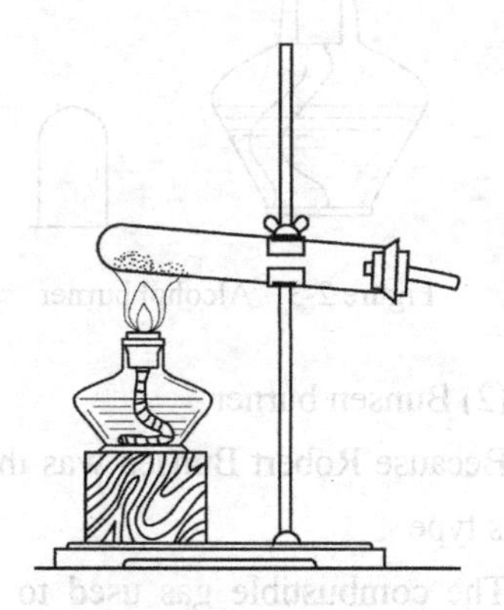

图 2-10　加热试管中的固体

（4）灼烧

当需要在高温加热固体时，可以把固体放在坩埚中灼烧（图2-11）。首先用小火预热坩埚，再用灯的氧化焰灼烧。应避免让还原焰接触埚底，以免在锅底结成炭黑。

要夹取高温下的坩埚时，必须使用干净的坩埚钳，用前先把坩埚钳放在火焰上预热片刻。坩埚钳用后应平放在石棉网上。

当加热较多固体时，可把固体放在蒸发皿中进行。但应注意充分搅拌，使固体受热均匀。

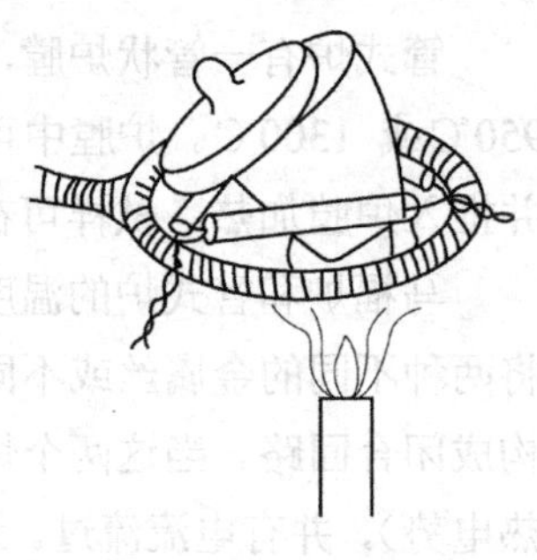

图 2-11　灼烧坩埚

2.2 Heating

2.2.1 Heating devices and methods

(1) Alcohol burner

An alcohol burner is used for experiments with moderate temperature requirements. It produces a temperature of about 400～500℃.

Alcohol burners consist of a cover, a wick with a porcelain tube and a lamp kettle for alcohol storage. Pay attention to the following when using an alcohol burner.

① The volume of alcohol in the lamp kettle should be less than 2/3 of the total volume.

② Lift the porcelain tube to let out the alcohol vapor before lighting.

③ Light an alcohol burner with matches (Figure 2-3). Never light an alcohol burner with another burning alcohol burner.

④ Never blow out the flame by mouth. Use the cover to extinguish the flame.

⑤ When adding alcohol, first extinguish the flame, and then add the alcohol with a funnel (Figure 2-4).

Figure 2-3　Alcohol burner

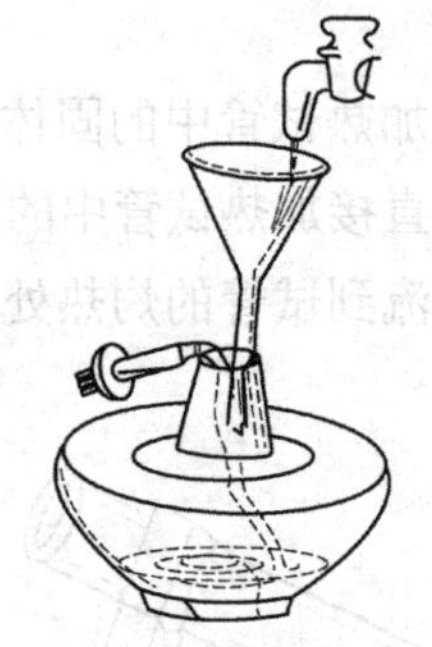

Figure 2-4　Adding alcohol to an alcohol burner

(2) Bunsen burner

Because Robert Bunsen was the first to design this burner, his name is given to most burners of this type.

The combustible gas used to supply the fuel for the Bunsen burner (Figure 2-5), in most laboratories, is natural gas. Natural gas is a mixture of gaseous hydrocarbons, but primarily

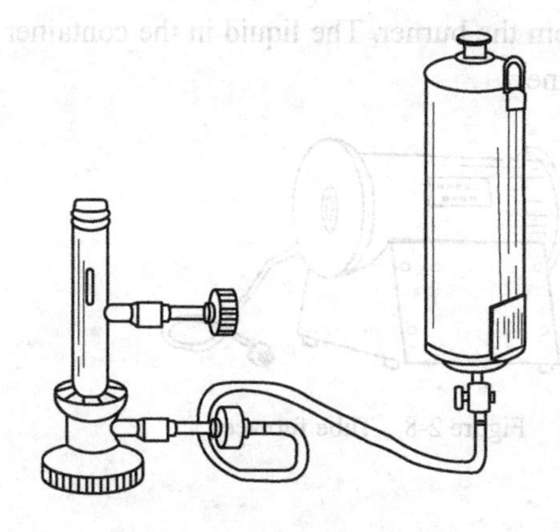
Figure 2-5 Bunsen burner

methane. If sufficient oxygen is supplied, methane burns with a blue, nonluminous flame.

Light a Bunsen burner properly using the following sequence of steps:

① Attach the tubing from the Bunsen burner to the gas outlet. Close the gas control valve on the Bunsen burner and fully open the gas valve at the outlet. Close the air holes at the base of the Bunsen burner and slightly open the gas control valve.

② Bring a lighted match or striker up the outside of the Bunsen burner barrel until the escaping gas at the top ignites.

③ After the gas ignites, adjust the gas control valve until the flame is pale blue and has two or more distinct cones. Slowly open the air control valve until you hear a slight buzzing sound. This sound is characteristic of the hottest flame from the Bunsen burner. Too much air may blow the flame out. When the best adjustment is reached, three distinct cones are visible. Temperatures within the second (inner) cone of a nonluminous blue flame approach 1500℃.

(3) Water bath, oil bath and sand bath

A water bath is used when a temperature of less than 100℃ is needed. An electric thermostatically-controlled water bath is available in modern chemistry laboratories. A beaker filled with water on a hot plate can also be used as a water bath.

Pay attention to the following when using a water bath:

① Fill with the appropriate amount of water.

② To hold and heat various sizes of containers, a stainless steel lid with a series of tight-fitting rings is used.

③ The reaction containers, such as a beaker and a flask, should be submerged in the water, but should not touch the wall or bottom of the water bath.

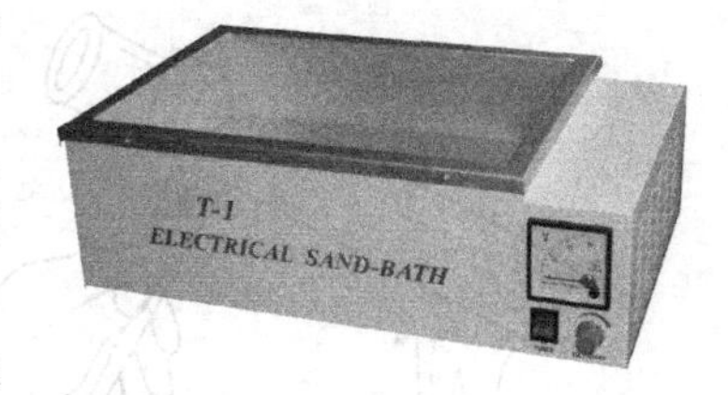

Figure 2-6 Electrical sand bath

An oil bath and a sand bath (Figure 2-6) can be used when a temperature of more than 100℃ is needed. An electrically heated oil bath is often used to heat small or irregularly shaped containers. A sand bath provides another method for heating microscale reactions.

(4) Electric heated devices

Hot plates, muffle furnaces (Figure 2-7), and tube furnaces (Figure 2-8)are often used heating devices in the laboratory. Hot plates are used for heating containers such as beakers and Erlenmeyer flasks. Tube furnaces and muffle furnaces can be heated to a temperature as high as 950℃ or 1300℃.

The thermometer used for tube furnace and muffle furnace is not an ordinary thermometer made of glass, but a thermocouple pyrometer. The electrodes of thermocouple are made of two different materials. The temperature differences between the measuring end and the reference end result in pyroelectric potential. The meter indicates the corresponding temperature to the pyroelectric potential.

2.2.2 Heating operations

(1) Heating a liquid in a beaker or a flask

When heating a liquid in a beaker or a flask with a burner, the container should be set on the wire gauze or asbestos gauze to evenly spread the heat from the burner. The liquid in the container should be no more than half the total volume of the container.

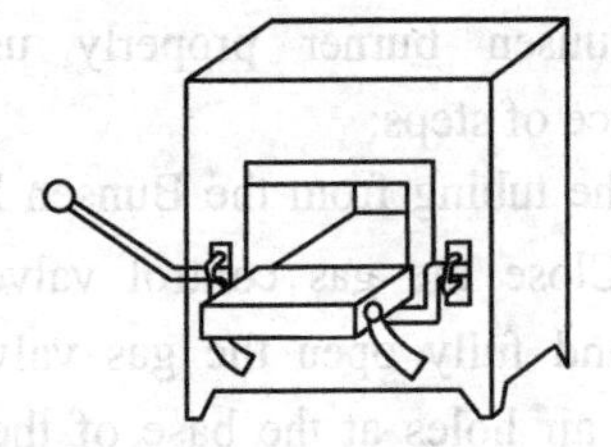

Figure 2-7 Muffle furnace

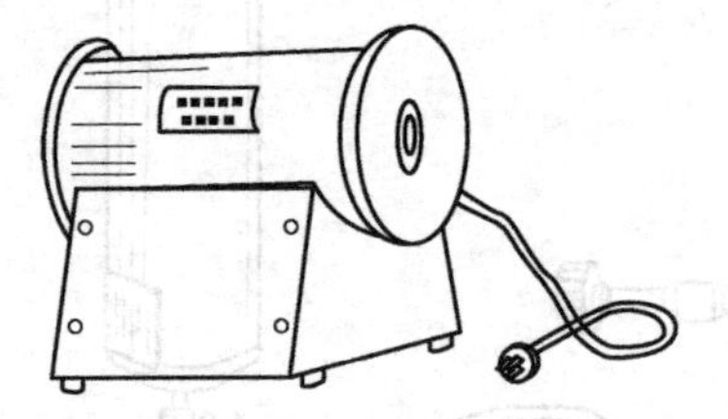

Figure 2-8 Tube furnace

(2) Heating a liquid in a test tube

① When heating a liquid in a test tube with a burner, the test tube should be held with a test tube holder at an angle of about 45 degrees to the flame (Figure 2-9).

② Never fix the position of the flame at the base of the test tube, and never point the test tube at anyone. The contents may be ejected violently if the test tube is not heated properly.

③ The test tube should be less than one-third full of liquid.

④ Move the test tube in and out of the cool flame, heating from top to bottom, mostly near the top of the liquid.

(3) Heating a solid in a test tube

When heating a solid in a test tube with a burner, the opening of the test tube should be slightly tilted down so that condensed water can not settle to the bottom of the test tube and break it (Figure 2-10).

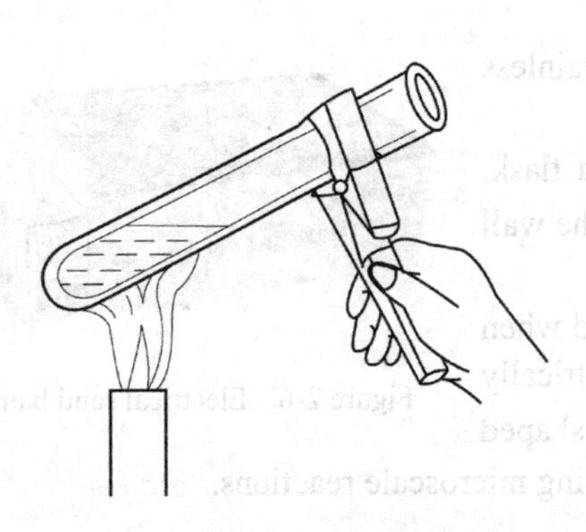

Figure 2-9 Heating a liquid in a test tube

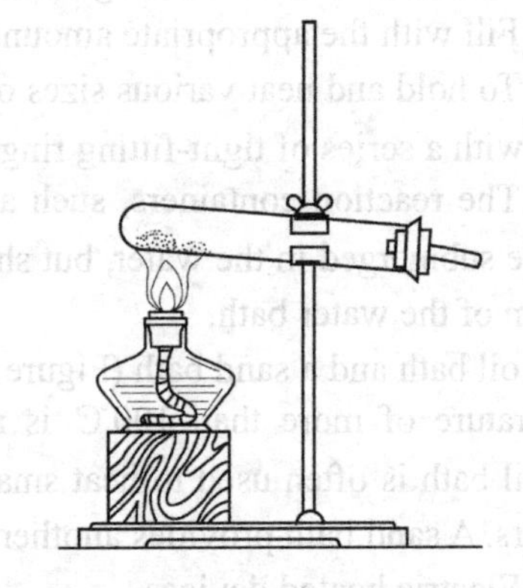

Figure 2-10 Heating a solid in a test tube

(4) High temperature combustion

A porcelain crucible is used to heat a solid sample to extremely high temperatures (Figure 2-11). To avoid contamination of the solid sample, thoroughly clean the crucible prior to use by supporting the crucible and lid on a clay triangle and heating in a hot flame until the bottom of the crucible glows a dull red. This complete firing of the crucible can eliminate impurities in the crucible. Handle the crucible and lid with the crucible tongs.

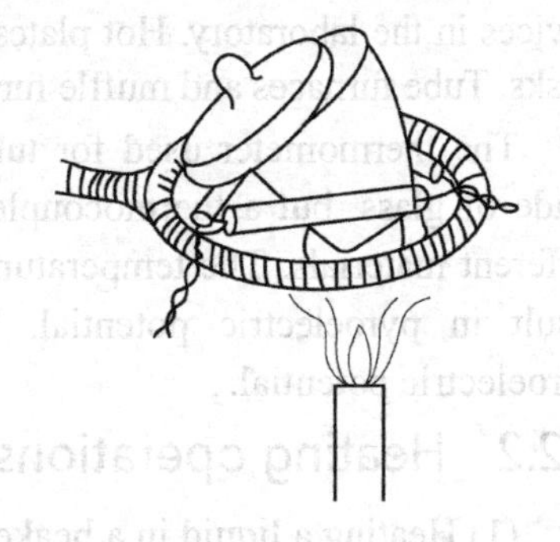

Figure 2-11 Heating a porcelain crucible

2.3 容量仪器及其使用

实验中用于量度液体体积常用以下几种容量仪器。

（1）量筒

常用于液体体积的一般量度。量取液体时要按图 2-12 所示，使视线与量筒内液体的弯月面的最低处保持水平，偏高或偏低都会读不准而造成较大的误差。

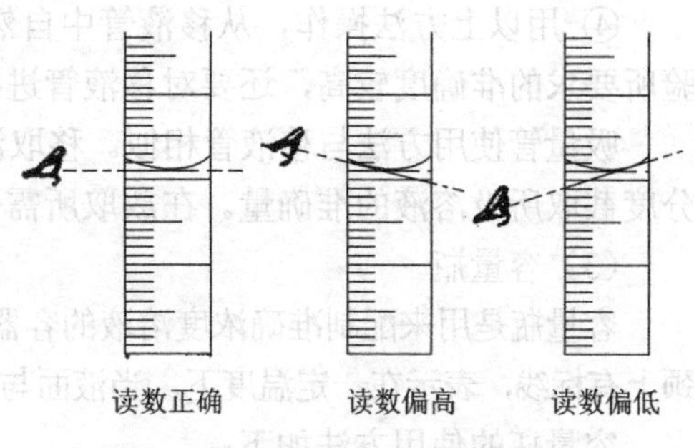

图 2-12 观察量筒内液体的体积

（2）移液管

要求准确地移取一定体积的液体时，可用各种不同容量的移液管[图 2-13(a)]。每支移液管上都标有它的容量和使用温度。另外还有一种带有分刻度的移液管用于移取非整数或 10mL 以下的液体。这种移液管称为吸量管[图 2-13(b)]。

移液管的使用方法如下（图 2-14）：

① 依次用自来水、洗液、自来水洗涤移液管（可用洗耳球将洗液等吸入移液管内进行洗涤），直至管内壁不挂水珠。然后用蒸馏水荡洗 3 次，用滤纸将移液管下端内外的水吸去，再用被移取的液体洗 3 次（每次用量不必太多，吸液体至刚进球部即可），以确保被移取的溶液的浓度不变。

② 把移液管下端伸入所吸溶液液面下约 1cm 处(伸入太深，管外壁附着溶液过多，伸入太浅，则易吸入空气)，右手拇指及中指拿住管颈标线以上的地方，左手拿洗耳球(先挤出空气)，在移液管的上口将液体吸入管内至标线以上（若管内有气泡必须排除）。移开洗耳球，用右手的食指按住管口，将移液管下口提出液面，靠在容器壁上，然后稍微放松食指，或轻轻转动移液管使液面缓慢平稳地下降，直到液体弯月面最低点与标线水平相切时立即按紧食指，使溶液不再流出，取出移液管。

(a) (b)

图 2-13 移液管（a）和吸量管（b）

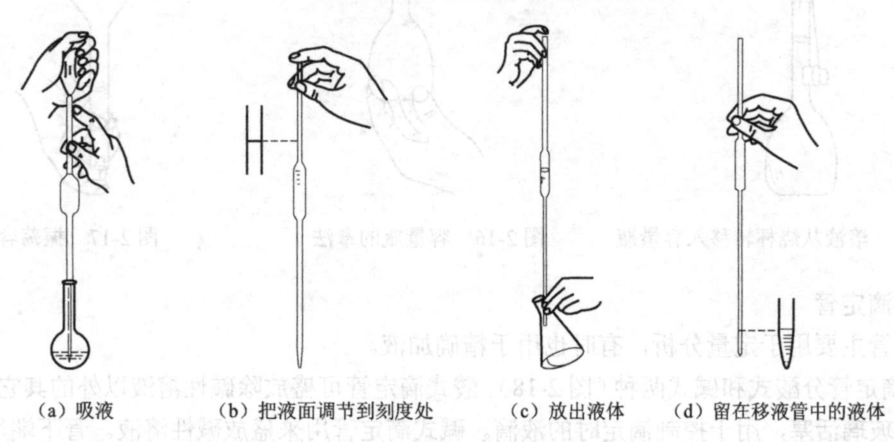

图 2-14 移液管的使用方法

③ 把移液管的尖端靠在接收容器的内壁上放松食指，令液体自然流出，这时应使接收容器倾斜而使移液管直立。等液体不再流出时，还要稍等片刻（约 15s），再移开移液管（使管内壁附着的液膜每次厚薄一致，以保证量度的准确性），但不要把留在移液管尖端的残留液吹出，因为在标定移液管的体积时，并未把这部分液体计算在内。

④ 用以上方法操作，从移液管中自然流出的液体正好是移液管上标明的体积。如果实验所要求的准确度较高，还要对移液管进行校正。

吸量管使用方法与移液管相似。移取溶液时，应尽量避免使用尖端处的刻度。可以使用分度截取所吸溶液的准确量。在放取所需体积的液体后即按住管口，使液体停止流出。

（3）容量瓶

容量瓶是用来配制准确浓度溶液的容器。它是一个细颈梨形的平底瓶，并带有磨口瓶塞。颈上有标线，表示在一定温度下，当液面与标线相切时，液体体积恰好与瓶上所标的体积相等。

容量瓶的使用方法如下：

① 使用前应检查是否漏水，为此，在瓶内加水，塞好瓶塞，右手拿瓶，左手顶住瓶塞，将瓶倒立，观察瓶塞周围是否有水漏出。如不漏，把塞子旋转 180°，塞紧，倒置，再试验这个方向是否漏水。合适的瓶塞要用小绳系在瓶颈上，以免打碎或遗失。

② 用固体物质配制溶液，要先在烧杯里把固体溶解，再把溶液转移到容量瓶中（图 2-15），然后用蒸馏水洗涤烧杯和玻璃棒 3～4 次，洗液一并转入容量瓶中。用洗瓶慢慢加入蒸馏水，至离标线约 1cm 处，稍等片刻，让附在瓶颈上的水流下，改用滴管滴加水至标线（小心勿过标线）。加水时，视线要平行标线。然后塞好塞子，用食指顶住瓶塞，用另一只手指顶住瓶底，如图 2-16 所示（较小的容量瓶不必用手指顶住瓶底），将瓶倒转和摇动多次，使溶液混合均匀（图 2-17）。

③ 热溶液要冷至室温才能倾入容量瓶中，否则溶液的体积会有误差。

④ 必要时，容量瓶的体积也应进行校正。

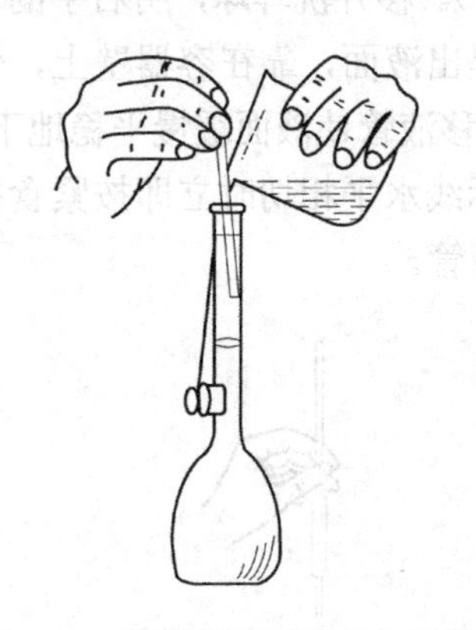

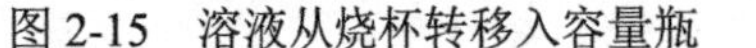

图 2-15　溶液从烧杯转移入容量瓶

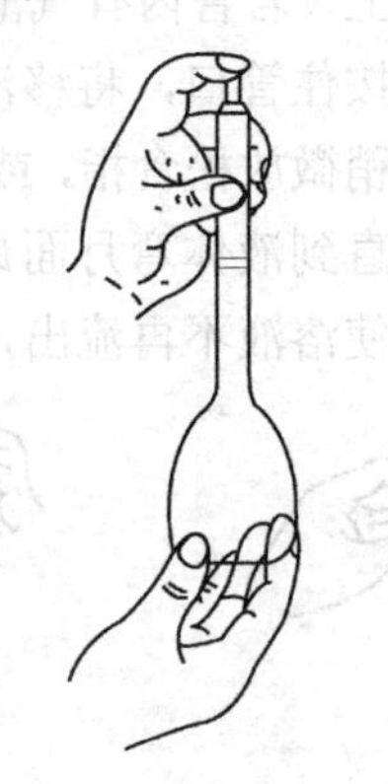

图 2-16　容量瓶的拿法

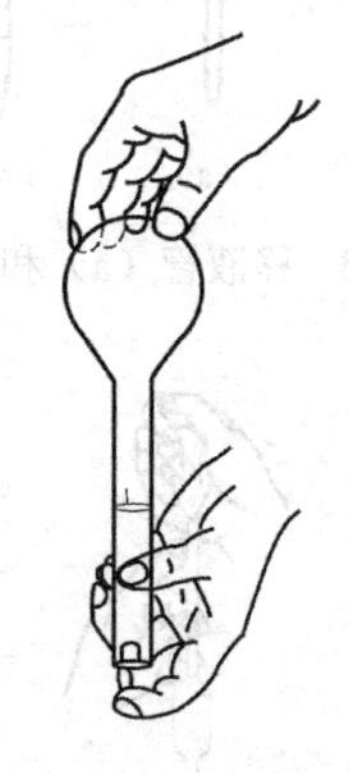

图 2-17　振荡容量瓶

（4）滴定管

滴定管主要用于定量分析，有时也用于精确加液。

① 滴定管分酸式和碱式两种（图 2-18）。酸式滴定管可盛放除碱性溶液以外的其它溶液。管下有一玻璃活塞，用于控制滴定时的液滴。碱式滴定管用来盛放碱性溶液。管下端用橡皮管连接一个一端有尖嘴的小玻璃管，橡皮管内装一个玻璃圆球以代替玻璃活塞（图 2-19）。

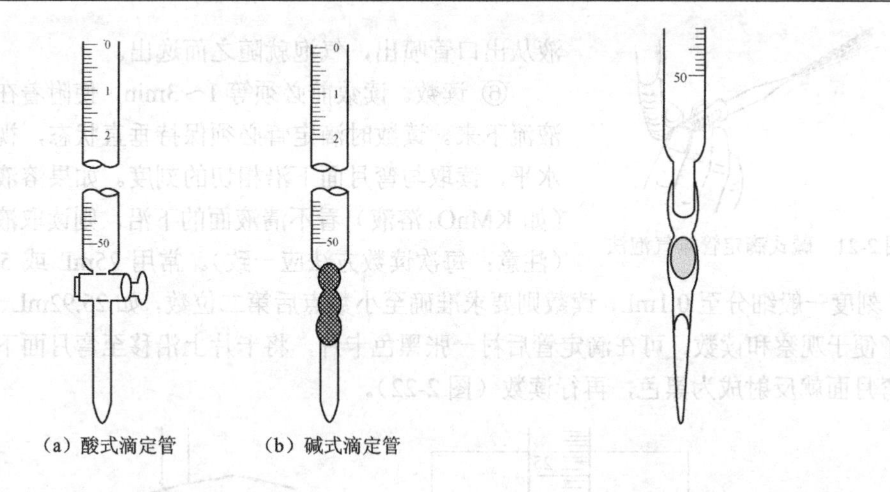

（a）酸式滴定管　（b）碱式滴定管

图 2-18　滴定管

图 2-19　碱式滴定管下端的结构

② 活塞涂油的方法。滴定管洗涤前必须检查是否漏水，活塞转动是否灵活。若活塞渗漏或转动不灵活，可将活塞取下，用滤纸或布擦净活塞和塞槽，然后在活塞两头涂一薄层凡士林。注意不要涂在活塞孔所在的那一圈，以免塞住塞孔。将活塞一直插入槽内，向同一方向转动，直到活塞与塞槽接触的地方透明为止（图 2-20）。

如果碱式滴定管漏水，则更换玻璃球或橡皮管。

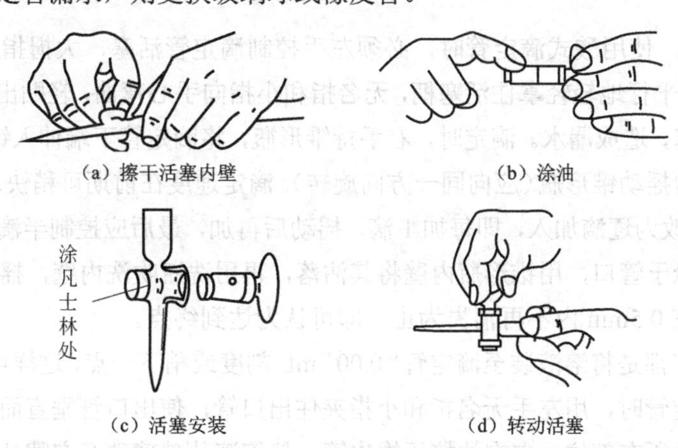

（a）擦干活塞内壁　（b）涂油

（c）活塞安装　（d）转动活塞

图 2-20　活塞擦干、涂油、安装、转动操作

③ 清除活塞孔或出口管孔中凡士林的方法。如果活塞孔堵塞，取下活塞用细铜丝捅出即可。若是出口孔堵塞，则将水充满全管，将出口管浸在热水中，温热片刻后，打开活塞使管内水突然冲下，将熔化的凡士林带出。如此法不能奏效，则用四氯化碳浸溶或用一根细铜丝捅通。

④ 洗涤方法。滴定管在使用前应按常规操作洗涤，洗净后的滴定管内壁不应附有液滴。经由蒸馏水荡洗后再用滴定用的溶液洗 3 遍（每次约 5mL），以确保溶液浓度不变。

⑤ 装溶液。将溶液加到滴定管内至刻度“0.00”以上，开启活塞或挤压玻璃球，使管下端充满溶液后，并调节液面在“0.00”刻度处。

必须注意滴定管下端不得留有气泡。如有气泡，必须排除，排除的方法是将酸式滴定管倾斜约 30° 角，左手迅速打开活塞使溶液冲出，管中气泡随之逐出。碱式滴定管可按图 2-21 所示的方法，把橡皮管向上弯曲，出口斜向上方，用手指挤压玻璃球稍上边的橡皮管，使溶

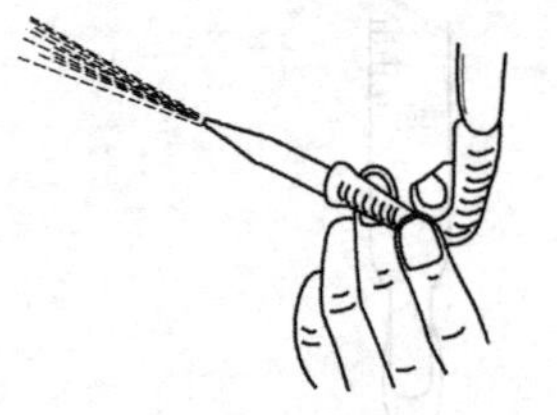

图 2-21　碱式滴定管排气泡法

液从出口管喷出，气泡就随之而逸出。

⑥ 读数。读数前必须等 1～3min，使附着在内壁上的溶液流下来。读数时滴定管必须保持垂直状态，视线应与液面水平，读取与弯月面下沿相切的刻度。如果溶液的颜色太深（如 $KMnO_4$ 溶液）看不清液面的下沿，则读取液面的最高点（注意：每次读数方法应一致）。常用 25mL 或 50mL 的滴定管，刻度一般细分至 0.1mL，读数则要求准确至小数点后第二位数，如 25.92mL、1.30mL 等。为了便于观察和读数，可在滴定管后衬一张黑色卡片，将卡片上沿移至弯月面下 1mm 左右，则弯月面就反射成为黑色，再行读数（图 2-22）。

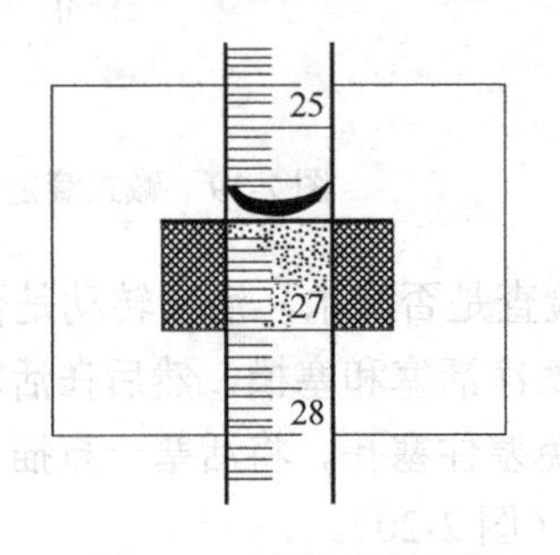

图 2-22　读数卡的使用

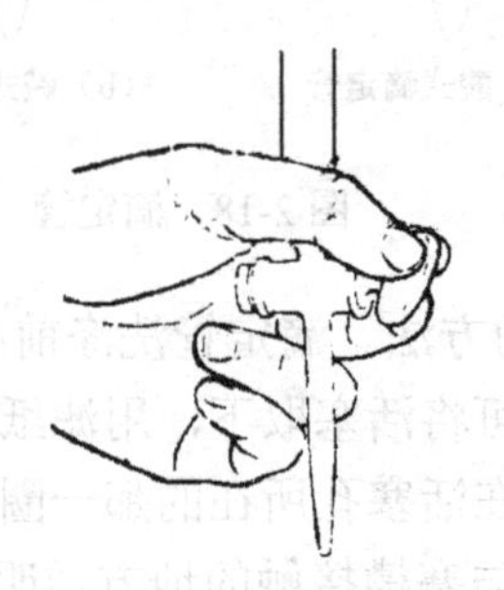

图 2-23　左手旋转活塞

⑦ 滴定操作。使用酸式滴定管时，必须左手控制滴定管活塞，大拇指在管前，食指和中指在管后，三指平行地轻轻拿住活塞柄，无名指和小指向手心弯曲，轻贴出口管（图 2-23），注意不要顶出活塞，造成漏水。滴定时，右手持锥形瓶，将滴定管下端伸入锥形瓶口约 1cm，然后边滴加溶液边摇动锥形瓶（应向同一方向旋转）。滴定速度在前期可稍快。但不能滴成“水线”。接近终点时改为逐滴加入，即每加 1 滴，摇动后再加，最后应控制半滴加入，将活塞稍稍转动，使半滴悬于管口，用锥形瓶内壁将其沾落，再用洗瓶吹洗内壁，摇匀。如此重复操作直到颜色变化在 0.5min 内不再消失为止，即可认为达到终点。

每次滴定最好都是将溶液装至滴定管“0.00”mL 刻度或稍下一点，这样可减少滴定误差。

使用碱式滴定管时，用左手无名指和小指夹住出口管，使出口管垂直而不摆动。用拇指和食指捏住玻璃珠所在部位，向右边挤压橡皮管，使溶液从玻璃珠旁空隙处流出。注意不要用力捏玻璃珠，也不能使玻璃珠上下移动。

2.3 Volumetric Glassware

(1) Graduated cylinder

To measure a liquid in a graduated cylinder, the volume of the liquid is read at the bottom of its meniscus (Figure 2-12). Position the eye horizontally at the bottom of the meniscus to read the level of the liquid.

(2) Pipet

Pipets measure volumes of liquids accurately. Each pipet is marked with the volume and temperature. There are mono-graduated pipets with a big bell [Figure 2-13 (a)], and graduated tube pipets [Figure 2-13 (b)].

Use a pipet properly according to the following procedures (Figure 2-14).

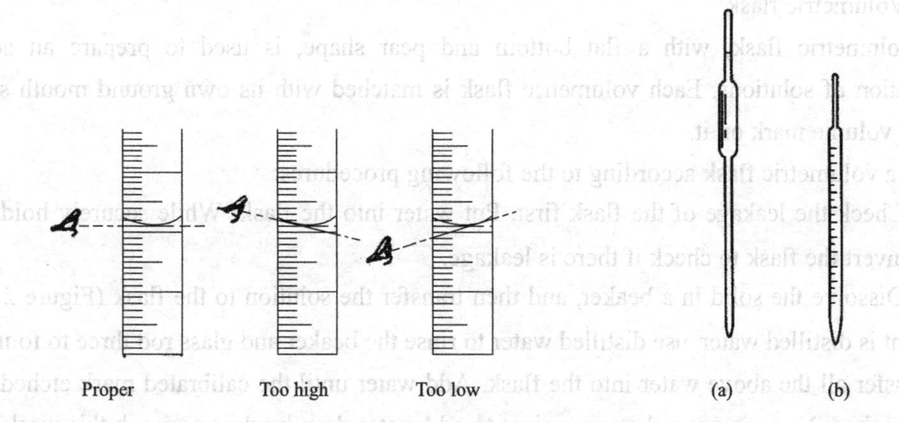

Figure 2-12 Reading the volume of a liquid at the bottom of its meniscus Figure 2-13 Pipets

① A pipet should be first cleaned with tap water, detergent, and tap water again, until there are no water droplets adhering to its inner wall. Then rinse the pipet 3 times with distilled water. Transfer the liquid that you intend to pipet from the reagent bottle into a clean, dry container. Use a rubber pipet bulb to draw a small amount of the liquid into the pipet as a rinse. Rinse 3 times, and deliver each rinse through the pipet tip into a waste container.

② Place the pipet tip well below the surface of the liquid. Using the suction from the rubber pipet bulb, draw the liquid into the pipet until the level is above the desired mark on the pipet. Remove the bulb and quickly cover the top of the pipet with your forefinger. Remove the tip from the liquid, and place the tip against the wall of the container. Control the delivery of the excess liquid until the level is at the desired mark of the pipet.

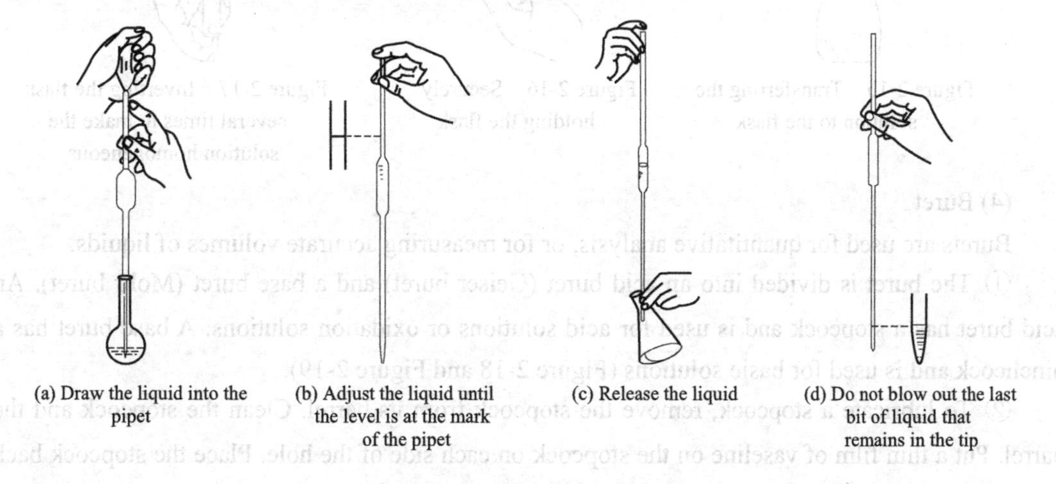

Figure 2-14 Proper use of pipets

③ Place the pipet tip against the wall of the receiving container. Deliver the liquid to the container by releasing the forefinger from the top of the pipet. Wait a little while (about 15 seconds) before removing the pipet. Do not blow out the last bit of liquid that remains in the tip because this liquid has been included in the calibration of the pipet.

④ Calibrate the pipet if necessary.

(3) Volumetric flask

A volumetric flask, with a flat bottom and pear shape, is used to prepare an accurate concentration of solutions. Each volumetric flask is matched with its own ground mouth stopper, and has a volume mark on it.

Use a volumetric flask according to the following procedures:

① Check the leakage of the flask first. Put water into the flask. While securely holding the stopper, invert the flask to check if there is leakage.

② Dissolve the solid in a beaker, and then transfer the solution to the flask (Figure 2-15). If the solvent is distilled water, use distilled water to rinse the beaker and glass rod three to four times. Then transfer all the above water into the flask. Add water until the calibrated mark etched on the flask is reached. You can use a dropping pipet to add water drop by drop to reach this mark. While securely holding the stopper, invert the flask several times to ensure that the solution is homogeneous (Figure 2-16 and Figure 2-17).

③ If the solution is hot, cool it to room temperature before transferring it into the flask.

④ Calibrate the volume of the volumetric flask if necessary.

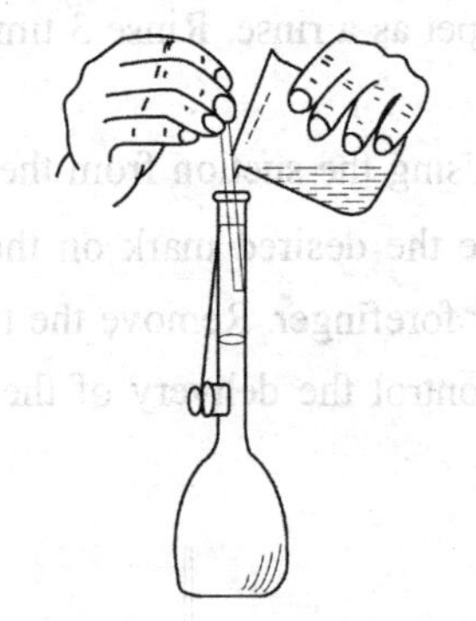

Figure 2-15 Transferring the solution to the flask

Figure 2-16 Securely holding the flask

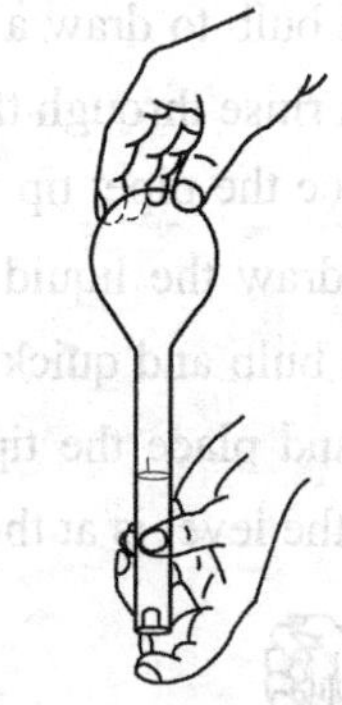

Figure 2-17 Inverting the flask several times to make the solution homogeneous

(4) Buret

Burets are used for quantitative analysis, or for measuring accurate volumes of liquids.

① The buret is divided into an acid buret (Geiser buret) and a base buret (Mohr buret). An acid buret has a stopcock and is used for acid solutions or oxidation solutions. A base buret has a pinchcock and is used for basic solutions (Figure 2-18 and Figure 2-19).

② To lubricate a stopcock, remove the stopcock from its barrel. Clean the stopcock and the barrel. Put a thin film of vaseline on the stopcock on each side of the hole. Place the stopcock back into the barrel and roll the stopcock until both the barrel and the stopcock are lubricated (Figure 2-20).

Change the pinchcock if a base buret leaks.

③ If there is vaseline in the buret tip, fill the buret with water and soak the tip in hot water.

Then open the stopcock and flush the melted vaseline out.

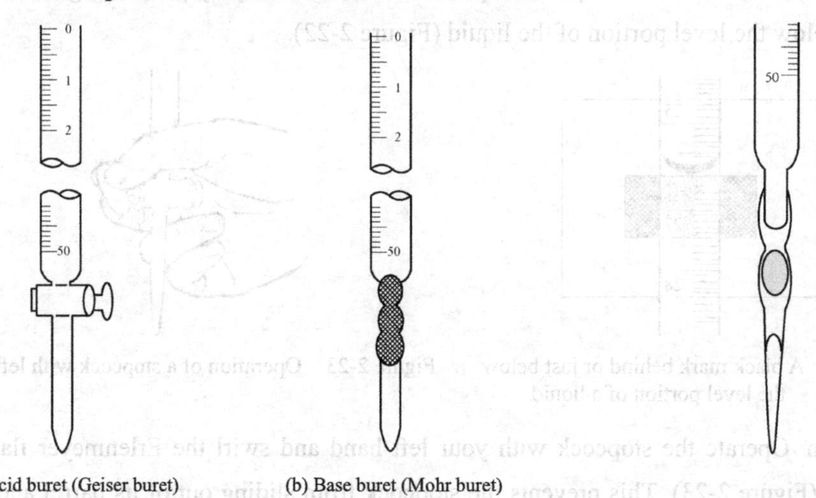

(a) Acid buret (Geiser buret)　　(b) Base buret (Mohr buret)

Figure 2-18 Burets　　Figure 2-19 Bottom tip of a base buret

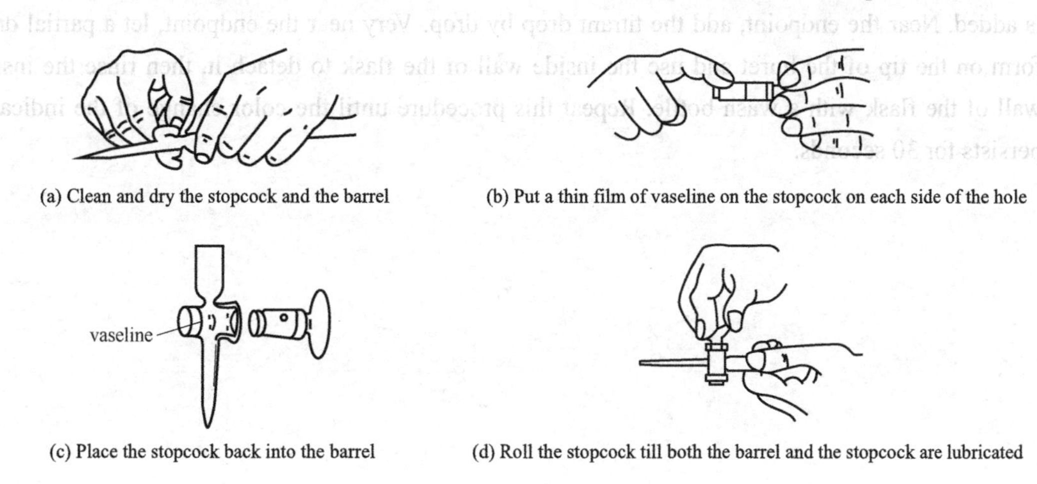

(a) Clean and dry the stopcock and the barrel

(b) Put a thin film of vaseline on the stopcock on each side of the hole

(c) Place the stopcock back into the barrel

(d) Roll the stopcock till both the barrel and the stopcock are lubricated

Figure 2-20 Operations of lubricating a stopcock

④ A clean buret should have no water droplets adhering to its inner wall. Rinse the buret with distilled water three times, and then rinse with titrant three times.

⑤ Fill the buret with the titrant to above the zero mark. Open the stopcock to adjust the level to the mark.

If there are air bubbles in the tip of the acid buret, open the stopcock quickly to remove any air bubbles in the tip. To remove the air bubbles in the base buret, aim the tip upward and allow a little titrant to flow through (Figure 2-21).

Figure 2-21 Removal of air bubbles in a base buret

⑥ To read the volume of the titrant in the buret, position the eye horizontally at the bottom of the meniscus. Record the data using all certain digits plus one uncertain digit. The minimum mark on a 25mL or 50mL buret is 0.1mL. So the reading should be recorded to 0.01mL. For example:

25.92mL, or 1.30mL. A clear or transparent liquid is read more easily by positioning a black mark behind or just below the level portion of the liquid (Figure 2-22).

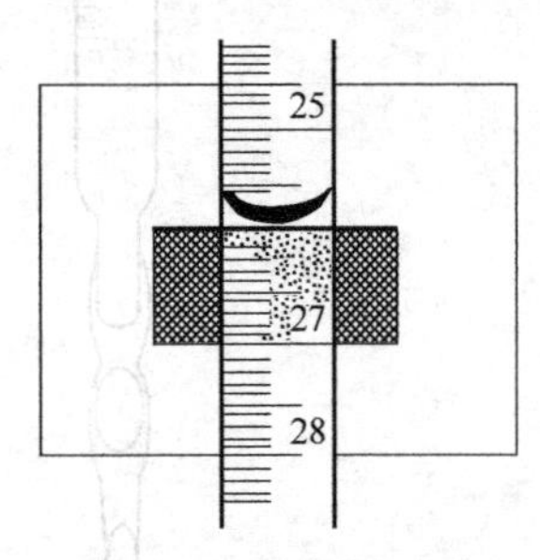

Figure 2-22 A black mark behind or just below the level portion of a liquid

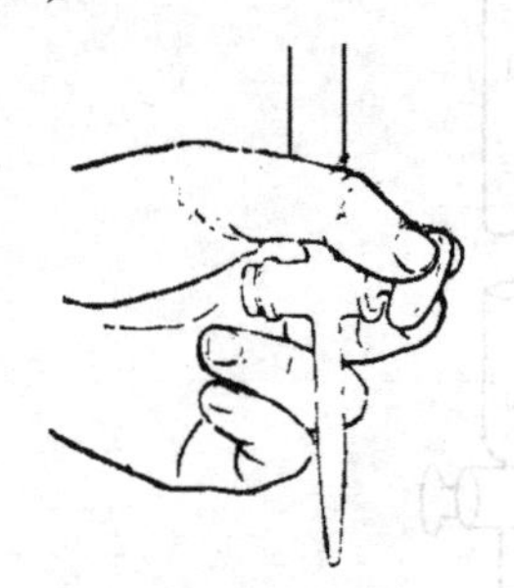

Figure 2-23 Operation of a stopcock with left hand

⑦ Titration. Operate the stopcock with your left hand and swirl the Erlenmeyer flask with your right hand (Figure 2-23). This prevents the stopcock from sliding out of its barrel and allows you to maintain a normal, constant swirling motion of the reaction mixture in the flask as the titrant is added. Near the endpoint, add the titrant drop by drop. Very near the endpoint, let a partial drop form on the tip of the buret and use the inside wall of the flask to detach it, then rinse the inside wall of the flask with a wash bottle. Repeat this procedure until the color change of the indicator persists for 30 seconds.

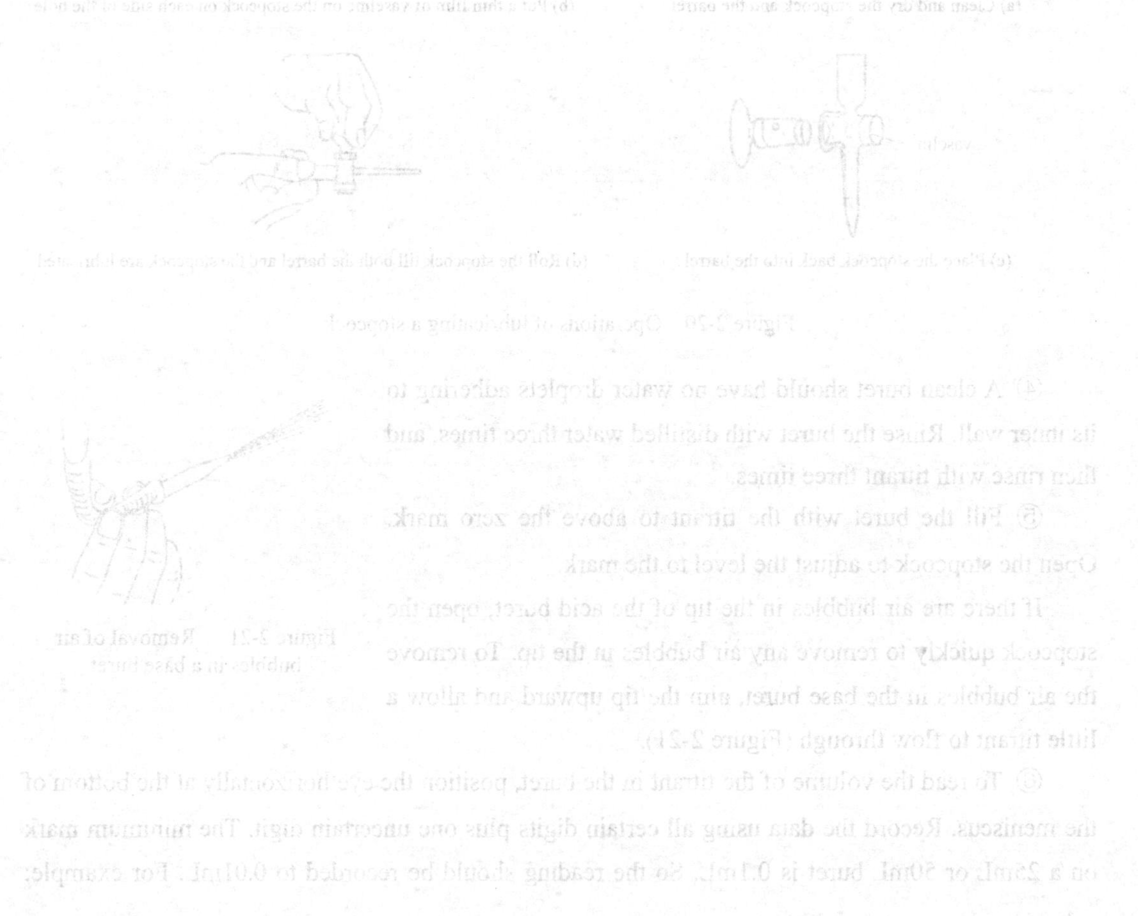

2.4 化学试剂及其取用

按照试剂中杂质含量的多少，我国把常用试剂分为实验试剂（LR，四级）、化学纯试剂（CP，三级）、分析纯试剂（AR，二级）和优质纯试剂（GR，一级）4 种规格。应根据实验要求，分别选用不同规格的试剂。

在实验室分装试剂时，一般把固体试剂装在广口瓶内；液体试剂盛在细口瓶或滴瓶中；见光易分解的试剂（如硝酸银、高锰酸钾）则应装在棕色瓶内。装碱液的瓶塞要用软木塞或橡皮塞。每个试剂瓶上都应贴有标签，写明试剂的名称和规格，液体试剂应注明浓度和配制日期。

（1）液体试剂的取用

① 取滴瓶中的液体试剂时，要用滴瓶中的专用滴管，滴头不要与接收容器的器壁接触，更不应把滴管伸入到其它液体中。滴管不能平握或倒置，以免试剂倒灌入橡皮帽。放回滴管时，管内试剂要排空。

② 用倾注法取液体试剂时，应将瓶塞反放在桌上，以免沾污。用手心握住瓶上贴有标签的一面，倒出的试剂应沿一干净的玻璃棒流入容器（图 2-24），或沿试管壁流下。取出所需的量后，慢慢竖起试剂瓶，把瓶口剩余的那滴试剂“碰”到容器内，以免液滴沿瓶子外壁流下。已取出的试剂不能再倒回试剂瓶。倒入容器的液体不应超过容器容量的 2/3。

图 2-24 液体试剂倒入烧杯

（2）固体试剂的取用

① 要用干净的药勺取试剂。用过的药勺必须洗净和擦干后才能再使用，以免沾污试剂。

② 取出试剂后，应立即盖紧瓶盖，不要盖错盖子。

③ 注意不要多取。取多的药品不能倒回原瓶，可放在指定容器中供他人使用。

④ 要求取用一定质量的固体试剂时，应把试剂放在称量纸上称量。具有腐蚀性或易潮解的试剂必须放在表面皿或玻璃容器内称量。

⑤ 往试管（特别是湿试管）中加入粉末状固体试剂时，可用药匙或将取出的药品放在对折的纸片上，伸进平放的试管中约 2 / 3 处，然后把试管竖直，让试剂滑下去。

⑥ 加入块状固体试剂时，应将试管倾斜使其沿管壁慢慢滑下，不得垂直悬空投入，以免击破管底。

⑦ 固体试剂的颗粒较大时，可在洁净而干燥的研钵中研碎，然后取用。

⑧ 有毒的药品要在教师指导下取用。

2.4 Transferring Chemical Reagents

Chemical reagents are usually divided into four grades in China, i.e., Lab Reagent (LR, the fourth grade), Chemically Pure Reagent (CP, the third grade), Analytical Reagent (AR, the second grade), and Guaranteed Reagent (GR, the first grade).

Solid reagents should be stored in wide-mouth bottles. Liquid reagents should be stored in dropping bottles or narrow-mouth bottles. Light-sensitive reagents, such as silver nitrate and potassium permanganate, should be stored in brown or red bottles to avoid visible light, ultraviolet and infrared radiation which may alter or break the reagents down. Alkaline solutions should not be stored in bottles with glass stoppers. All bottles should be clearly labeled with the reagent name, the grade, the concentration, and the date prepared, etc.

(1) Transferring liquids and solutions

① When a liquid is to be transferred from a dropping bottle, never hold a filled dropper upside down. Furthermore, the dropper should not touch any surface outside of the dropping bottle itself.

② When a liquid is to be transferred from a reagent bottle, remove the stopper and put it top side down on the laboratory bench if the stopper is hollow. Hold a stirring rod against the lip of the bottle and pour the liquid down the stirring rod, which, in turn, should touch the inner wall of the receiving container (Figure 2-24). Return the stopper to the reagent bottle. Do not return any excess or unused liquid to the original reagent bottle. The receiving container should be less than two-third full of liquid.

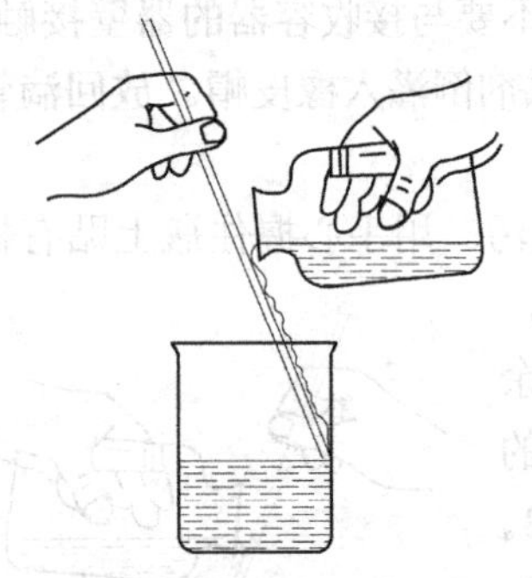

Figure 2-24 Transferring liquids and solutions to a beaker

(2) Transferring solids

① Use clean weighing spoons to transfer solids.

② Return the stopper to the reagent bottle immediately after finishing.

③ Try not to transfer any more reagent than is necessary for the experiment. Do not return any excess reagent to the reagent bottle.

④ Use weighing paper, when necessary, to weigh solids. Corrosive and deliquescent solids should be weighed by using a watch glass or glass container.

⑤ When you transfer solids to a test tube, use a weighing spoon or weighing paper.

⑥ If transferring a lump of solid into a test tube, slowly slide the solid down to the bottom of the test tube.

⑦ A clean mortar can be used to grind solids before transferring.

⑧ Consult your laboratory instructor when using poisonous reagents.

2.5 溶液的配制

（1）饱和溶液的配制

若配制硫化氢、氯气等气体的饱和溶液，只要在常温下把新制备出来的硫化氢、氯气等气体通入蒸馏水一段时间即可。若配制某固体试剂的饱和溶液时，先按试剂在室温下的溶解度数据计算出配制时所需的试剂量和蒸馏水量。称量出比计算量稍多的固体试剂。若颗粒较大，则用研钵磨碎。溶解时，常用搅拌、加热等方法来加快溶解。搅拌时，应该手持搅拌棒并转动手腕，使搅拌棒在液体中均匀转圈。转动速度不要太快，也不要使搅拌棒碰到器壁和器底，以免打碎烧杯。加热到一定温度而固体残留不再溶解为止，冷却至室温后又有一些固体析出，这种溶液就是饱和溶液。

（2）易水解盐溶液的配制

氯化锡（Ⅱ）、氯化锑（Ⅲ）、硝酸铋（Ⅲ）等盐，极易水解成氢氧化物或碱式盐，所以配制它们的水溶液时，必须把它们溶解在相应的稀酸溶液中，以抑制水解才能得到透明的溶液。

2.5 Preparing Solutions

(1) Preparing saturated solutions

To prepare a saturated solution of H_2S or Cl_2 gas, insert freshly prepared H_2S or Cl_2 gas into distilled water for an appropriate time. To prepared a saturated solution of a solid, calculate the amount of solid and water according to the solubility first. Then weigh an amount of solid that is slightly more than the calculated amount. Dissolve the solid in water with a stirring rod. Heat until the solid is dissolved. Cool to the room temperature, and precipitation will occur. This is a saturated solution.

(2) Preparing solutions which are easily hydrolized

Some salts, such as $SnCl_2$, $SbCl_3$ and $Bi(NO_3)_3$, should be dissolved in acid first to avoid precipitation by hydrolysis. Then dilute the obtained solution with water to get the appropriate concentration.

2.6 气体的发生、净化、干燥和收集

（1）气体的发生

实验室中常用启普发生器（图 2-25）来制备氢气、二氧化碳和硫化氢等气体。

启普发生器由一个葫芦状的玻璃容器和球形漏斗组成。固体药品放在中间圆球内，可在固体下面放些玻璃丝或有孔橡皮块来承受固体，以免固体掉至下部球内。酸从球形漏斗加入。使用时只要打开活塞，由于压力差，酸液自动下降而进入中间球内，与固体接触而产生气体。要停止使用时，只要关闭活塞，继续发生的气体会把酸从中间球内压入下球及球形漏斗内，使酸液与固体不再接触而停止反应。下次使用时，只要重新打开活塞，又会产生气体。

发生器中的酸液长久使用后会变稀。此时，可从下球侧口倒掉废酸，再向球形漏斗中加入新的酸液。若固体需要更换时，先倒出酸液或在酸与固体脱离接触的情况下，用一橡皮塞将球形漏斗上口塞紧，再从中间球侧口将固体残渣取出，更换新的固体。

启普发生器不能加热，装入的固体反应物又必须是较大的块粒，不适用于小颗粒或是粉末的固体反应物，所以氯化氢、氯气、二氧化硫等气体就不能使用启普发生器制备，但可改用图 2-26 所示的气体发生装置。

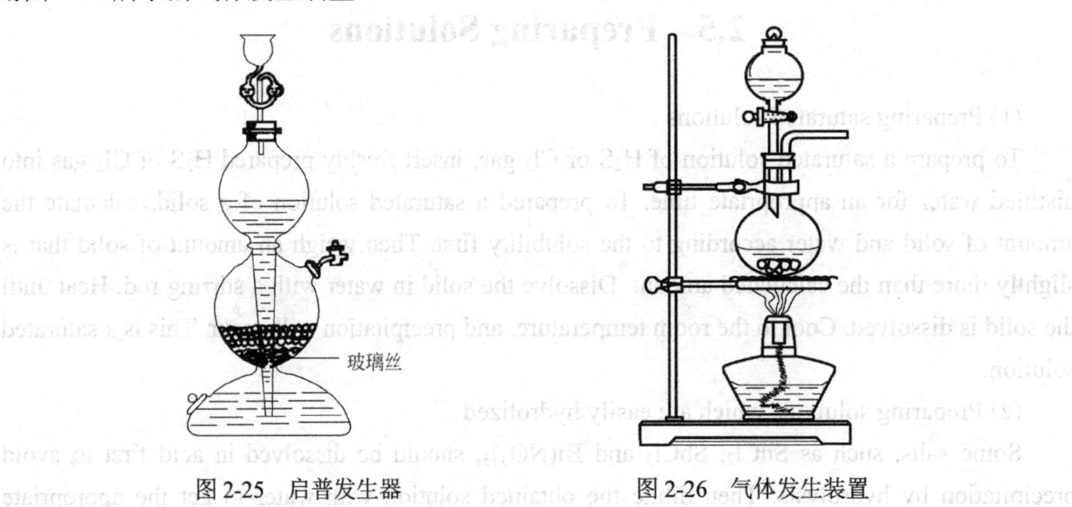

图 2-25 启普发生器　　图 2-26 气体发生装置

把固体加在蒸馏瓶内，酸液装在分液漏斗中。使用时，打开分液漏斗下面的活塞，使酸液均匀地滴加在固体上，就产生气体。当反应缓慢或不产生气体时，可以微微加热。如加热后仍不起反应，则需更换固体药品。

在实验室，还可以使用气体钢瓶直接获得各种气体。气体钢瓶是储存压缩气体特制的耐压钢瓶。使用时，通过减压器（气压表）有控制地放出。由于钢瓶的内压很大（有的高达 150×10^5Pa），而且有些气体易燃或有毒，所以在使用钢瓶时一定要注意安全，操作要特别小心。

使用钢瓶时的注意事项：

① 钢瓶应存放在阴凉、干燥、远离热源（如阳光、暖气、炉火）的地方。可燃性气体钢瓶必须与氧气钢瓶分开存放。

② 绝对不可使油或其它易燃性有机物沾在气瓶上（特别是气门嘴和减压器）。也不得用

棉、麻等物堵漏，以防止燃烧而引起事故。

③ 使用钢瓶中的气体时，要用减压器（气压表）。可燃性气体的钢瓶，其气门螺纹是反扣的（如氢气、乙炔气）。不燃或助燃性气体钢瓶，其气门螺纹是正扣的。各种气体的气压表不得混用。

④ 钢瓶内的气体绝不能全部用完，一定要保留 0.5kg（0.05MPa）以上的残留压力（表压）。可燃性气体如乙炔剩余压力应为 2～3kg（0.2～0.3MPa）。

⑤ 为了避免把各种气瓶混淆而用错气体（这样会发生很大事故），通常在气瓶外面涂以特定的颜色以便区别，并在瓶上写明瓶内气体的名称。表 2-2 为我国气瓶常用的标记。

表 2-2　我国的气瓶标记

气体类别	瓶身颜色	标字颜色
氮气	黑	黄
氧气	天蓝	黑
氢气	深绿	红
空气	黑	白
氨气	黄	黑
二氧化碳	黑	黄
氯气	黄绿	黄
乙炔	白	红
其它一切可燃气体	红	白
其它一切不可燃气体	黑	黄

（2）气体的干燥和净化

实验室中发生的气体都带有酸雾和水汽，有时需要进行净化和干燥。酸雾可用水或玻璃棉除去；水汽可用浓硫酸、无水氯化钙或硅胶吸收。一般情况下使用洗气瓶（图 2-27）、干燥塔（图 2-28）或 U 形管（图 2-29）等仪器进行净化。液体（如水、浓硫酸）装在洗气瓶内，无水氯化钙和硅胶装在干燥塔或 U 形管内，玻璃棉装在 U 形管内。气体中如还有其它杂质，应根据具体情况分别用不同的洗涤液或固体吸收。

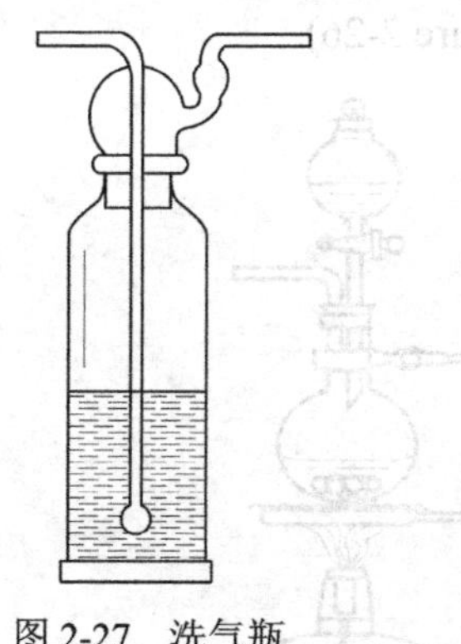

图 2-27　洗气瓶

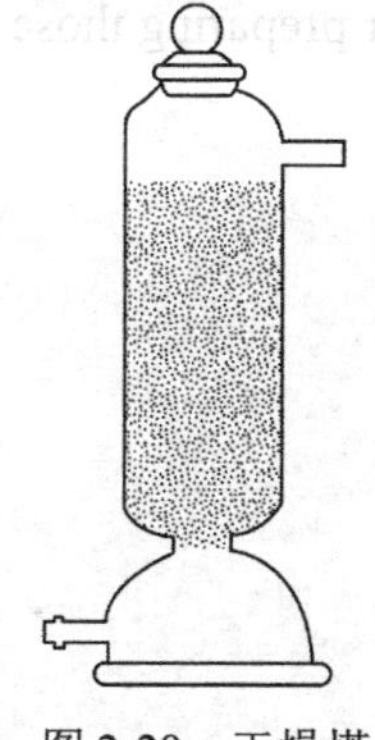

图 2-28　干燥塔

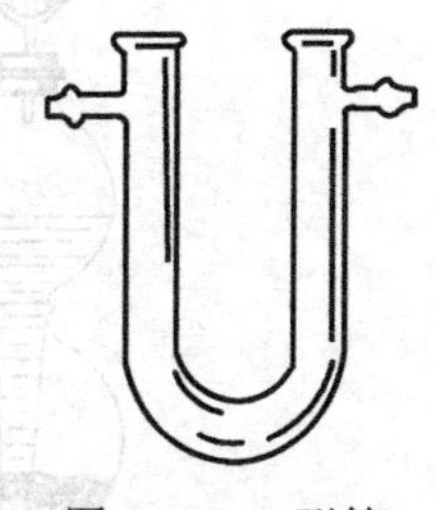

图 2-29　U 形管

具有还原性或碱性的气体如硫化氢、氨气等，不能用浓硫酸来干燥，可分别用氯化钙（硫化氢）或氢氧化钠固体（氨）来干燥。注意：氨气不能用无水氯化钙来干燥。

（3）气体的收集

① 在水中溶解度很小的气体（如氢气、氧气）可用排水集气法收集（图 2-30）。

② 易溶于水而比空气轻的气体（如氨）可用瓶口向下的排气集气法收集（图 2-31）。

③ 易溶于水而比空气重的气体（如氯气和二氧化碳）可用瓶口向上的排气集气法收集

（图 2-32）。

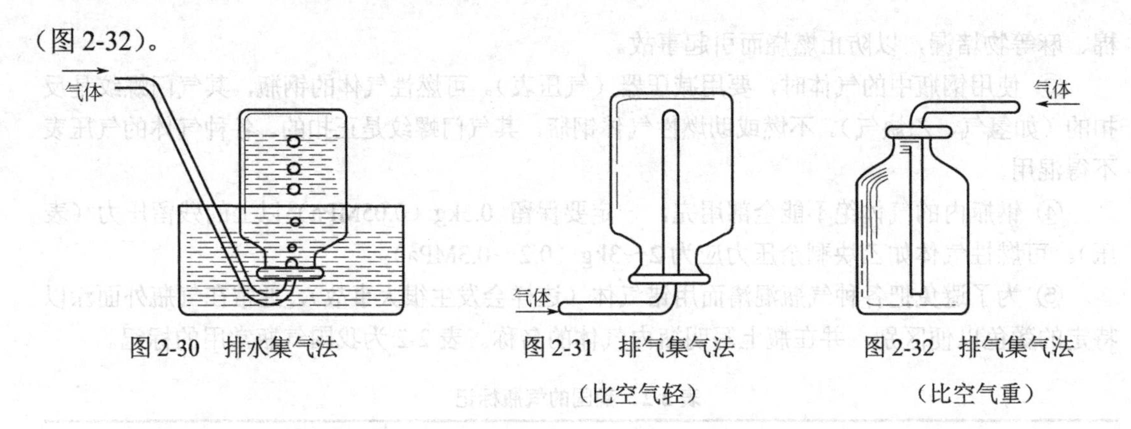

图 2-30 排水集气法　　图 2-31 排气集气法（比空气轻）　　图 2-32 排气集气法（比空气重）

2.6 Preparing, Purifying, Drying and Collecting Gases

(1) Preparing gases

Kipp's apparatus (Figure 2-25), also called Kipp generator, is an apparatus designed for preparation of hydrogen, carbon dioxide, and hydrogen sulfide gases in the laboratory.

This apparatus make up by three vertically stacked cylinders. The solid is placed into the middle cylinder, and acid is put into the top spherical cylinder. A tube extends from the top spherical cylinder into the bottom cylinder. The middle cylinder has a tube with a stopcock attached, which is used to draw off the evolved gas. When the stopcock is closed, the pressure of the gas in the middle cylinder rises and expels the acid back into the top cylinder until it is not in contact with the solid anymore, and the chemical reaction stops. Turn on the stopcock, and the reaction will continue.

When changing the acid, pour out the old acid from the opening on the bottom cylinder. Add new acid from the top cylinder.

Kipp's apparatus can not be heated. The solid should be large pieces or lumps. Therefore, it can not be used to prepare hydrogen chloride, chlorine, and sulfur dioxide gases. A separating funnel with a distilling flask can be used for preparing those gases (Figure 2-26).

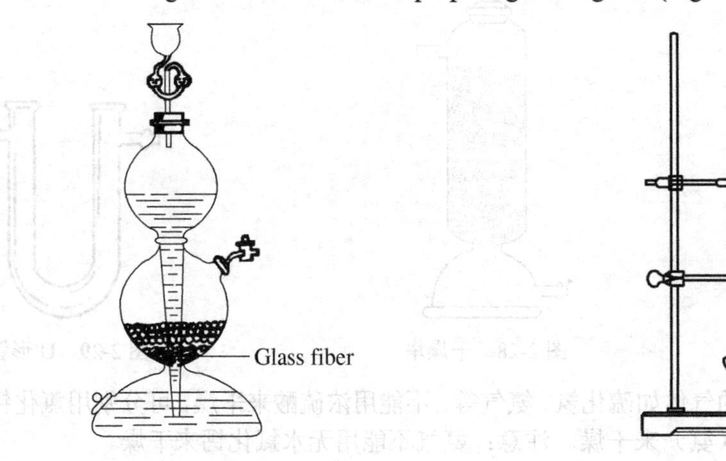

Figure 2-25 Kipp's apparatus　　Figure 2-26 Gas generator

The solid is placed into the distilling flask, and the acid is put into the separating funnel. Turn on the stopcock of the separating funnel, and the acid will be added to the solid to produce gas.

A gas cylinder or tank is a pressure vessel used to store gases at above atmospheric pressure. Because the contents are under pressure and are sometimes hazardous, there are special safety regulations for handling gas cylinders.

① Gas cylinders should be put in dry and cool places. Keep away from heat (the sun, heater, stove). Gas cylinders, which store flammable gases, should never be put proximity to oxygen gas cylinders.

② Keep the gas cylinders clean. Never stain gas cylinders with oil or flammable organic compounds. Never try to stop a leak by using cotton etc.

③ When gases are supplied in gas cylinders, the cylinders have a stop angle valve at the end on top. When the gas is not in use, a cap may be screwed over the protruding valve to protect it from damage. Instead of a cap, cylinders commonly have a protective collar or neck ring around the service valve assembly.

When the gas in the cylinder is ready to be used, the cap is taken off and a pressure-regulating assembly is attached to the stop valve. This attachment typically has a pressure regulator with upstream (inlet) and downstream (outlet) pressure gauges and a further downstream needle valve and outlet connection. For gases that remain gaseous under ambient storage conditions, the upstream pressure gauge can be used to estimate how much gas is left in the cylinder according to pressure. For gases that are liquid under storage, such as propane, the outlet pressure depends on the vapor pressure of the gas, and does not fall until the cylinder is nearly exhausted. The regulator can be adjusted to control the flow of gas out of the cylinder according to the pressure shown by the downstream gauge.

④ Do not use up all the gas in the cylinder.

⑤ Gas cylinders with different colors store different gases. Table 2-2 lists the color of the gas cylinders and the gases they represent in China.

Table 2-2 Gas cylinders in China

Gas	Color of the gas cylinder	Color of Chinese characters on the gas cylinder
nitrogen	black	yellow
oxygen	blue	black
hydrogen	dark green	red
air	black	white
ammonia	yellow	black
carbon dioxide	black	yellow
chlorine	yellowish green	yellow
acetylene	white	red
all other flammable gases	red	white
all other nonflammable gases	black	yellow

(2) Purifying and drying gases

Gases prepared in the laboratory should be purified and dried due to the possible contamination with acid or water vapor. Acid vapor can be removed by water or glass wool. Water vapor can be absorbed by concentrated sulfuric acid, anhydrous calcium chloride, or silica gel. Gas bottles (Figure 2-27), drying towers (Figure 2-28), and U tubes (Figure 2-29) are apparatuses used for purifying gases. Liquids, such as water and concentrated sulfuric acid, should be placed in a gas bottle. Anhydrous calcium chloride or silica gel should be put in a drying tower or U tube. Glass wool should be put in the U tube.

Concentrated sulfuric acid can not be used to dry reducing agents (e.g. hydrogen sulfide) and alkaline gases (e.g. ammonia). Hydrogen sulfide can be dried with anhydrous calcium chloride. Ammonia can be dried with solid sodium hydroxide, not with anhydrous calcium chloride.

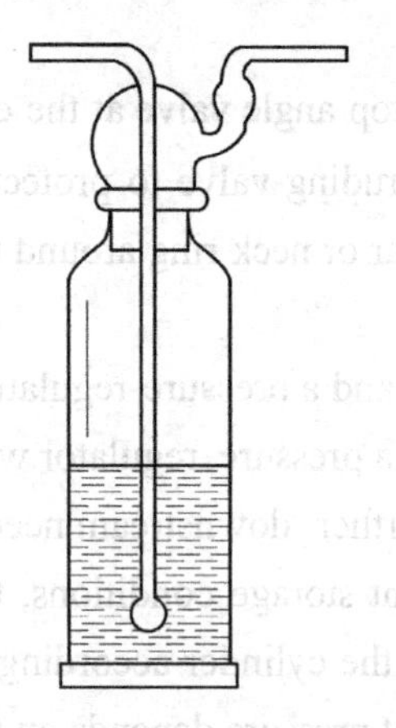

Figure 2-27 Gas bottle

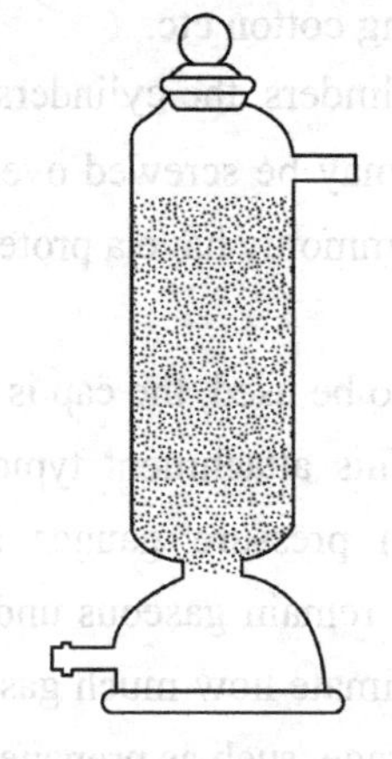

Figure 2-28 Drying tower

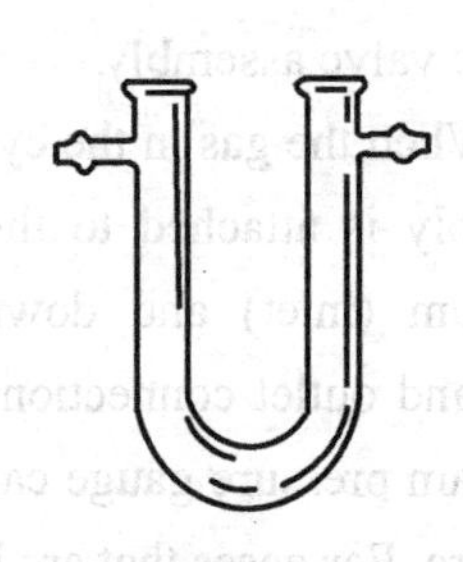

Figure 2-29 U tube

(3) Collecting gases

① Some gases, such as hydrogen and oxygen, are relatively insoluble in water and are collected by water displacement (Figure 2-30).

② Water-soluble gases less dense than air are collected by air displacement. The less dense gas pushes the more dense air down and out of the gas bottle (Figure 2-31).

③ Water-soluble gases more dense than air are also collected by air displacement except that, the more dense gas pushes the less dense air up and out of the gas bottle (Figure 2-32).

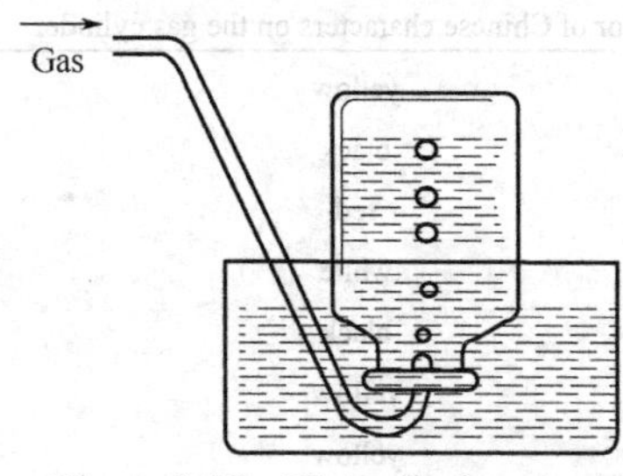

Figure 2-30 Water displacement

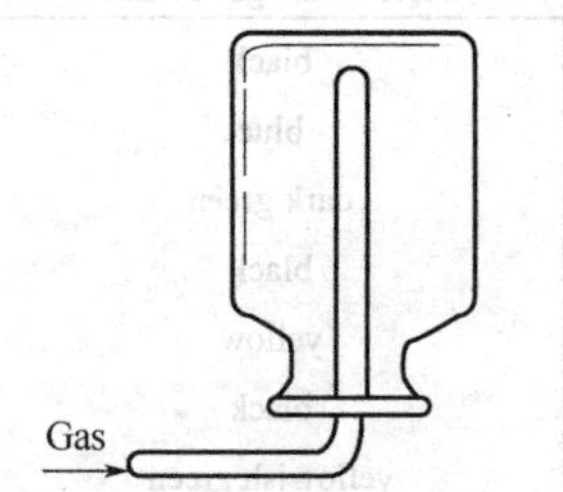

Figure 2-31 Air displacement (less dense than air)

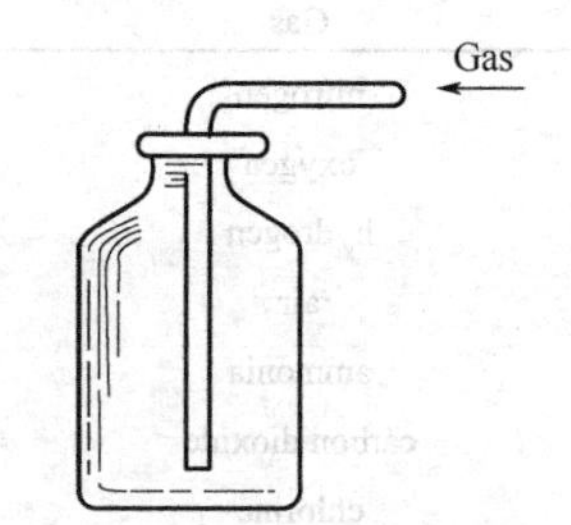

Figure 2-32 Air displacement (more dense than air)

2.7 蒸发（浓缩）、结晶（重结晶）

（1）蒸发（浓缩）

当物质的溶解度很大、溶液很稀时，要使溶质结晶析出，必须通过加热，使溶剂蒸发，溶液浓缩。蒸发到一定程度后冷却，即可析出晶体。

蒸发通常是在蒸发皿中进行（有时也可在烧杯中加热蒸发）。蒸发皿所盛的溶液不要超过容量的 2/3。如果物质对热较稳定，溶剂又不易燃烧，可将蒸发皿放在石棉网上用火直接加热，否则用水浴间接加热蒸发。

蒸发浓缩的程度，根据溶质溶解度的大小和结晶时对浓度的要求而定。但不得蒸至干涸。如果溶质的溶解度较小或其溶解度随温度变化较大，则蒸发到一定程度即可停止。如果溶质的溶解度较大，则应蒸发到溶液表面出现晶膜为止。如果结晶时希望得到较大的晶体，溶液就不能浓缩得太浓。

（2）结晶（重结晶）

结晶是指当溶质超过其溶解度时，晶体从溶液中析出的过程。通常采用蒸发减少溶剂，改变溶剂或改变温度等方法，使溶液变成过饱和状态而析出结晶。

析出晶体颗粒的大小与条件有关。如果溶液浓度较高，溶质的溶解度较小，快速冷却，并不时搅拌溶液、摩擦器壁，则析出的晶体就较小。如果溶液的浓度不高，投入一小粒晶体后，静置溶液，缓慢冷却（如放在温水浴上冷却），这样就可得到较大的结晶。

晶体颗粒的大小要适当。颗粒较大且均匀的晶体挟带母液较少，易于洗涤，有利于提高产品的纯度。晶体太小且大小不匀时，能形成稠厚的糊状物，挟带母液较多，不易洗净，影响产品的纯度。只得到几粒大晶体时，母液中剩余的溶质较多，损失较大。所以结晶颗粒大小适宜且较均匀有利于物质的纯度。

当第一次结晶物质的纯度不符合要求时，可在加热的情况下，用尽可能少的溶剂重新溶解成为饱和溶液，趁热滤去不溶性杂质，冷却后，溶液呈过饱和状态，析出溶质的晶体，而可溶性杂质含量少仍留在母液中。这种操作方法称为重结晶。重结晶后的产品纯度较高，产品产率低。

2.7 Evaporation (Concentration) and Crystallization (Recrystallization)

(1) Evaporation (concentration)

To obtain crystals from a diluted solution, evaporation or concentration needs to be performed.

A nonflammable liquid can be evaporated either with a direct flame or over a steam bath. If a direct flame is used, place the liquid in an evaporating dish centered on a wire/asbestos gauze. The evaporating dish should be less than two-thirds full of liquid.

Evaporate and concentrate the solution according to the solubility of the solute and the concentration of the solution. Do not evaporate the solution to dry. If the solubility of the solute is low and the solubility changes dramatically with the temperature, evaporate to a limited extent. If the solubility of the solute is high, evaporate till a film of crystal forms on the surface. Do not concentrate the solution too much if bigger crystals are needed.

(2) Crystallization (recrystallization)

Crystallization is based on the principles of solubility. Solutes tend to be more soluble in hot solvents than they are in cold solvents. If a saturated hot solution is allowed to cool, the solute is no longer soluble in the solvent and forms crystals of pure compound. Impurities are excluded from the growing crystals and the pure solid crystals can be separated from the dissolved impurities by filtration.

Recrystallization is used to purify solid crystals. Use an appropriate amount of solvent to dissolve crystals by heating. Filter the solution if there are impurities. Cool the solution and crystals will form from the hypersaturated solution. The impurities will be retained in the solution. Enhanced purity will be obtained after recrystallization with low yield.

2.8 结晶（沉淀）的分离和洗涤

2.8.1 倾析法

当结晶的颗粒较大或沉淀的密度较大，静置后能沉降至容器底部时，可用倾析法分离和洗涤。

按图 2-33 所示，把沉淀上部的溶液倾入另一容器内，然后往盛有沉淀的容器内加入少量洗涤液（如蒸馏水）充分搅拌后，沉降，倾去洗涤液。如此重复操作 3 遍以上，即可把沉淀洗净。

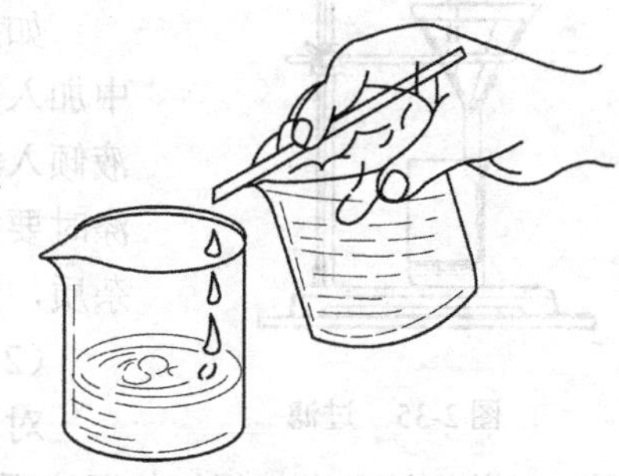

图 2-33 倾析法洗涤

2.8.2 过滤法

过滤是最常用的固液分离方法之一。当溶液和结晶（沉淀）的混合物通过过滤器（如滤纸）时，结晶（沉淀）就留在过滤器上，溶液则通过过滤器进入承接器中，过滤后所得的溶液为滤液。

溶液的温度、黏度、过滤时的压力、过滤器的孔隙大小和沉淀物的状态等都会影响过滤的速度。热溶液比冷溶液易过滤，溶液黏度愈大，过滤愈慢。减压过滤比常压过滤快。过滤器的孔隙要选择合适，太大沉淀会通过，太小则被沉淀堵塞，使过滤难于进行。呈胶状的沉淀物必须用加热的办法来破坏它，否则会透过滤纸。总之，要考虑上述各因素来选用不同的过滤方法。常用的过滤方法有下列 3 种。

（1）常压过滤

使用普通玻璃漏斗和滤纸进行过滤。玻璃漏斗锥体角度应为 60°，但有的稍有偏差，使用时应注意。

滤纸分定性滤纸和定量滤纸两种。按照孔隙大小，滤纸又可分为“快速”“中速”和“慢速”3 种。应该根据实际需要，选用不同规格的滤纸（注意，在使用滤纸前，应把手洗净、擦干）。

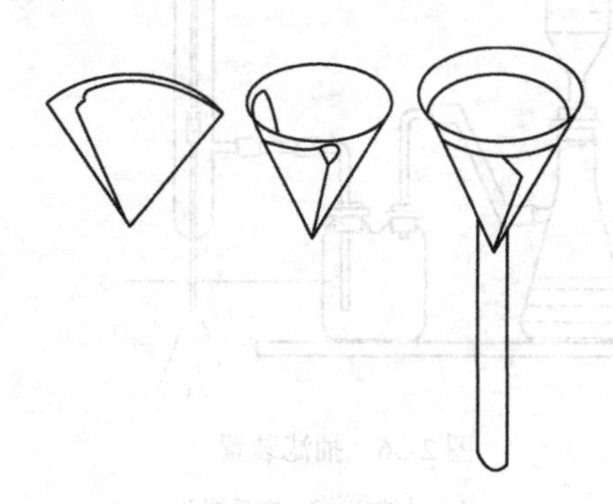

图 2-34 滤纸的折叠方法与安放

过滤时，先按图 2-34 所示，把圆形滤纸或方形滤纸折叠成 4 层（方形滤纸还要剪成扇形），然后在 3 层厚的外层滤纸折角处撕去一角，把滤纸展开成锥形，用食指把滤纸按在玻璃漏斗的内壁上，再用水润湿滤纸，并使它紧贴在玻璃漏斗的内壁上。滤纸的边缘应略低于漏斗的边缘。有的玻璃漏斗的锥角略大或略小于 60°，则在折叠滤纸时要做相应的调整。如果滤纸贴在漏斗壁后，两者之间有气泡，应该用手指轻压滤纸，把气泡赶掉。在这种情况下过滤时，漏斗颈内可充满滤液，滤液以本身的重力曳引漏斗内液体下漏，使过滤大为加速。

过滤时应注意：

① 漏斗应放在漏斗架上，漏斗颈紧靠在接收容器的内壁上，使滤液顺着容器壁流下，不致溅开来（图 2-35）。

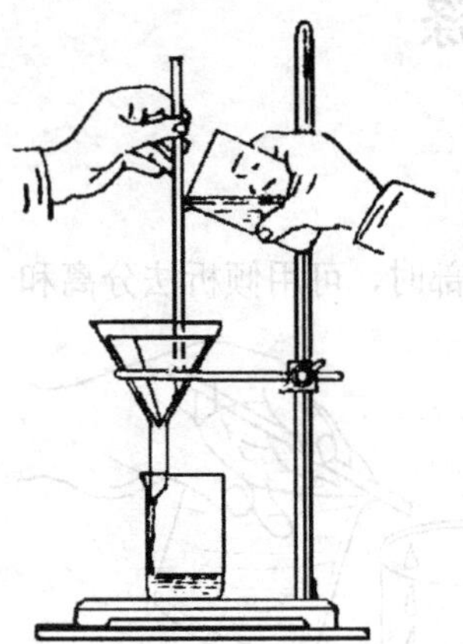

图 2-35　过滤

② 用倾析法过滤，先转移溶液，后转移沉淀，以免沉淀堵塞滤纸的孔隙而减慢过滤的速度。

③ 转移溶液时，应借助玻璃棒引流，把溶液滴在 3 层滤纸处。

④ 每次加入漏斗中的溶液不要超过滤纸高度的 2 / 3。

如果需要洗涤沉淀，则等溶液转移完毕后，往盛沉淀的容器中加入少量洗涤剂，充分搅拌并静置，待沉淀下沉后，把上层清液倾入漏斗内，如此重复操作两三遍，再转移沉淀到滤纸上。洗涤时要按照少量多次的原则，才能提高洗涤效率。检查滤液中的杂质，判断沉淀是否已经洗净。

（2）减压过滤（抽滤或吸滤）

对于大颗粒的沉淀或欲使沉淀较干燥，可采用抽滤以加快过滤。胶状沉淀和颗粒很细的沉淀不宜用减压过滤。减压过滤的装置如图 2-36 所示，它由布氏漏斗、抽滤瓶、安全瓶和水泵组成。

布氏漏斗（或称瓷孔漏斗）：上面有很多瓷孔，下端颈部装有橡皮塞，借以与抽滤瓶相连。橡皮塞塞入抽滤瓶的部分一般不得超过橡皮塞高度的 1 / 2。

抽滤瓶：用来承受被过滤下来的液体，并有支管与安全瓶短管相连。安全瓶的长管与水泵相连。

安全瓶：当因减压过滤完毕而关闭水龙头时，或者当水的流量突然加大后又变小时，都会由于抽滤瓶内的压力低于外界的压力而使自来水溢入抽滤瓶内（这一现象称为反吸），所以操作时要在抽滤瓶和水泵之间装上一安全瓶，作为缓冲。

水泵：减压用。在泵内有一窄口，当水急剧流至窄口时，水即把空气带出，而使与水泵相连的仪器减压。

抽滤操作与注意事项：

① 按图 2-36 所示，安装抽滤装置。

② 布氏漏斗的斜口应与抽滤瓶的支管相对。

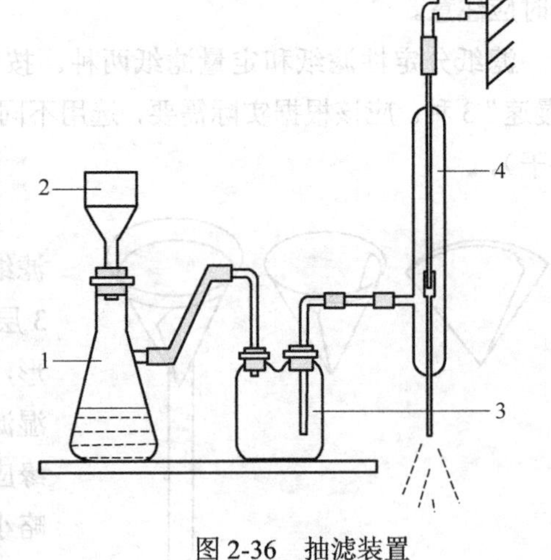

图 2-36　抽滤装置

1—抽滤瓶；2—布氏漏斗；3—安全瓶；4—水泵

③ 滤纸应略小于布氏漏斗的内径，以盖住瓷板上的小孔为宜。先用少量蒸馏水润湿滤纸，微启水泵使滤纸紧贴瓷板。

④ 用倾析法转移液体，加入的溶液不要超过漏斗容积的 2 / 3，待溶液漏完后，再将沉淀移入滤纸的中间。

⑤ 过滤时，抽滤瓶内的液面应低于支管的位置，否则滤液会被水泵抽出。当液面快升到支管时应拔出橡皮管，取下漏斗，从抽滤瓶的上口倒出滤液，此时应注意其支管必须向上。

⑥ 在布氏漏斗内洗涤沉淀时，应停止抽滤，待洗涤剂通过沉淀后再继续抽滤，但过滤速度不宜太快。

⑦ 在抽滤过程中，不得突然关闭水泵以防倒吸。若需暂时停止抽滤或过滤完毕，应先

拆下抽滤瓶上的橡皮管，再关闭水泵。

⑧ 取出沉淀时，将漏斗的颈口朝上，轻轻敲打漏斗边缘或在漏斗口用力一吹，沉淀即可脱离漏斗。也可用玻璃棒轻轻揭起滤纸边，以取下滤纸和沉淀。

瓶内溶液从抽滤瓶上口倒出，不得从瓶的支管口倒出。

⑨ 有些强酸性、强碱性或强氧化性的溶液过滤时不宜用滤纸。因为溶液会和滤纸作用而破坏滤纸。这时就需要在布氏漏斗上铺石棉纤维或用尼龙布来代替滤纸。先将石棉纤维在水中浸泡一段时间后，把它搅匀，倾入布氏漏斗内，再减压使石棉纤维紧贴在漏斗上。石棉纤维要铺得均匀些，不要太厚。但由于过滤后沉淀会夹杂有石棉纤维，所以此法适用于过滤后只要滤液的情况。

⑩ 如果过滤后既要滤液又要沉淀，可用玻璃砂漏斗，它不适用于过滤强碱性的溶液。过滤作用是通过熔接在漏斗中部具有微孔的烧结玻璃片上进行的。根据玻璃片的孔径大小，可分 1 号、2 号、3 号、4 号共 4 种规格，1 号孔径最大。可根据不同需要分别选用。过滤操作与减压过滤方法相同。

玻璃砂漏斗不宜用硫酸、盐酸或洗液去洗涤。可能生成不溶性的硫酸盐和氯化物把烧结玻璃片的微孔堵塞。通常用水洗去可溶物，然后在 6mol·L^{-1} 硝酸溶液中浸泡一段时间，再用水冲洗干净。

（3）热过滤

如果溶液中溶质在冷却后析出，而又不希望这些溶质在过滤过程中析出而留在滤纸上，这时就需要趁热过滤。为此，在过滤前把漏斗放在水浴上用水蒸气加热（抽滤时抽滤瓶也需加热），然后再进行过滤。

常压过滤时，可把玻璃漏斗放在铜质的热漏斗内（图 2-37）。

热漏斗内装有热水，以维持溶液的温度。另外选用的玻璃漏斗的颈部愈短愈好，不至于滤液在颈内停留过久、因散热降温而析出晶体，使颈部堵塞。

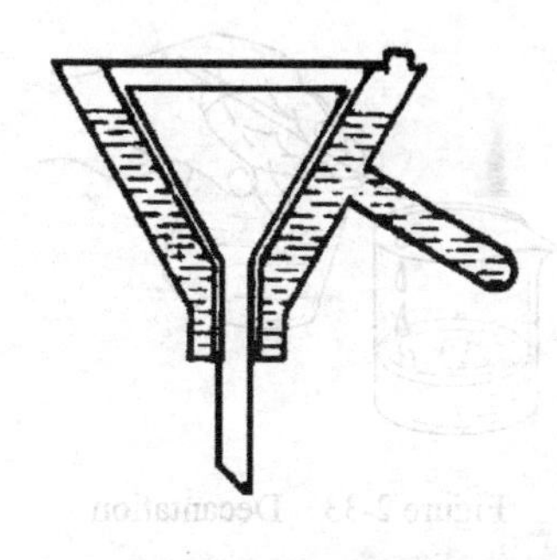

图 2-37　热过滤用漏斗

2.8.3　离心分离法

当被分离的沉淀量很少时，可用离心分离法。将盛有沉淀和溶液的离心试管放在离心机中高速旋转，沉淀受到离心力的作用，向离心管的底部移动，因此沉淀聚集在管底尖端，上面是澄清溶液。

实验室常用的离心仪器是电动离心机（图 2-38）。使用时，将装试样的离心试管放在离心机的套管中，管底垫一点棉花。为使离心机旋转保持平衡，几个离心试管要放在对称的位置上。如果只有一个试样，则在对称的位置也要放一支离心试管，管内装等质量的水。

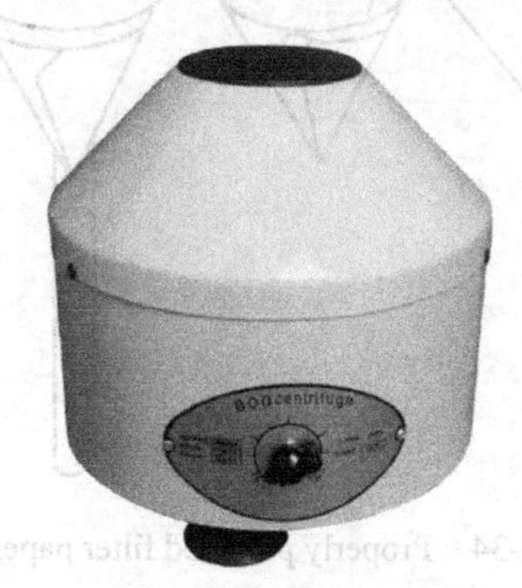

图 2-38　电动离心机

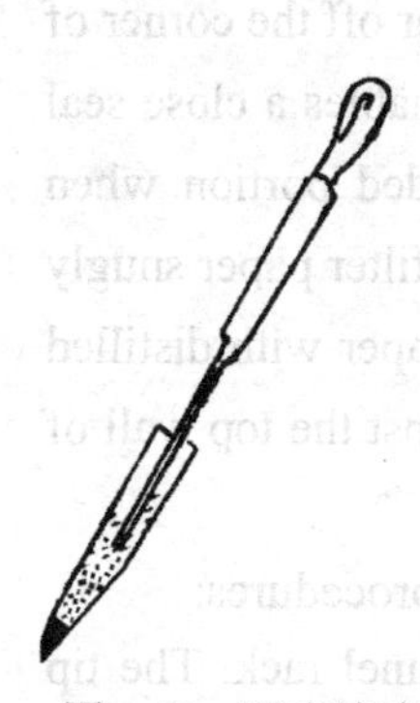

图 2-39　吸出溶液

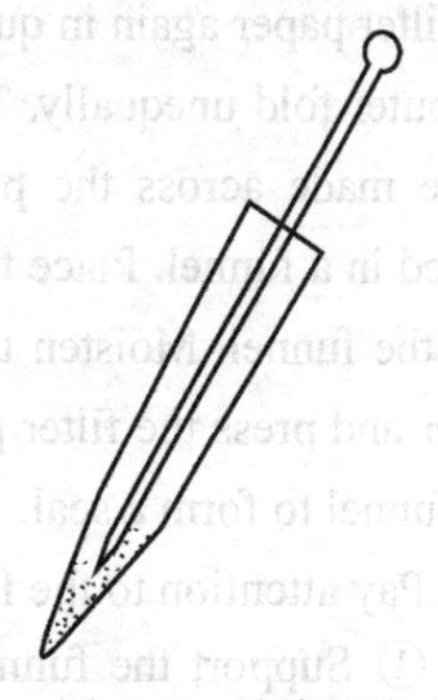

图 2-40　洗涤沉淀

电动离心机转动极快，要注意安全。放好离心试管后，应把盖盖上。开始时，把变速器放在最低挡，然后逐渐加速。停止时，任其自然停下，决不可以用手强制它停止转动。

离心沉降后，欲将沉淀和溶液分离，可用左手斜持离心管，右手拿毛细吸管，用手捏紧吸管上橡皮乳头以排除其中的空气，然后按图 2-39 所示，把毛细吸管伸入离心管，直到毛细吸管的末端恰好进入液面为止。这时慢慢减小手对橡皮乳头上的挤压力量，清液即进入毛细吸管。随着离心管中清液的减少，毛细吸管应逐渐下移，至全部清液吸入毛细吸管为止。

沉淀和溶液分离后，沉淀表面仍含有少量溶液，必须洗涤，往盛沉淀的离心管中加入适量的蒸馏水或其它洗涤剂，如图 2-40 所示，用玻璃棒充分搅拌后，进行离心沉降。用毛细吸管将上层清液吸出，再用上法操作 2～3 次即可。

2.8 Separating a Liquid or Solution From a Solid

2.8.1 Decantation

When large crystals are formed or the density of precipitates is heavy, the liquid can be decanted from the solid.

Allow the solid precipitate to settle to the bottom of the beaker. Transfer the liquid (supernatant) with the aid of a clean stirring rod (Figure 2-33). Do this slowly so as not to disturb the precipitate. Then flush the precipitate using a wash bottle. Stir with a stirring rod. Decant the supernatant. Repeat washing the precipitate for more than three times to make it clean.

Figure 2-33 Decantation

2.8.2 Filtration

Filtration is one of the common methods to separate a liquid from a solid. The precipitate or crystal will be on the filter paper after filtration. The filtrate, which is the liquid after filtration, will be in the receiving container.

(1) Gravity filtration

A funnel and filter paper will be used for gravity filtration. The filter paper must be properly prepared according to the following sequence(Figure 2-34). First fold the filter paper in half, and then fold the filter paper again in quarters. Tear off the corner of the outer fold unequally. The tear enables a close seal to be made across the paper's folded portion when placed in a funnel. Place the folded filter paper snugly into the funnel. Moisten the filter paper with distilled water and press the filter paper against the top wall of the funnel to form a seal.

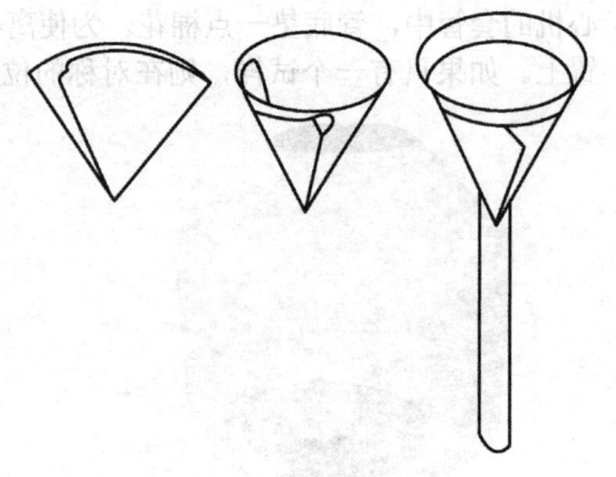

Figure 2-34 Properly prepared filter paper

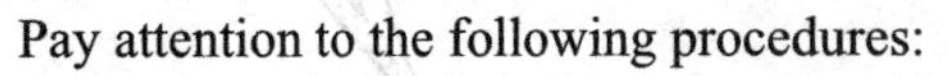

Pay attention to the following procedures:

① Support the funnel in a funnel rack. The tip of the funnel should touch the wall of the receiving beaker to reduce any splashing of the filtrate

(Figure 2-35).

② In order to maintain an adequate filtration rate, transfer the liquid into the funnel first, and then transfer the precipitate.

③ A stirring rod should be used, and the stirring rod should be against the folded portion of the filter paper.

④ Fill the funnel with the mixture until it is less than two-thirds full.

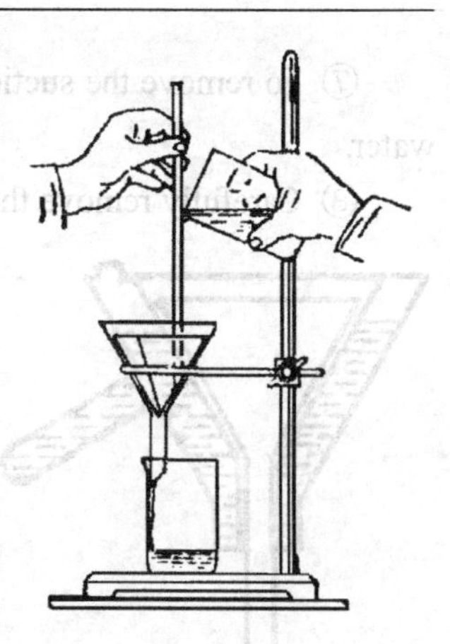

Figure 2-35 Gravity filtration

(2) Vacuum filtration

Vacuum filtration is faster than gravity filtration, because the solution and air are forced through the filter paper by the application of reduced pressure. A Büchner funnel set into a filter flask connected to a safety flask and a water aspirator is the apparatus normally used for vacuum filtration (Figure 2-36).

Büchner funnel: there are many holes on the bottom of the funnel. The neck of the funnel is surrounded by a rubber stopper which is connected to the filter flask.

Filter flask: it is to receive filtrate. The filter flask is attached to the safety flask.

Safety flask (trap): when the operation of the vacuum filtration is over and the water aspirator is turned off, the pressure in the filter flask is lower than the atmospheric pressure, and the reverse suction will occur. The water will enter the filter flask. So a safety flask should be used as a buffer between the filter flask and the water aspirator.

Water aspirator: the water aspirator or vacuum pump absorbs the air in the filter flask, so the pressure in the filter flask is reduced.

Pay attention to the following procedures when using vacuum filtration:

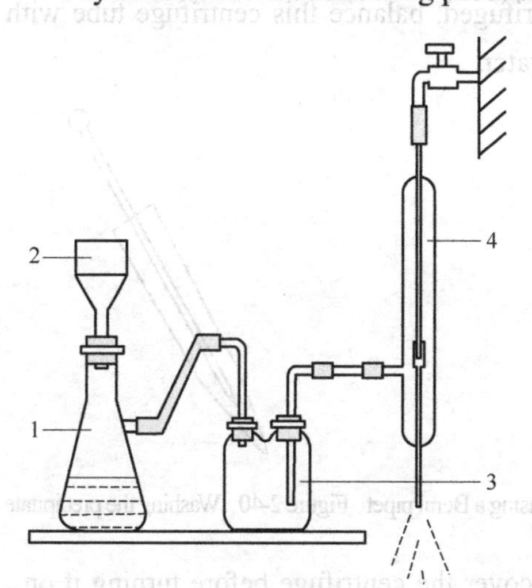

Figure 2-36 Vacuum filtration apparatus

1—Filter flask; 2—Büchner funnel;
3—Safety flask; 4—Water aspirator

① Set up the vacuum filtration apparatus properly (Figure 2-36).

② The bevel tip of the Büchner funnel should face the branch of the filter flask.

③ The diameter of the filter paper should be smaller than that of the Büchner funnel, and cover all the holes. Seal the disk of filter paper onto the bottom of the Büchner funnel by applying a light suction to the filter paper while adding a small amount of distilled water.

④ Fill the funnel with the mixture until it is less than two-thirds full.

⑤ The filtrate in the filter flask should be lower than the branch. Decant the filtrate from the top opening of the filter flask.

⑥ Wash the precipitate in the Büchner funnel with an appropriate solvent to help remove impurities.

⑦ To remove the suction, first disconnect the hose from the filter flask, and then turn off the water.

⑧ Carefully remove the filter paper and the precipitate from the Büchner funnel with the help of a stirring rod.

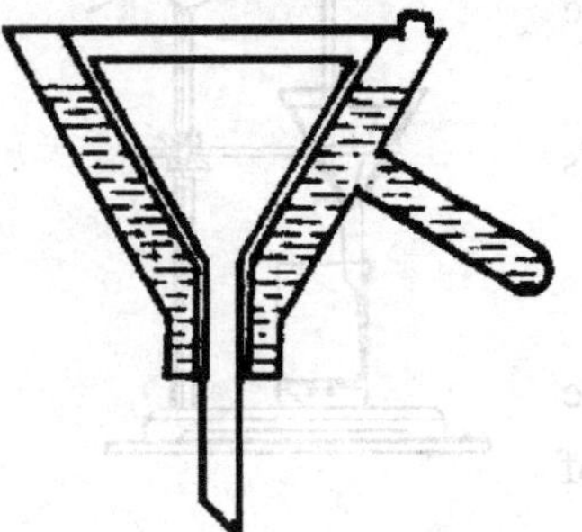

Figure 2-37 Hot gravity filtration

⑨ The filter paper can not be used for strong acids, strong alkali, or strong oxidants.

⑩ A glass sand funnel can be used in the vacuum filtration. The glass sand funnel can not be used for hydrofluoric acid, concentrated alkali, or activated carbon because they will block and contaminate the pores.

(3) Hot gravity filtration

Hot gravity filtration is used to separate insoluble impurities from a hot solution. Hot filtration requires keeping the apparatus warm (Figure 2-37). The neck of the glass funnel should be short to prevent the formation of precipitates.

2.8.3 Centrifugation

A centrifuge (Figure 2-38) is commonly used in the laboratory. A solid/liquid mixture in a centrifuge tube or a test tube is placed into a sleeve of the rotor of the centrifuge. By centrifugal force the solid is forced to the bottom of the tube. The supernatant, the clear liquid, is then easily decanted.

To keep the centrifuge balanced is very important. Always operate the centrifuge with an even number of centrifuge tubes containing equal volumes of liquid placed opposite one another in the centrifuge. If only one centrifuge tube needs to be centrifuged, balance this centrifuge tube with another centrifuge tube containing the same volume of water.

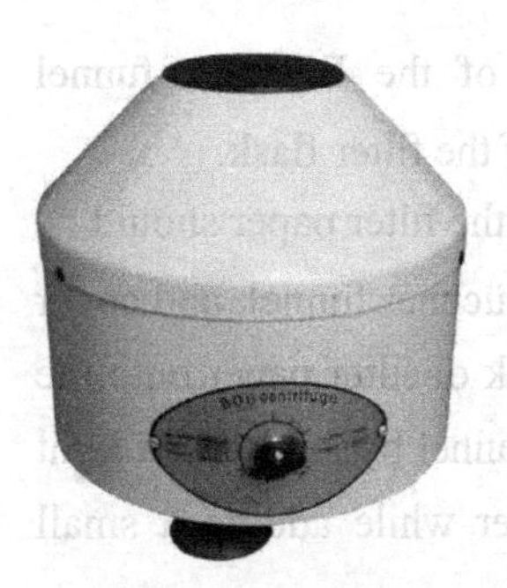

Figure 2-38 Centrifuge

Figure 2-39 Separating the supernatant by using a Beral pipet

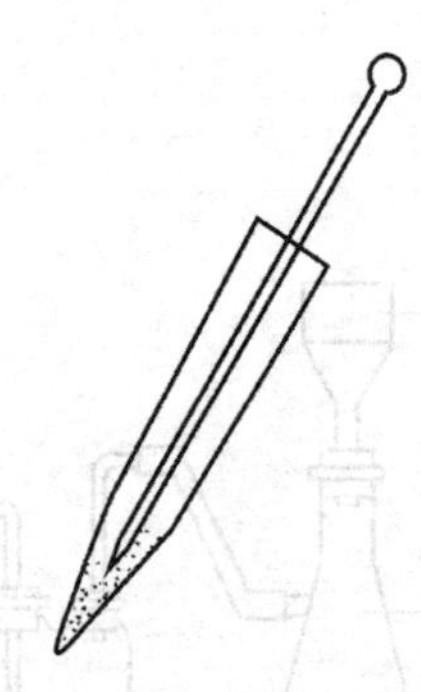

Figure 2-40 Washing the precipitate

A centrifuge spins at a very high velocity. Always cover the centrifuge before turning it on . When the centrifuge is turned off, let the rotor come to rest on its own. Never attempt to manually stop a centrifuge.

After centrifugation, the supernatant in the centrifuge tube can be separated by using a Beral pipet (Figure 2-39). Hold the centrifuge tube at an angle of about 30 degrees, removing the supernatant by slowly drawing it into the capillary pipet. The tip of the pipet should be just below the surface of the liquid. As the pressure on the bulb is slowly released, the supernatant will arise in the pipet, and lower the pipet into the centrifuge tube until all of the supernatant is removed.

To wash the precipitate, add an appropriate solvent to the centrifuge tube, and stir with a stirring rod (Figure 2-40). Decant the supernatant after the centrifugation. Repeat washing two or three times.

2.9 试纸的使用

（1）几种常用的试纸

试纸的作用是通过其颜色变化来测试溶液的性质，主要用于定性或定量的分析，其特点是简易、方便、快速，并具有一定的精确度。目前我国生产的各种用途的试纸已达几十种。在无机实验室中常用的有酚酞试纸、红色或蓝色石蕊试纸、pH 试纸、醋酸铅试纸以及碘化钾淀粉试纸等。

其中酚酞试纸在碱性溶液中变红，而红色石蕊试纸在碱性溶液中变蓝，蓝色石蕊试纸在酸性溶液中变红。

pH 试纸分广泛 pH 试纸和精密 pH 试纸。广泛 pH 试纸用以粗略地检验溶液的 pH 值，而精密 pH 试纸能较精细地检验溶液的 pH 值。

醋酸铅试纸用以检查痕量的硫化氢，作用时，试纸由白色变黑色。

碘化钾淀粉试纸用以检查氧化剂（特别是游离卤素以及亚硝酸和臭氧等）。作用时变蓝色（有时某些气体的氧化性很强且浓度很大时，可将 I_2 继续氧化为 IO_3^- 而使蓝色的试纸褪色）。

（2）试纸的使用方法及其注意事项

① 用石蕊试纸或酚酞试纸检验溶液的酸碱性时，可先将试纸剪成小块，放于干燥洁净的表面皿上，再用玻璃棒蘸取待测的溶液，滴在试纸上，在 0.5min 内观察试纸颜色的变化。不得将试纸投入溶液中进行试验。

② 使用 pH 试纸的方法与①同，差别在于当 pH 试纸显色后 0.5min 内，需将所显示的颜色与标准色标相比较，方能知其具体 pH 值。广泛 pH 试纸的色阶变化为“1”个 pH 值单位。精密 pH 试纸的色阶变化小于“1”个 pH 值单位。

③ 检查挥发性物质时，可将所用的试纸用蒸馏水润湿，然后悬空放在气体的出口处，观察试纸颜色的变化，检查挥发性物质的酸碱性，或确定某种物质的存在。

④ 试纸应密闭保存，不要用沾有酸性或碱性的湿手去取试纸，以免变色。

⑤ 注意节约，尽量将试纸剪成小块用。

2.9 Using Test Paper

(1) Test paper

The color change of a test paper can tell us the property of a solution. Test paper enables easy, convenient and quick qualitative analysis in the laboratory. The frequently used test papers include phenolphthalein test paper, red or blue litmus test paper, pH test paper, lead acetate test paper, and potassium iodide-starch test paper.

Basic solutions turn phenolphthalein test paper red. For litmus test paper, basic solutions turn red litmus blue; acidic solutions turn blue litmus red.

There are wide and short range pH test strips. Wide range pH test strips provide a distinct color for each pH unit from 1 to 14. Short range pH test strips provide more precise pH values than the wide range test strips.

H_2S turns white lead acetate test paper black.

Oxidants turn the potassium iodide-starch test paper blue.

(2) Methods to use the test paper

① To test the acidity or basicity of a solution with test paper, insert a stirring rod into the solution, withdraw it, and touch it to the test paper. Never place the test paper directly into the solution.

② The approximate pH of a solution can be obtained by comparing the color of the pH test paper to the standard colors.

③ To test a volatile substance, moisten the test paper with distilled water before use.

④ The test paper should be stored sealed and dry.

2.10 误差与数据处理

化学是一门实验的科学，要进行许多定量的测量，如常数的测定、物质组成的测定以及溶液浓度的测定等。这些测定，有的是直接进行的，有的则是根据实验数据推演计算得出的。在处理实验数据以及研究这些测定与计算结果的准确性时，都会遇到误差等有关问题。因此，树立正确的误差和有效数字的概念、掌握分析和处理实验数据的科学方法是十分必要的。下面仅就有关问题介绍一些基础知识。

2.10.1 准确度和误差

在定量的分析测定中，对于实验结果的准确度都有一定的要求。可是，绝对准确是没有的。在实际过程中，即使是技术很熟练的人，用最好的测定方法和仪器，测定出的数值与真实值之间总会产生一定的差值。这种差值越小，实验结果的准确度就越高；反之，则准确度就越低。所以，准确度是表示实验结果与真实值接近的程度。

准确度的高低常用误差来表示。误差有绝对误差和相对误差两种。

绝对误差（$\mathit{\Delta}$）是测量值与真实值之差。例如：称得某 3 个试样的质量分别为 2.3657g、1.5628g、0.2364g，而它们的真实质量分别为 2.3658g、1.5627g 和 0.2365g。则其绝对误差$\mathit{\Delta}$分别为：

$$\mathit{\Delta}_1 = 2.3657 - 2.3658 = -0.0001\ (\mathrm{g})$$

$$\mathit{\Delta}_2 = 1.5628 - 1.5627 = +0.0001\ (\mathrm{g})$$

$$\mathit{\Delta}_3 = 0.2364 - 0.2365 = -0.0001\ (\mathrm{g})$$

由此可见，当测定值小于真实值时，绝对误差为负值，表示测定结果偏低；反之，若测定值大于真实值时，则绝对误差为正值，表示测定结果偏高。其中 1、3 两个物体的真实质量相差近 10 倍，而绝对误差却一样，为−0.0001g。可见绝对误差并没有表明测量误差在真实值中所占的比重。为此引入相对误差的概念。

相对误差（δ）表示绝对误差与真实值之比，即误差在真实值中所占的百分率。故上述 1、3 两次测量的相对误差为：

$$\delta_1 = \frac{-0.0001}{2.3658} \times 100\% = -0.004\%$$

$$\delta_3 = \frac{-0.0001}{0.2365} \times 100\% = -0.04\%$$

由此看出，尽管测量的绝对误差相同，但由于被测量的量的大小不同，其相对误差也不同。被测量的量较大时，相对误差较小，测量的准确度较高。

2.10.2 精密度和偏差

在实际工作中，由于真实值不知道，通常是在同一条件下进行多次正确测量，求出其算术平均值代替真实值。或者以公认的手册上的数据作为真实值。

在多次测量中，如果每次测量结果的数值比较接近，就说明测定结果的精密度比较高。可见精密度是表示各次测量结果相互接近的程度。

精密度的高低用偏差表示。偏差愈小，精密度愈高。偏差有不同表示方法。

绝对偏差（d）为测得值与平均值之差：

$$d_i = x_i - \bar{x}$$

x_i指个别一次测得结果；d_i指它的绝对偏差；$\bar{x}$指平均值。

相对偏差

$$相对偏差 = \frac{d_i}{\bar{x}} \times 100\%$$

相对偏差指第 i 个测得结果中所包含的偏差占平均值的百分数。由于所需要的是整组数据对平均值的离散度，所以常用平均偏差（$\bar{d}$）或相对平均偏差来表示。

平均偏差（$\bar{d}$）

$$\bar{d} = \frac{|d_1| + |d_2| + \cdots + |d_n|}{n}$$

式中，$|d_1|$、$|d_2|\cdots|d_n|$分别代表各次测定结果的绝对偏差的绝对值；n 代表一共测量的次数。

相对平均偏差

$$相对平均偏差 = \frac{\bar{d}}{\bar{x}} \times 100\%$$

由于相对平均偏差能代表一组结果偏离平均值程度在所得结果中的影响程度，所以常用以表示精密度。

【例 2-1】

x	$\lvert d \rvert$
1.234	0.002
1.238	0.002
1.239	0.003
1.234	0.002
1.235	0.001
$\bar{x}$:1.236	$\bar{d}$:0.002

$$相对平均偏差 = \frac{0.002}{1.236} \times 100\% = 0.16\%$$

【例 2-2】

x	$\lvert d \rvert$
1.226	0.010
1.234	0.002
1.245	0.009
1.239	0.003
1.236	0
$\bar{x}$:1.236	$\bar{d}$:0.005

$$相对平均偏差 = \frac{0.005}{1.236} \times 100\% = 0.40\%$$

由以上两例可见，虽然两组结果的平均值相同，但后一组的精密度不如前一组高。

精密度和准确度的概念不同。测量的精密度高，其准确度不一定高；同样，测量的准确度高，其精密度不一定也高。

精密度高是保证准确度高的前提。精密度低的测量，其准确度是不可信的。因为数据波动大，说明在每次测量过程中，引起误差的因素在起变化。或者说，测量条件并非完全一致。

例如，甲、乙、丙、丁4位同学用同种方法测量同一样品中某物质的质量分数（设真实质量分数为50.37%），测量结果如下：

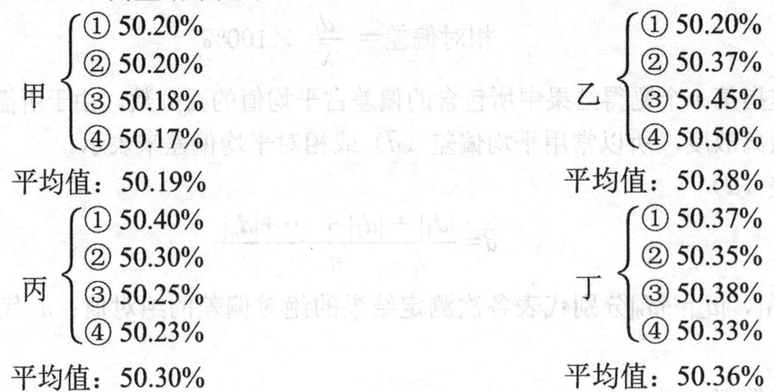

所得结果绘图如图2-41所示。

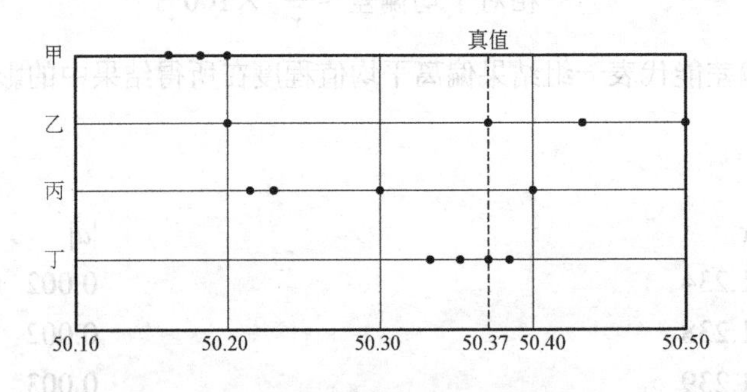

图2-41　四位同学测量同一物体的结果图

由图可见，甲测量的精密度很高，但平均值与真实值相差很大，说明准确度低；乙测量的准确度高，但不可信，因为精密度太差；丙测量的精密度和准确度都不高；只有丁的测量结果两全其美，即精密度和准确度都高。

必须说明的是，误差和偏差的含义也不同。误差以真实值为基准，偏差以平均值为基准。但严格说来，任何量的真实值都是未知的。通常所说的“真实值”，是采用各种方法进行多次平行测量所得到的相对可靠的平均值。因此，用这一平均值代替真实值计算误差，实际得到的也有一种偏差。故有时不严格区分误差和偏差。

2.10.3　误差产生的原因

误差产生的原因很多，一般分为系统误差和偶然误差两大类。

（1）系统误差

系统误差是测量过程中某种经常性的原因所引起的。这些误差对测量结果的影响比较固定，会在同一条件下的多次测量中反复显示出来，使测量结果系统偏高或偏低。例如，用未经校正的砝码称量时，由于砝码的值不准确，故在多次测量中使误差重复出现且误差大小不变。另外，在观测条件改变时，误差按某一确定的规律变化，这种观测误差也称为系统误差，它对测量结果的影响并不是固定的。例如，标准溶液会因其体积随温度变化而改变，而且有确定的规律，因此，对于浓度的变化，可以进行适当的校正。总之，对于系统误差，不论是固定或不固定的，如能找到其来源，就可设法加以控制或消除。

系统误差主要来源于以下几个方面。

① 方法误差　由测量方法引入的误差。例如，借助于从溶液中形成沉淀的方法确定某元素的含量时，由于沉淀物有一定的溶解度而造成的损失所带来的误差，就是方法误差。

② 仪器误差　仪器本身的制造精度有限而产生的误差。

③ 试剂误差　由试剂或试液（包括常用溶剂水）引入一些对测量有干扰的杂质所造成的误差。

④ 人员误差　由试验人员的生理缺陷和生理定势所引入的误差。例如，有的人对颜色的变化不甚敏感，在比色或光度的测量中引起的误差。测量者个人的习惯和偏向引起的误差。例如，有的人读数常偏高或偏低。

（2）偶然误差

偶然误差是由某些难于觉察的偶然原因所造成的误差。例如，外界条件（温度、湿度、振动和气压等）波动引起瞬间微小变化，或者实验仪器性能（灵敏度）的微小变化，以及实验者对各份试样处理的微小差别等。由于引起误差的原因是偶然性的，就单个误差值的出现情况而言是可变的，有时大、有时小，有时为正、有时为负，因此，既不可预料也没有确定的规律，它随具体的偶然因素的不同而不同。但是在相同条件下，对同一个量进行大量重复的测量而得到的一系列偶然误差来说，则显示出如下的统计分布规律：

① 绝对值相等的正误差和负误差出现的机会相等；

② 绝对值小的误差比绝对值大的误差出现的机会大；

③ 误差超出一定范围的机会很小。

根据上述特点可知，在同一测量条件下，随着测量次数的增加，偶然误差的算术平均值将趋近于零。也就是说，多次测量结果的算术平均值更接近于真实值。

从以上的分析可以看出，系统误差和偶然误差对测量结果所产生的影响是不同的。精密度是反映偶然误差大小的程度，而准确度是反映系统误差大小的程度。但在实际测量中，两类误差对测量结果往往都有影响。因此，采用一个新的概念来反映系统误差和偶然误差总和大小的程度，称为精确度。

此外，还可能由于实验人员粗枝大叶、不遵守操作规程，以致造成不应有的过失。例如，器皿未洗净、试液丢失、试剂误用、记录及运算错误等。如果已发现有错误的测量，应该取消该数据，不能参与总测量结果的计算。因此，对于初学者来说，从一开始就必须严格遵守操作规程，养成一丝不苟的良好科学实验的习惯。

2.10.4　提高实验精确度的方法

虽然误差在定量实验中总是客观存在的，但必须设法尽量减小。减小误差的方法有以下几方面。

① 选择合适的测量方法　各种测量方法的相对误差和灵敏度是不同的（灵敏度是在测量的条件下所能测得的最小值）。例如，测量元素含量时可用重量法和仪器测量法，前者相对误差比较小（一般为±0.2%），但灵敏度低；后者相对误差比较大（一般为±2%），但灵敏度高。因此，测量含量较高的元素可用重量法，而测量含量低的元素时，重量法的灵敏度一般达不到要求，应采用灵敏度高的仪器测量法。

② 减少测量误差　应根据不同的方法，不同的仪器和不同的要求确定待测量的最小实验量。在重量法中，主要操作是称量。由于一般电子天平称量的绝对误差为±0.0002g。如果使测量时的相对误差在0.1%以下，试样的质量就不能太小。因为：

$$相对误差=\frac{绝对误差}{试样质量}\times 100\%$$

$$试样质量=\frac{绝对误差}{相对误差}\times 100\%$$

$$=\frac{0.0002}{0.1\%}\times 100\%=0.2(g)$$

此时要求试样质量必须在0.2g以上。

不同的测量任务要求的准确度不同。如用仪器测量法，称取试样0.5g，试样的称量误差不大于0.5g×2%＝0.0l（g）即行，不必要强调称准到0.0001g。

③ 减小偶然误差　在系统误差很小的情况下，平行测量的次数越多，所得的平均值就越接近真实值。偶然误差对平均值的影响也就越小。通常要求平行测量2~4次以上，以获得较准确的测量结果。

④ 改进实验方法　为消除固定性的系统误差（其数值和符号在测量时总是保持恒定的），常采用交换抵消法，即进行两次测量，在这两次测量中将测量中的某些条件（例如被测物所处的位置等）相互交换，使产生系统误差的原因对两次测量的结果起相反的作用，从而使系统误差抵消。例如，通过交换被测物与砝码的位置，取两次称量结果的平均值作为被测物的质量，可以抵消等臂天平由于实际不等臂而引起的系统误差。

⑤ 对照实验　用标准件、标准样品做对照实验。校测后以修正值的方式加入测量值中，以消除系统误差。

也常用空白实验的方法来减小系统误差。即从试样测量结果中扣除空白值，就得到比较可靠的结果。

此外，对于仪器不准所引起的系统误差，可通过校准仪器来减小其影响。例如，砝码、滴定管和移液管等的校正。

2.10 Errors and Treatment of Data

Chemistry is an experimental science. Chemists do a lot of quantitative analyses such as determination of a constant, analysis of the composition of a compound, and determination of the molar concentration of a solution. Some data obtained from the quantitative analysis are directly from the experiment, but other data are deduced and calculated. A scientist must be concerned with the accuracy of data and the errors. It is important to understand the source of errors, the significant figures, and the treatment of data.

2.10.1 Accuracy and error

The accuracy of data is very important in the quantitative analysis. To obtain accurate data, we must use instruments that are carefully calibrated with known standards of measurement. But even with accurate measuring devices, there are differences between measured values and true values. The smaller the differences are in these values translates to higher accuracy of the data. The accuracy of the data means how closely the measured values lie to the true vales.

Errors are used to show the accuracy of the data. There are absolute errors and relative errors.

Absolute error (Δ) is the difference between measured value and true value. For example, the measured values of three samples are 2.3657g, 1.5628g, 0.2364g, respectively; the true values are 2.3658g, 1.5627g, and 0.2365g, respectively. The absolute errors (Δ) for three measurements are:

$$\Delta_1 = 2.3657 - 2.3658 = -0.0001 \text{ (g)}$$
$$\Delta_2 = 1.5628 - 1.5627 = +0.0001 \text{ (g)}$$
$$\Delta_3 = 0.2364 - 0.2365 = -0.0001 \text{ (g)}$$

The absolute error can be positive or negative. The mass of compound 1 is 10 times that of compound 3, but their measurements have the same absolute errors, which is –0.0001g. Therefore, the relative error (δ) is used to show the percentage of error in the true value. The relative error is the ratio of absolute error to the true value. The relative errors for the mass measurements of compound 1 and 3 are:

$$\delta_1 = \frac{-0.0001}{2.3658} \times 100\% = -0.004\%$$
$$\delta_3 = \frac{-0.0001}{0.2365} \times 100\% = -0.04\%$$

From the example, it is clear that the absolute errors are the same, but the relative errors are different because these two compounds have different values of mass. So a higher accuracy and lower relative error will occur when determining a compound with a bigger value of mass.

2.10.2 Precision and deviation

In the laboratory, we usually don't know the true values. So we repeat the measurements and calculate the average or mean value by dividing the sum of all the measured values by the total number of values. Scientists check the accuracy of their measurements by comparing their results with values that are well established and are considered accepted values. Many reference books are used to check a result against an accepted value.

When all measurements are close to one another, we say the data are of high precision. The precision of the data shows how closely the measured values lie to one another.

Deviation is used to show the precision of the data. The lower the deviation, the higher the precision.

The absolute deviation (d) is the difference between the average and each measured value (x_i means each measured value; d_i is the deviation for the measured value x_i; $\overline{x}$ means the average value.):

$$d_i = x_i - \overline{x}$$

The relative deviation (d%) is

$$d_i\% = \frac{d_i}{x} \times 100\%$$

$d_i\%$ is the percentage of deviation for the measured value x_i in the average value.

The average deviation ($\bar{d}$) is

$$\bar{d}=\frac{|d_1|+|d_2|+\cdots+|d_n|}{n}$$

In the above formula, $|d|$ is the absolute value of the deviation for each measured value; n is the number of measurements.

The relative average deviation ($\bar{d}\%$) is

$$\bar{d}\%=\frac{\bar{d}}{\bar{x}}\times 100\%$$

The relative average deviation is often used to show the precision of data.

Example 1

x	$\|d\|$
1.234	0.002
1.238	0.002
1.239	0.003
1.234	0.002
1.235	0.001
$\bar{x}$:1.236	$\bar{d}$:0.002

$$\bar{d}\%=\frac{0.002}{1.236}\times 100\%=0.16\%$$

Example 2

x	$\|d\|$
1.226	0.010
1.234	0.002
1.245	0.009
1.239	0.003
1.236	0
$\bar{x}$:1.236	$\bar{d}$:0.005

$$\bar{d}\%=\frac{0.005}{1.236}\times 100\%=0.40\%$$

The two examples show that the average values are the same, but Example 1 has higher precision than Example 2.

Precision and accuracy are two different terms. If the result is of high precision, it may not be of high accuracy, and vise versa.

High precision is the prerequisite of high accuracy. If the precision of an analysis is low, confidence may not be placed in the results. The reason is the data may vary greatly due to the variability and conditions under which measurements are obtained.

For example, four students analyzed the percent mass of a sample and their data were shown as follows. Suppose the true value is 50.37%.

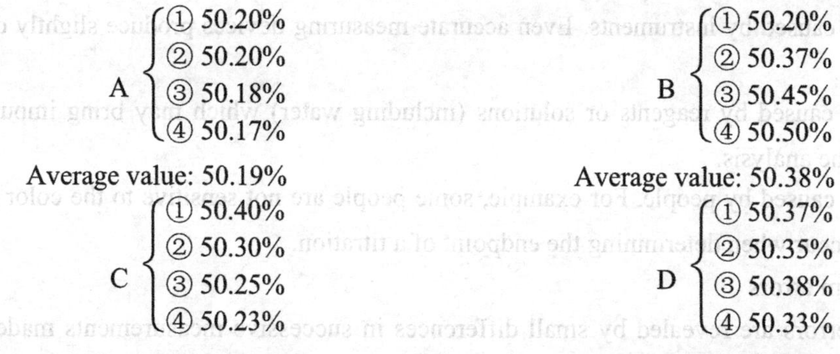

The above data are shown in Figure 2-41.

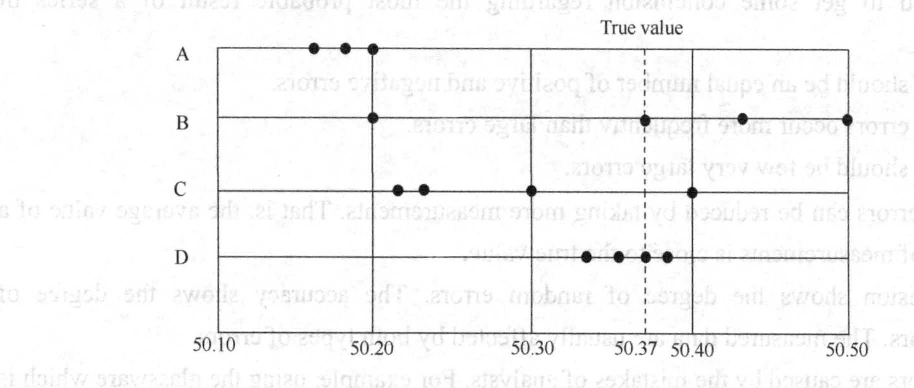

Figure 2-41 Data from four students

A is of high precision, but low accuracy; B is of low precision, so the result has no confidence; C is of low precision and low accuracy; D is of high precision and high accuracy.

The error and the deviation are two different terms. The error is based on the true value, but the deviation is based on the average value. The true value is usually obtained by the relatively reliable average value from a large number of measurements. So sometimes the error and deviation have no big differences.

2.10.3 Reasons for errors

There are many reasons for errors. The errors are divided into systematic errors and random errors.

(1) Systematic errors

Repeated measurements of the same quantity produce slightly different numerical values. These errors are called systematic errors. All experimental measurements are subject to systematic errors. For example, if a scale is not accurate, the error will occur whenever we use the scale.

The reasons for systematic errors are as follows:

① Errors caused by experimental methods. For example, when determining the concentration of a metal ion by precipitating it from solution, errors will occur because the precipitate will dissolve slightly.

② Errors caused by instruments. Even accurate measuring devices produce slightly different values.

③ Errors caused by reagents or solutions (including water) which may bring impurities to interfere with the analysis.

④ Errors caused by people. For example, some people are not sensitive to the color change, so errors will occur when determining the endpoint of a titration.

(2) Random errors

Random errors are revealed by small differences in successive measurements made by the same analyst under virtually identical conditions, and they can not be predicted or estimated. These accidental errors will follow a random distribution, so mathematical laws of probability can be applied to get some conclusion regarding the most probable result of a series of measurements.

① There should be an equal number of positive and negative errors.

② Small errors occur more frequently than large errors.

③ There should be few very large errors.

Random errors can be reduced by taking more measurements. That is, the average value of a large number of measurements is close to the true value.

The precision shows the degree of random errors. The accuracy shows the degree of systematic errors. The measured data are usually affected by both types of errors.

Some errors are caused by the mistakes of analysts. For example, using the glassware which is not clean, using the wrong reagent, misreading the data, etc. If you are aware of a mistake during an experiment, you should not use the data obtained from the experiment, and you should redo the experiment. In the laboratory, always perform your experiments strictly.

2.10.4 Methods to improve precision and accuracy

① Choose the proper experimental method. Different methods have different sensitivities and different relative errors.

② Reduce measuring errors. Measuring mass by using electronic balance is common in the laboratory. The absolute error for the electronic balance is generally ± 0.0002g. If we want the relative error to be less than 0.1%, the least mass of a sample to be weighed is:

$$\text{Relative error} = \frac{\text{absolute error}}{\text{mass of sample}} \times 100\%$$

$$\text{Mass of sample} = \frac{\text{absolute error}}{\text{relative error}} \times 100\%$$

$$= \frac{0.0002}{0.1\%} \times 100\% = 0.2(\text{g})$$

Therefore, the mass of the sample should be more than 0.2 g.

③ Reduce random errors. Random errors can be reduced by taking more measurements. That is, the average value of a large number of measurements is close to the true value. Usually we repeat doing experiments 2 to 4 times to obtain a more accurate result.

④ Improve the experimental methods.

⑤ Use control experiments. A standard sample is usually used as a control.

Blank experiments are often used to reduce systematic errors.

To obtain accurate data we must use instruments that are carefully calibrated with known standards of measurement. For example, a balance, a buret, and a pipet, should be periodically calibrated.

2.11 有效数字

在定量实验中，为了使结果可靠，要准确地测量，正确地读数、记录和计算数据。

（1）有效数字的概念

在实验中，使用的仪器所标出的刻度的精确程度总是有限的。例如，50mL 量筒，最小刻度为 1mL，在两刻度间可再估计一位，所以，实际测量读数能读至 0.1mL，如 34.5mL 等。若为 50mL 滴定管，最小刻度为 0.1mL，再估计一位，可读至 0.01mL，如 24.78mL 等。总之，在 34.5mL 与 24.78mL 这两个数字中，最后一位是估计出来的，是不准确的。通常把只保留最后一位不准确数字，而其余数字均为准确数字的这种数字称为有效数字。也就是说，有效数字实际上是能测出的数字。

由上述可知，有效数字与数学上的数有着不同的含义。数学上的数只表示大小。有效数字则不仅表示量的大小，而且反映了所用仪器的准确程度。例如，“取 NaCl 6.5g”，这不仅说明 NaCl 质量为 6.5g，而且表明用灵敏度 0.1g（或 0.5g）的天平称就可以了。若是“取 NaCl 6.5000g”，则表明一定要在灵敏度为 0.0001g 的电子天平上称。

这样的有效数字还表明了称量误差。对灵敏度 0.1g 的天平，称 6.5g NaCl，绝对误差为 0.1g，相对误差为：

$$\frac{0.1}{6.5}\times 100\%=2\%$$

对灵敏度为 0.0001g 的电子天平称 6.5000g NaCl，绝对误差为 0.0001g，相对误差为：

$$\frac{0.0001}{6.5000}\times 100\%=0.002\%$$

所以，记录测量数据时，不能随便乱写，不然便会夸大或缩小了准确度。例如，用电子天平称 6.5000g NaCl 后，若记成 6.50g，则相对误差就由$\frac{0.0001}{6.5000}\times 100\%=0.002\%$夸大到$\frac{0.01}{6.5000}\times 100\%=0.2\%$。

因此，有效数字的位数与仪器的准确度有关。高于或低于仪器的准确度都是不恰当的。

“0”的作用是值得注意的，“0”在数字中起的作用是不同的，有时是有效数字，有时则不是，这与“0”在数字中的位置有关。

下面一组数据的有效数字如下：

有效数字	0.0275	2.0065	6.5000	0.0030	54000
有效数字位数	三位	五位	五位	两位	不确定

“0”在数字前，仅表示小数点的位置，不属于有效数字。小数点的位置与测量所用的单位有关。如测得某物质的质量为 27.5g，若要换算为以千克为单位，则得 0.0275kg，单位的变换并不引起有效数字的增减，因 0.0275 仍为三位有效数字。

“0”在数字中间是有效数字。故 2.0065 为五位有效数字。

“0”在小数的数字后也是有效数字。如 6.5000 中 3 个 0 都是有效数字，而 0.0030 中“3”之前 3 个 0 不是有效数字，“3”后面的 0 为有效数字，所以 6.5000 为五位有效数字，而 0.0030 只有两位有效数字。以“0”结尾的正整数，有效数字的位数不确定。如 54000 可能是两位、三位、四位、五位，这种数应根据有效数字情况改写为指数形式。如为两位，则写成 5.4×

10^4；如为三位，则写成 5.40×10^4。

此外，常数不算作有效数字。在计算过程中，常数所取的位数不应小于参加运算的各数据中的最小位数。

总之，要能正确判别与书写有效数字。下面列出了一些数字，并指出了它们的有效数字位数：

6.5000	46009	五位有效数字
23.14	0.06010%	四位有效数字
0.0173	1.56×10^{-10}	三位有效数字
48	0.000050	两位有效数字
0.002	5×10^5	一位有效数字
54000	100	有效数字位数不定

（2）有效数字的运算规则

① 加法和减法　在计算几个数字相加或相减时，所得的和或差的有效数字位数，应以小数点后位数最小的数为准。

例如：将 2.0113，31.25 及 0.357 三个数相加时，见下式（可疑数以 ? 表示）：

$$\begin{array}{r} 2.011\mathit{3} \\ 31.2\mathit{5} \\ +0.35\mathit{7} \\ \hline 33.6\mathit{183} \rightarrow 33.62 \\ ??? \end{array}$$

可见，小数点后位数最少的数 31.25 中的 5 已是可疑，相加后使得和 33.6183 中的 1 也可疑。所以再多保留几位已无意义，也不符合有效数字只保留一位可疑数字的原则，这样相加后，按“四舍六入五取双”的规则处理，结果应是 33.62。

为了看清加减法应保留的位数，上例采用了先运算后取舍的方法。但是，在一般情况下，也可先取舍后运算。即：

$$\begin{array}{rcr} 2.0113 & \rightarrow & 2.01 \\ 31.25 & \rightarrow & 31.25 \\ +0.357 & \rightarrow & 0.36 \\ \hline & & 33.62 \end{array}$$

② 乘法和除法　在计算几个数相乘或相除时，其积或商的有效数字位数应以有效数字位数最少的数为准。如 1.312 与 23 相乘时：

$$\begin{array}{r} 1.312 \\ \times\ \mathit{23} \\ \hline \mathit{3936} \end{array}$$

$$\begin{array}{r} 2624 \\ \hline 30.176 \end{array}$$

显然，由于 23 中的 3 是可疑的，就使得积 30.176 中的 0 也可疑。所以保留两位即可，其余按：“四舍六入五取双”处理，结果就是 30。

同加减法一样，也可先取舍后运算。即：

$$\begin{array}{lcr} 1.312 & \rightarrow & 1.3 \\ 23 & \rightarrow & \times\ 23 \\ \hline & & 39 \\ & & 26\ \ \\ \hline & & 29.9 \rightarrow 30 \end{array}$$

另外，若第一位数等于或大于 8，则有效数字的总位数可多算一位。例如，9.15 虽然只有三位数字，但第一位数大于 8，所以运算时可看作四位。

③ 对数　进行对数运算时，对数值的有效数字只由尾数部分的位数决定。首数部分为 10 的幂数，不是有效数字。

例如：2345 为四位有效数字，其对数 lg2345＝3.3701，尾数部分仍保留四位，首数“3”不是有效数字。不能记成 lg2345＝3.370，这只是三位有效数字，就与原数 2345 的有效数字位数不一致了。

在化学中对数运算很多，如 pH 值的计算，若$[H^+]=4.9\times10^{-11}$，这是两位有效数字，所以 $pH=-lg[H^+]=10.31$，有效数字仍只有两位。反过来，由 pH＝10.31 计算$[H^+]$时，也只能记作$[H^+]=4.9\times10^{-11}$，而不能记成 4.989×10^{-11}。

④ 使用计算器进行运算时，虽然计算器能显示出许多位数，但在运算结果取数时，必须注意保留适当的有效数字的位数。这是因为测量结果数值计算的准确度不应该超过测量的准确度。

2.11 Significant Figures

Scientists use significant figures to clearly express the accuracy of measurements.

(1) The concept of significant figures

The number of significant figures used to express the measurement is determined by the specific instrument used to make the measurement. For example, the minimum mark of a 50mL graduated cylinder is 1mL. We can estimate a figure between the minimum marks. So the reading should be to 0.1mL, such as 34.5mL. The last figure “5” is uncertain. The minimum mark of 50mL buret is 0.1mL, plus one estimated figure, the reading is 24.78mL. The last figure “8” is uncertain. Therefore, the number of significant figures in a measurement equals the number of figures that are certain in the measurement plus one additional figure that expresses uncertainty.

The significant figures reflect the reliability of instruments and equipment used to make the measurements. For example, weigh NaCl 6.5g, which means the mass of NaCl should be measured on a balance that has an accuracy of 0.1g. If weigh NaCl 6.5000g, which means the mass of NaCl should be measured on a balance that has an accuracy of 0.0001g.

The significant figures also reflect the measuring errors. For a balance that has an accuracy of 0.1g, the absolute error is 0.1g. To weigh 6.5g NaCl, the relative error is

$$\frac{0.1}{6.5}\times 100\% = 2\%$$

For a balance that has an accuracy of 0.0001g, the absolute error is 0.0001g. To weigh 6.5000g NaCl, the relative error is

$$\frac{0.0001}{6.5000}\times 100\% = 0.002\%$$

Therefore, record the significant figures correctly. The significant figures clarify the accuracy of the measurements.

Zeros at the front end of a measurement are not significant. Zeros at the end of a measurement of data may or may not be significant.

For example:

Significant figures	0.0275	2.0065	6.5000	0.0030	54000
Number of significant figures	3	5	5	2	uncertain

Zeros at the front end of a measurement are not significant because they only represent the place of decimal point. Therefore, 0.0275 has 3 significant figures.

Zeros between figures are significant. So 2.0065 has 5 significant figures.

Zeros at the end of a decimal are significant. For example, 6.5000 has 5 significant figures, and 0.0030 has 2 significant figures. The number 54000 may have 2, 3, 4, or 5 significant figures. It has 2 significant figures if expressing the measurement as 5.4×10^4. It has 3 significant figures if expressing the measurement as 5.40×10^4. Expressing measurements in scientific notation often simplifies the recording of measurements with the correct number of significant figures.

For example:

6.5000	46009	5 significant figures
23.14	0.06010%	4 significant figures
0.0173	1.56×10^{-10}	3 significant figures
48	0.000050	2 significant figures
0.002	5×10^5	1 significant figure
54000	100	uncertain

(2) Mathematical operations of significant figures

① In addition and substraction, the answer has no more digits to the right of the decimal point than are found in the measurement with the fewest digits to the right of the decimal point.

For example (italic figure means the uncertain digit),

2. 011*3*

31. 2*5*

$$\begin{array}{r} +0.357 \\ \hline 33.6183 \rightarrow 33.62 \end{array}$$

For 31.25, the last figure "5" is uncertain. So for 33.6183, the figure "1" is also uncertain, and 33.6183 is rounded to 33.62.

Sometimes, round off the numbers first, and then calculate. For example,

$$\begin{array}{rcr} 2.0113 & \rightarrow & 2.01 \\ 31.25 & \rightarrow & 31.25 \\ +0.357 & \rightarrow & 0.36 \\ \hline & & 33.62 \end{array}$$

② In multiplication and division, the number of significant figures in the answer is the same as the number of significant figures in the measurement that contains the fewest significant figures. For example,

$$\begin{array}{r} 1.312 \\ \times \quad 23 \\ \hline 3936 \\ 2624 \\ \hline 30.176 \end{array}$$

The answer "30" is rounded to two significant figures because the measurement with the fewest significant figures "23" contains only two significant figures.

Sometimes, round off the numbers first, then calculate. For example,

$$\begin{array}{rcr} 1.312 & \rightarrow & 1.3 \\ 23 & \rightarrow & \times \quad 23 \\ \hline & & 39 \\ & & 26 \\ \hline & & 29.9 \rightarrow 30 \end{array}$$

③ In changing from logarithms to antilogarithms, and vice versa, the number being operated on and the logarithm mantissa have the same number of significant figures.

For example, the number 2345 has four significant figures, so lg 2345=3.3701. The mantissa has four significant figures also. It can not be written as lg 2345=3.370, because the mantissa has three significant figures.

For the calculation of pH, if $[H^+]=4.9\times10^{-11}$, it has two significant figures. Then $pH=-lg[H^+]=10.31$, it also has two significant figures. In reverse, to calculate the concentration of hydrogen ion with a $pH=10.31$, we should write $[H^+]=4.9\times10^{-11}$, instead of 4.989×10^{-11}.

④ When doing mathematical operations by using a calculator, be careful to record the proper significant figures. The accuracy of calculations should not be higher than the measurement.

实 验 部 分
Experiments

实验 1　滴定操作练习

【实验目的】

1．掌握酸碱滴定的原理。

2．掌握滴定操作。

3．学会正确判断滴定终点。

【实验提要】

如果酸（A）与碱（B）的中和反应为

$$aA + bB = cC + dH_2O$$

当反应达到化学计量点时，则 A 的物质的量 n_A 与 B 的物质的量 n_B 之比为：

$$\frac{n_A}{n_B} = \frac{a}{b} \text{ 或 } n_A = \frac{a}{b} n_B$$

又因为

$$n_A = c_A V_A \quad n_B = c_B V_B$$

故

$$c_A V_A = \frac{a}{b} c_B V_B$$

式中，c_A、c_B 分别为 A，B 的物质的量浓度，$mol \cdot L^{-1}$；V_A、V_B 分别为 A，B 的体积，L 或 mL。

由此可见，酸碱溶液通过滴定，确定它们中和反应达到终点时所需的体积比，即可确定它们的浓度比。如果其中一溶液的浓度已确定，则另一溶液的浓度可求出。

本实验以酚酞为指示剂，用 NaOH 溶液分别滴定 HCl 和 HAc，当指示剂由无色变为淡粉红色时，即表示已达到终点。

【仪器和药品】

碱式滴定管，25mL 移液管。

$0.1mol \cdot L^{-1}$ HCl 标准溶液（准确浓度已知），$0.1mol \cdot L^{-1}$ NaOH 溶液（浓度待标定），$0.1mol \cdot L^{-1}$ HAc 溶液（浓度待标定），酚酞溶液（质量分数为 1%）。

【实验内容】

1．NaOH 溶液浓度的标定

用 $0.1mol \cdot L^{-1}$ NaOH 操作液荡洗已洗净的碱式滴定管，每次 10mL 左右，荡洗液从滴定管两端分别流出弃去，共洗 3 次。然后再装满滴定管，赶出滴定管下端的气泡。调节滴定管内溶液的弯月面在“0”刻度以下。静置 1min，准确读数，并记录在报告本上。

将已洗净的 25mL 移液管用 $0.10mol \cdot L^{-1}$ HCl 标准溶液荡洗 3 次后（每次用 5～6mL 溶液），准确移取 25.00mL 的 HCl 标准溶液于 250mL 锥形瓶中。加酚酞指示剂 2 滴，此时溶液应无色。用已备好的 $0.1mol \cdot L^{-1}$ NaOH 操作液滴定酸液。近终点时，用洗瓶吹洗锥形瓶内壁，再继续滴定，直至溶液在加入半滴 NaOH 后，变为明显的淡粉红色，在 30s 内不褪色，此时即为终点。准确读取滴定管中 NaOH 的体积。终读数和初读数之差，即为与 HCl 中和所消耗的

NaOH 体积。

重新把碱式滴定管装满溶液（每次滴定最好用滴定管的相同部分），重新移取 25mL HCl，按上法再滴定两次。计算 NaOH 的浓度。3 次测定结果的相对平均偏差不应大于 0.2%。

2．HAc 溶液浓度的测定

用上面已测知浓度的 NaOH 溶液，按上法测定 HAc 溶液的浓度 3 次。3 次测定结果的相对平均偏差也不应大于 0.2%。

3．数据记录和结果处理

将 NaOH 溶液浓度的标定和 HAc 溶液浓度的测定的有关数据分别填入表 2-3 和表 2-4。

表 2-3　NaOH 溶液浓度的标定

数据记录与计算		1	2	3
HCl 标准溶液的浓度/$mol \cdot L^{-1}$				
HCl 标准溶液的净用量/mL		25.00	25.00	25.00
NaOH 溶液	终读数/mL			
	初读数/mL			
	净用量/mL			
NaOH 溶液的浓度/$mol \cdot L^{-1}$				
平均值/$mol \cdot L^{-1}$				
相对平均偏差				

表 2-4　HAc 溶液浓度的测定

数据记录与计算		1	2	3
NaOH 溶液的浓度 / $mol \cdot L^{-1}$				
HAc 溶液	终读数 / mL			
	初读数 / mL			
	净用量 / mL			
HAc 溶液净用量 / mL		25.00	25.00	25.00
HAc 溶液浓度 / $mol \cdot L^{-1}$				
平均值 / $mol \cdot L^{-1}$				
相对平均偏差				

【问题与讨论】

1．分别用 NaOH 滴定 HCl 和 HAc，当达化学计量点时，溶液 pH 值是否相同？

2．滴定管和移液管均需用待装溶液荡洗 3 次的原因何在？滴定用的锥形瓶也要用待装溶液荡洗吗？

3．如果取 10.00mL HAc 溶液，用 NaOH 溶液滴定其浓度，所得的结果与取 25.00mL HAc 溶液的相比，哪一个误差大？

4．以下情况对滴定结果有何影响：

① 滴定管中留有气泡；

② 近终点时，没有用蒸馏水冲洗锥形瓶的内壁；

③ 滴定完后，有液滴悬挂在滴定管的尖端处；

④ 滴定过程中，有一些滴定液自滴定管的活塞处渗漏出来。

Experiment 1 Titrimetric Analysis

【Objectives】

1. To understand the principles of acid-base titration.

2. To learn the titration procedure.

3. To properly determine the endpoint.

【Introduction】

The neutralization reaction of an acid (A) and a base (B) is

$$aA+bB=cC+dH_2O$$

When the reaction reaches stoichiometric point, the ratio of the number of moles between A and B is

$$\frac{n_A}{n_B}=\frac{a}{b} \text{ or } n_A=\frac{a}{b}n_B$$

The number of moles of a solution is calculated using the following equation

$$n_A=c_AV_A \quad n_B=c_BV_B$$

Therefore

$$c_AV_A=\frac{a}{b}c_BV_B$$

In the above equations, c_A and c_B represent the molar concentration of solution A and solution B; V_A and V_B (L or mL) represent the volume of solution A and solution B respectively.

From the stoichiometry of the reaction and the measured volumes of both acid and base in a titration procedure, the molar concentration of the acid can be calculated if the molar concentration of the base is already known.

In this experiment, NaOH solution is used to determine the molar concentration of HCl solution or HAc solution. The indicator is phenolphthalein. The endpoint is reached when the color of the indicator changes from colorless to pink.

【Apparatus and Chemicals】

Base buret, 25mL pipet.

$0.1mol \cdot L^{-1}$ HCl standard solution, $0.1mol \cdot L^{-1}$ NaOH solution, $0.1mol \cdot L^{-1}$ HAc solution, phenolphthalein solution (1%, wt/wt).

【Experimental procedure】

1. Standardization of a sodium hydroxide solution

Prepare a clean buret. Rinse the buret with three 10mL portions of the NaOH solution. Drain each rinse through the buret tip. Then, fill the buret with the NaOH solution. Remove the air bubbles in the tip. Adjust the level to below the zero mark. Read the volume by viewing the bottom of the meniscus after 1 min. Record this initial volume.

Rinse a clean 25mL pipet with three 5mL portions of the HCl standard solution. Accurately pipet 25.00mL of HCl standard solution into an Erlenmeyer flask. Add two drops of phenolphthalein. The solution should be colorless. Add the NaOH solution from the buret to the HCl standard solution. As the endpoint nears, occasionally rinse the wall of the flask with distilled water from your wash bottle. Continue adding the NaOH titrant until the endpoint is reached. The endpoint should be within one-half drop of a slight pink color, and the color should persist for 30 seconds. Accurately read the final volume of the NaOH solution in the buret.

Refill the buret and repeat the titration at least twice. The relative average deviation of the molar concentrations of NaOH from the three analyses should be less than 0.2%.

2. Determination of the molar concentration of HAc solution

The NaOH solution, now a solution with accurately known molar concentration, is used to determine the molar concentration of the HAc solution. Do the titration three times. The relative average deviation of the molar concentrations of HAc from the three analyses should be less than 0.2%.

3. Data and results

Fill in the Table 2-3 and Table 2-4 with proper data.

Table 2-3 The standardization of NaOH solution

Data		1	2	3
Molar concentration of HCl standard solution/mol·L^{-1}				
Volume of HCl standard solution /mL		25.00	25.00	25.00
NaOH solution	Final buret reading/mL			
	Initial buret reading/mL			
	Volume dispensed/mL			
Molar concentration of NaOH/mol·L^{-1}				
Average molar concentration of NaOH /mol·L^{-1}				
Relative average deviation				

Table 2-4 The molar concentration of HAc solution

Data		1	2	3
Molar concentration of NaOH / mol·L^{-1}				
HAc solution	Final buret reading/mL			
	Initial buret reading/mL			
	Volume dispensed/mL			
Volume of HAc / mL		25.00	25.00	25.00
Molar concentration of HAc / mol·L^{-1}				
Average molar concentration of HAc / mol·L^{-1}				
Relative average deviation				

【Questions and Discussion】

1. In this experiment, NaOH is used to titrate HCl and HAc. For these two acid-base titrations, do they have the same pH at the stoichiometric point?

2. Why do the buret and the pipet need to be rinsed three times with the solution that is going to be measured? Do you need to rinse the Erlenmeyer flask with the solution that is going to be

measured?

3. To determine the molar concentration of a HAc solution, we can pipet 10.00 mL HAc or 25.00mL HAc into the Erlenmeyer flask. Which has the possibility of a bigger error?

4. What effect will the following operation cause?

① There are air bubbles in the tip of the buret.

② As the endpoint nears, we don't rinse the wall of the flask with distilled water.

③ When the titration is over, there is a drop of solution hanging at the tip of the buret.

④ During the titration, there is solution leaking from the stopcock of the buret.

实验 2　酸 碱 滴 定

【实验目的】

1．巩固电子天平的称量操作，练习准确配制并转移样品溶液。

2．学习滴定操作，掌握中和滴定原理并测定酸碱溶液的物质的量浓度。

【实验提要】

酸碱滴定是利用酸碱的中和反应，测定酸或碱的物质的量浓度的一种定量分析方法。滴定反应的终点常借助于指示剂的颜色变化来确定。指示剂本身是一种弱酸或弱碱，它们在不同的 pH 值范围显示出不同的颜色。例如：酚酞变色范围在 pH 为 8.0～10.0，pH 在 8.0 以下为无色、10.0 以上显红色、8.0～10.0 之间显浅红色。又如甲基红，在 pH<4.4 时显红色、6.2 以上显黄色、4.4～6.2 之间显橙色（见附录 4）。

草酸是常用于标定碱的浓度的基准物质之一，它是固体弱酸，能溶于水，且 $K_a>10^{-7}$ ($K_{a_1}=5.9\times10^{-2}$，$K_{a_2}=6.4\times10^{-5}$)。故可在水溶液中用 NaOH 溶液进行滴定，反应产物为强碱弱酸盐，终点在弱碱性范围。所以选用酚酞为指示剂，滴定至溶液呈浅红色即为终点。反应方程式为：

$$H_2C_2O_4(aq)+2NaOH(aq)=\!=\!=Na_2C_2O_4(aq)+2H_2O(l) \tag{1}$$

根据等物质的量反应规则：

$$\begin{aligned} n(H_2C_2O_4) &= n(2NaOH) \\ &= \frac{1}{2}n(NaOH) \\ &= \frac{1}{2}c(NaOH)V(NaOH) \end{aligned}$$

因

$$\begin{aligned} n(H_2C_2O_4) &= n(H_2C_2O_4\cdot 2H_2O) \\ &= \frac{m(H_2C_2O_4\cdot 2H_2O)}{M(H_2C_2O_4\cdot 2H_2O)} \end{aligned}$$

由

$$c(NaOH)=\frac{2m(H_2C_2O_4\cdot 2H_2O)}{M(H_2C_2O_4\cdot 2H_2O)V(NaOH)}$$

准确称取一定量的分析纯草酸晶体 $H_2C_2O_4\cdot 2H_2O$，溶于适量的蒸馏水中，用 NaOH 溶液滴定，记录反应到达终点时所消耗的 NaOH 体积，便可以计算 NaOH 溶液的浓度 c(NaOH)。

【仪器和药品】

碱式滴定管（50mL），锥形瓶（250mL），移液管（25mL），容量瓶（100mL），电子天平，称量瓶。

草酸($H_2C_2O_4\cdot 2H_2O$) (AR)，NaOH、HCl 溶液（均为 $0.1mol\cdot L^{-1}$）。

【实验内容】

1．样品称量及溶液配制

用电子天平准确称取草酸晶体（0.6300～0.6500g）1 份，置于洁净的小烧杯中，加入适量的蒸馏水溶解，溶液转入 100mL 容量瓶中，用少量蒸馏水冲洗烧杯 3 次，冲洗液转入容量瓶，加水稀释至刻度，摇匀备用。

2．滴定练习

量取 10.00 mL HCl 溶液于锥形瓶中，加入 2 滴酚酞指示剂，用 NaOH 溶液滴定，观察终点变化，并练习终点的确定（滴定操作见无机化学实验操作有关内容）。

本部分练习可反复多次，直至能够较熟练地掌握滴定操作和正确判断终点，才开始做下面实验。

3．c (NaOH)的标定

用干净的移液管准确移取 20.00mL 草酸样品溶液于锥形瓶中，加入 2～3 滴酚酞指示剂，用 NaOH 溶液进行滴定，记录所消耗的 NaOH 溶液体积。

另吸取 20.00mL 草酸样品溶液进行滴定，两次滴定所用 NaOH 溶液的体积相差不超过±0.05mL。记录数据，计算 c (NaOH)。

【选做部分】

另准确称取两份草酸样品（每份的质量约 0.1500g），分别放入锥形瓶中，用适量蒸馏水溶解，加入指示剂，按上述方法进行滴定，记录数据，计算 c (NaOH)。

【问题与讨论】

1．滴定管装入溶液后没有将下端尖管内的气泡赶尽就开始滴定，对实验结果有何影响？

2．本实验中，当滴定至锥形瓶中溶液出现粉红色，振摇 0.5min 内粉红色不消失，但约 1min 后消失，这种情况可否认为是达到终点？

3．下列情况对实验结果各有什么影响？

① 滴定管的下端留有液滴；

② 溅在锥形瓶壁上的液滴没有用蒸馏水冲下。

Experiment 2　Acid-Base Titration

【Objectives】

1. To be familiar with the operation of electronic balance; to practice preparing and transferring solutions accurately.

2．To learn the titration operation, to understand the principle of acid-base titration, and to determine the molar concentration of a solution.

【Introduction】

Acid-base titration is a quantitative analysis to determine the molar concentration of an acid or a base by their neutralization reaction. The endpoint of the titration is detected by the color change of the indicator. The indicator itself is a weak acid or a weak base, and shows different colors at different pH values. For example, the pH range of color change for phenolphthalein is 8.0~10.0. Phenolphthalein is colorless when the pH is below 8.0, red when the pH is greater than 10.0, and pink when the pH is between 8.0 and 10.0. Methyl red is red when the pH is below 4.4, yellow when the pH is greater than 6.2, and orange when the pH is between 4.4 and 6.2 (See Appendix 4).

Oxalic acid is one of the primary standards used to find the molar concentration of a NaOH solution. It is a weak solid acid, and water-soluble with $K_a>10^{-7}$ ($K_{a_1}=5.9\times10^{-2}$，$K_{a_2}=6.4\times10^{-5}$). Therefore, the NaOH solution can be standardized with the oxalic acid. At the endpoint of the titration, the solution is basic. Therefore, phenolphthalein can be used as the indicator with the color change from colorless to pink at the endpoint. The reaction of oxalic acid with NaOH is

$$H_2C_2O_4\,(aq)+2NaOH\,(aq)=\!=\!=Na_2C_2O_4\,(aq)+2H_2O\,(l)$$

According to the reaction

$$n(H_2C_2O_4)=n(2NaOH)$$
$$=\frac{1}{2}n\,(NaOH)$$
$$=\frac{1}{2}c(NaOH)V(NaOH)$$

The number of moles of oxalic acid used for the analysis is

$$n(H_2C_2O_4)=n(H_2C_2O_4\cdot 2H_2O)$$
$$=\frac{m(H_2C_2O_4\cdot 2H_2O)}{M(H_2C_2O_4\cdot 2H_2O)}$$

The molar concentration of NaOH is calculated as

$$c(NaOH)=\frac{2m(H_2C_2O_4\cdot 2H_2O)}{M(H_2C_2O_4\cdot 2H_2O)V(NaOH)}$$

Accurately weigh a certain amount of $H_2C_2O_4\cdot 2H_2O$ crystal (AR). Dissolve the crystal in distilled water. Titrate the above solution with NaOH solution. Therefore, the molar concentration of NaOH can be calculated according to its volume used in the titration.

【Apparatus and Chemicals】

Base buret (50mL), Erlenmeyer flask (250mL), pipet (25mL), volumetric flask (100mL), electronic balance, weighing bottle.

$H_2C_2O_4\cdot 2H_2O$ (AR), NaOH solution（$0.1mol\cdot L^{-1}$）, HCl solution（$0.1mol\cdot L^{-1}$）.

【Experimental Procedure】

1. Preparation of oxalic acid

Accurately weigh a less than 0.6500g sample of oxalic acid crystal. Dissolve the crystal in a clean beaker with the appropriate amount of distilled water. Transfer the solution to a 100 mL volumetric flask, and rinse the beaker three times with distilled water. Transfer all the above water into the flask. Add water until the calibrated mark etched on the flask is reached. While securely holding the stopper, invert the flask slowly a few times to make the solution homogeneous.

2. Titration practice

Pipet 10.00mL of HCl solution into a clean Erlenmeyer flask. Add 2 drops of phenolphthalein. Titrate the HCl solution with NaOH solution. Practice the titration operation and observe the endpoint.

3. Standardization of NaOH solution

Pipet 20.00mL of oxalic acid solution to a clean Erlenmeyer flask. Add 2 to 3 drops of phenolphthalein. Titrate the oxalic acid solution with NaOH. Record the volume of the NaOH solution dispensed.

Repeat the titration. The difference in volumes of the NaOH solution dispensed for the two analyses should be less than ±0.05mL. Record your data, and calculate the molar concentration of NaOH.

【Questions and Discussion】

1. What effect will be caused if the air bubbles in the tip of the buret are not removed before titration?

2. In this experiment, when pink appears in the Erlenmeyer flask, and the color persists for 1 minute instead of 30 seconds, is this the endpoint?

3. What effect will the following operation cause?

① When the titration is over, there is a drop of solution hanging at the tip of the buret.

② A few drops of the solution on the wall of the flask are not rinsed with distilled water.

实验 3　EDTA 标准溶液的配制与标定

【实验目的】

1．掌握 EDTA 标准溶液的配制和标定方法。

2．学会判断配位滴定的终点。

3．了解缓冲溶液的应用。

【实验提要】

配位滴定中通常使用的配位剂是乙二胺四乙酸的二钠盐（$Na_2H_2Y \cdot 2H_2O$），其水溶液的 pH 值为 4.5 左右，若 pH 值偏低，应该用 NaOH 溶液中和到 pH 值为 5 左右，以免溶液配制后有乙二胺四乙酸析出。

EDTA 能与大多数金属离子形成 1∶1 的稳定配合物，因此可以用含有这些金属离子的基准物，在一定酸度下，选择适当的指示剂来标定 EDTA 的浓度。

常用的标定 EDTA 溶液的基准物有 Zn、Cu、Pb、$CaCO_3$、$MgSO_4 \cdot 7H_2O$ 等。用 Zn 作基准物可以用铬黑 T（EBT）作指示剂，在 NH_3-NH_4Cl 缓冲溶液（pH＝10）中进行标定，其反应如下。

滴定前：

$$\underset{\text{(纯蓝色)}}{Zn^{2+} + In^{2-}} = \underset{\text{(紫红色)}}{ZnIn}$$

式中，In^{2-}为金属指示剂。

滴定开始至终点前：

$$Zn^{2+} + Y^{4-} = ZnY^{2-}$$

终点时：

$$\underset{\text{(紫红色)}}{ZnIn} + Y^{4-} = ZnY^{2-} + \underset{\text{(纯蓝色)}}{In^{2-}}$$

所以，终点时溶液从紫红色变为纯蓝色。

用 Zn 作基准物也可用二甲酚橙作指示剂，六亚甲基四胺作缓冲剂，在 pH 为 5~6 进行标定。两种标定方法所得结果稍有差异。通常选用的标定条件应尽可能与被测物的测定条件相近，以减少误差。

【仪器和药品】

NH_3-NH_4Cl 缓冲溶液（pH＝10）：取 6.75g NH_4Cl 溶于 20mL 水中，加入 57mL 15mol·L^{-1} $NH_3 \cdot H_2O$，用水稀释到 100mL。

铬黑 T 指示剂，纯 Zn，EDTA 二钠盐(AR)。

【实验内容】

1．EDTA 溶液的配制

称取 1.9~2.0g EDTA 二钠盐，溶于 500mL 水中，必要时可温热以加快溶解（若有残渣可过滤除去）。

2．Zn^{2+}标准溶液的配制

取适量纯锌粒或锌片，用稀 HCl 稍加泡洗（时间不宜长），以除去表面的氧化物。再用水洗去HCl，然后，用酒精洗一下表面，晾干后于110℃下烘几分钟，置于干燥器中冷却。

准确称取纯锌0.1500～0.2000g，置于100mL小烧杯中，加5mL (1∶1) HCl，盖上表面皿，必要时稍温热（小心），使锌完全溶解。吹洗表面皿及杯壁，小心转移至250mL容量瓶中，用水稀释至标线，摇匀。计算Zn^{2+}标准溶液的浓度$c(Zn^{2+})$。

3．EDTA浓度的标定

用25mL移液管吸取Zn^{2+}标准溶液置于250mL锥形瓶中，逐滴加入6mol·L^{-1}$NH_3 \cdot H_2O$，同时不断摇动直至开始出现白色$Zn(OH)_2$沉淀。再加5mL NH_3-NH_4Cl缓冲溶液、50mL水和3滴铬黑T，用EDTA标准溶液滴定至溶液由紫红色变为纯蓝色即为终点。记下EDTA溶液的用量V(EDTA)。平行标定3次，计算EDTA的浓度c(EDTA)。

【问题与讨论】

1．在配位滴定中，指示剂应具备什么条件？

2．本实验用什么方法调节pH？

Experiment 3 Preparation and Standardization of an EDTA Solution

【Objectives】

1. To learn the preparation and standardization of an EDTA solution.

2. To learn how to detect the endpoint of a complexometric titration.

3. To understand the use of buffer solutions.

【Introduction】

The most important chelating agent in complexometric titration is ethylenediaminetetraacetic acid (EDTA). Disodium EDTA (often written as $Na_2H_2Y \cdot 2H_2O$), which is much more soluble than EDTA, is commonly used to standardize aqueous solutions of transition metal cations. The pH for the aqueous solution of disodium EDTA is about 4.5.

EDTA forms very stable complexes with many metal cations, and the ratio of EDTA to metal cation is 1∶1. Therefore, the molar concentration of an EDTA solution can be standardized with primary standards such as Zn, Cu, Pb, $CaCO_3$ and $MgSO_4 \cdot 7H_2O$.

In this experiment, an EDTA solution is standardized with a primary standard Zn using the indicator Eriochrome Black T (EBT) at pH=10. The buffer solution used in this experiment is NH_3-NH_4Cl.

Before titration (In^{2-} means the indicator)

$$Zn^{2+} + \underset{\text{(blue)}}{In^{2-}} = \underset{\text{(red-violet)}}{ZnIn}$$

Before the endpoint

$$Zn^{2+} + Y^{4-} = ZnY^{2-}$$

At the endpoint

$$\underset{\text{(red-violet)}}{ZnIn} + Y^{4-} = ZnY^{2-} + \underset{\text{(blue)}}{In^{2-}}$$

So the endpoint is detected by the color change from red-violet to blue.

An EDTA solution can also be standardized with primary standard Zn using the indicator xylenol orange at pH=5~6. The buffer solution used here is hexamethylene tetramine. The molar concentrations obtained from these two methods may be a little bit different.

【Apparatus and Chemicals】

NH_3-NH_4Cl buffer solution (pH=10): dissolve 6.75g NH_4Cl into 20mL of distilled water, adding 57mL of 15mol·L^{-1} $NH_3 \cdot H_2O$. Dilute the solution to 100mL with distilled water.

EBT indicator, pure zinc, disodium EDTA (AR).

【Experimental Procedure】

1. Preparation of an EDTA solution

Weigh about 1.9~2.0g of disodium EDTA, and dissolve it into 500mL of distilled water. Heat and filter if necessary.

2. Preparation of the Zn^{2+} standard solution

Remove the oxide on the surface of pure zinc with dilute HCl. Do not put zinc into HCl for a long time. Then use water and ethanol to wash the zinc. Dry and heat the zinc at 110℃ for a few minutes. Finally cool the zinc in a desiccator.

Accurately weigh the zinc. Make sure the weight is between 0.1500 and 0.2000g. Put the zinc sample in a 100mL beaker. Add 5mL of HCl (1∶1) to the beaker, and cover the beaker with a watch glass. If necessary, heat carefully to totally dissolve the zinc. Rinse the watch glass and the wall of the beaker. Transfer all the above solution to a 250mL volumetric flask. Add water until the mark on the flask is reached. Invert the flask slowly a few times to make the solution homogeneous. Calculate the molar concentration of Zn^{2+} standard solution.

3. Standardization of an EDTA solution

Pipet 25mL of Zn^{2+} standard solution to an Erlenmeyer flask. Add 6mol·L^{-1} ammonia drop by drop until a white precipitate $Zn(OH)_2$ appears. Then, add 5mL of NH_3-NH_4Cl buffer, 50mL of distilled water and three drops of EBT. Titrate the Zn^{2+} standard solution with the EDTA solution until the color changes from red-violet to blue. Record the volume of the EDTA solution used for the titration. Repeat the titration 3 times. Calculate the molar concentration of the EDTA solution.

【Questions and Discussion】

1. What characteristics should the indicator possess in complexometric titrations?
2. How should the pH in this experiment be adjusted?

实验 4　水中钙、镁含量的测定（配位滴定法）

【实验目的】

1．掌握配位滴定的基本原理、方法和计算。

2．掌握铬黑 T、钙指示剂的使用条件和终点变化。

【实验提要】

用 EDTA 测定 Ca^{2+}、Mg^{2+}时，通常在两等份溶液中分别测定 Ca^{2+}量以及 Ca^{2+}和 Mg^{2+}的总量，Mg^{2+}量则从两者所用 EDTA 量的差数求出。

在测定 Ca^{2+}时，先用 NaOH 调节溶液到 pH＝12～13，使 Mg^{2+}生成难溶的 $Mg(OH)_2$ 沉淀。加入钙指示剂与 Ca^{2+}配位呈红色。滴定时，EDTA 先与游离 Ca^{2+}配位，然后夺取已和指示剂配位的 Ca^{2+}，使溶液的红色变成蓝色为终点。从 EDTA 标准溶液用量可计算 Ca^{2+}的含量。

测定 Ca^{2+}、Mg^{2+}总量时，在 pH＝10 的缓冲溶液中，以铬黑 T 为指示剂，用 EDTA 滴定。因稳定性 CaY^{2-}＞MgY^{2-}＞MgIn＞CaIn，铬黑 T 先与部分 Mg^{2+}配位为 MgIn（紫红色）。而当 EDTA 滴入时，EDTA 首先与 Ca^{2+}和 Mg^{2+}配位，然后再夺取 MgIn 中的 Mg^{2+}，使铬黑 T 游离，因此到达终点时，溶液由紫红色变为天蓝色，从 EDTA 标准溶液的用量，即可计算样品中的钙镁总量，然后换算为相应的硬度单位。

各国对水的硬度的表示方法各有不同。其中德国硬度是较早的一种，也是被我国较普遍采用的硬度单位之一，以度数计，1° 表示 1L 水中含 10mg CaO。为方便起见，我国也常以 $mol{\cdot}L^{-1}$ 或 $mmol{\cdot}L^{-1}$ 来表示。也有些国家采用 $CaCO_3$ 的含量来表示硬度，单位为 $mol{\cdot}L^{-1}$。

【仪器和药品】

$6mol{\cdot}L^{-1}$NaOH，NH_3-NH_4Cl 缓冲溶液（pH＝10），EDTA 标准溶液，铬黑 T 指示剂，钙指示剂。

【实验内容】

1．Ca^{2+}的测定

用移液管准确吸取水样 25mL 于 250mL 锥形瓶中，加 25mL 蒸馏水、2mL $6mol{\cdot}L^{-1}$ NaOH (pH＝12~13)、4~5 滴钙指示剂。用 EDTA 标准溶液滴定，不断摇动锥形瓶，当溶液变为纯蓝色时，即为终点❶。记下所用体积 V_1。用同样方法平行测定 3 份。

2．Ca^{2+}、Mg^{2+}总量的测定

准确吸取水样 25mL 于 250mL 锥形瓶中，加入 25mL 蒸馏水、5mL NH_3-NH_4Cl 缓冲溶液、3 滴铬黑 T 指示剂。用 EDTA 标准溶液滴定，当溶液由紫红色变为纯蓝色时，即为终点。记下所用体积 V_2。用同样方法平行测定 3 份。

按下式分别计算 Ca^{2+}、Mg^{2+}总量（以 CaO 的质量浓度表示，单位为 $mg{\cdot}L^{-1}$）及 Ca^{2+}和 Mg^{2+}的质量浓度（单位为 $mg{\cdot}L^{-1}$）。

$$\text{CaO 质量浓度}=\frac{c\bar{V}_2 M(\text{CaO})}{25}\times 1000$$

❶ 当试液中 Mg^{2+}的含量较高时，加入 NaOH 后，产生 $Mg(OH)_2$ 沉淀，使结果偏低或终点不明显（因沉淀吸附指示剂之故），可将溶液稀释后测定。

$$Ca^{2+}质量浓度=\frac{c\overline{V}_1M(Ca)}{25}\times1000$$

$$Mg^{2+}质量浓度=\frac{c\left(\overline{V}_2-\overline{V}_1\right)M(Mg)}{25}\times1000$$

式中，c 为 EDTA 的浓度，$mol\cdot L^{-1}$；$\overline{V}_1$ 为 3 次滴定 Ca^{2+} 所消耗 EDTA 的平均体积，mL；$\overline{V}_2$ 为 3 次滴定 Ca^{2+}、Mg^{2+} 总量所消耗 EDTA 的平均体积，mL。

【问题与讨论】

1．如果只有铬黑 T 指示剂，能否测定 Ca^{2+} 的含量？如何测定？

2．Ca^{2+}、Mg^{2+} 与 EDTA 的配合物，哪个稳定？为什么滴定 Mg^{2+} 时要控制 pH＝10，而测定 Ca^{2+} 则需控制 pH＝12～13？

3．测定的水样中若含有少量 Fe^{3+}、Cu^{2+} 时，对终点会有什么影响？如何消除其影响？

4．若在 pH＞13 的溶液中测定 Ca^{2+} 时会怎么样？

Experiment 4　An Analysis of Concentrations of Calcium and Magnesium Ions in a Water Sample

【Objectives】

1. To learn the principle, the method, and the calculation of complexometric titrations.

2. To learn the use and the color change of EBT indicator and calconcarboxylic acid.

【Introduction】

First, we will determine the Ca^{2+} concentration in a water sample with EDTA. Second, we will determine the total concentration of Ca^{2+} and Mg^{2+} in the water sample with EDTA. Finally, we can calculate the Mg^{2+} concentration in the water sample by the different volumes of EDTA used in the two experiments.

To determine the concentration of calcium ions, adjust the pH of the solution to 12~13 with NaOH to produce $Mg(OH)_2$ precipitate. Then add calconcarboxylic acid, to react with calcium ions, to produce red complexes. Next, EDTA is used to titrate the calcium ions. EDTA will firstly react with free calcium ions to form complexes, and then react with the calcium ions which have already formed complexes with the calconcarboxylic acid. The endpoint of the titration is detected by the color change from red to blue. The concentration of calcium ions can be calculated by the volume of EDTA used in the titration.

To determine the total concentration of calcium and magnesium ions, EDTA will be used with EBT indicator in a buffer solution with a pH of 10. The stability of different complexes is CaY^{2-}> MgY^{2-}>MgIn>CaIn. So EBT will react with magnesium ions to form MgIn complexes (red-violet). When EDTA is added, EDTA will firstly react with both calcium and magnesium ions, and then react with magnesium ions which are in the complexes MgIn. Now the EBT indicator is free, and the color changes from red-violet to blue. This is the endpoint of the titration. The total concentration of calcium and magnesium ions can be calculated by the volume of EDTA used in the titration. Also, we will know the hardness of the water sample.

The total water hardness is the sum of the molar concentrations of Ca^{2+} and Mg^{2+}, in $mol \cdot L^{-1}$ or $mmol \cdot L^{-1}$ units. Water hardness is often not expressed as a molar concentration, but rather in various units, such as German degrees (°), parts per million ($mg \cdot L^{-1}$, or American degrees), grains per gallon (gpg), English degrees, or French degrees. We often use the German degree, one degree (1°) means 10 mg CaO per liter of water. Some countries use $CaCO_3$ instead of CaO to express the water hardness.

【Apparatus and Chemicals】

$6mol \cdot L^{-1}$ NaOH, NH_3-NH_4Cl buffer solution (pH=10), EBT indicator, calconcarboxylic acid.

【Experimental Procedure】

1. Determine the concentration of Ca^{2+}

Pipet a water sample of 25mL to an Erlenmeyer flask. Add 25mL of distilled water, 2mL of $6mol \cdot L^{-1}$ NaOH (pH=12~13) and four drops of calconcarboxylic acid. Titrate the water sample with EDTA until the solution turns blue. Record the volume (V_1) of EDTA. Repeat the titration three times.

2. Determine the total concentration of Ca^{2+} and Mg^{2+}

Pipet the water sample of 25mL to an Erlenmeyer flask. Add 25mL of distilled water, 5mL of NH_3-NH_4Cl buffer solution and three drops of EBT. Titrate the water sample with EDTA until the solution turns from red-violet to blue. Record the volume (V_2) of EDTA. Repeat the titration three times.

Calculate the total concentration of Ca^{2+} and Mg^{2+} (expressed as the mass concentration of CaO with the unit of $mg \cdot L^{-1}$), and the mass concentrations of Ca^{2+} and Mg^{2+} ($mg \cdot L^{-1}$) according to the following formulae.

$$\text{Mass concentration of CaO} = \frac{c\overline{V}_2 M(\text{CaO})}{25} \times 1000$$

$$\text{Mass concentration of Ca}^{2+} = \frac{c\overline{V}_1 M(\text{Ca})}{25} \times 1000$$

$$\text{Mass concentration of Mg}^{2+} = \frac{c(\overline{V}_2 - \overline{V}_1) M(\text{Mg})}{25} \times 1000$$

In these formulae, c is the molar concentration ($mol \cdot L^{-1}$) of EDTA; $\overline{V}_1$ is the average volume (mL) of EDTA used for the three titrations of Ca^{2+}; $\overline{V}_2$ is the average volume (mL) of EDTA used for the three titrations of total Ca^{2+} and Mg^{2+}.

【Questions and Discussion】

1. Can we determine the concentration of Ca^{2+} only with the EBT indicator? How would we do it if possible?

2. Which complex is more stable, CaEDTA or MgEDTA? When titrating Mg^{2+}, the pH should be about 10. But for Ca^{2+}, the pH should be 12~13. Why?

3. If there are small amounts of Fe^{3+} or Cu^{2+} in the water sample, how will this affect the endpoint? How to alleviate this effect?

4. If we determine the concentration of Ca^{2+} at pH＞13, what will happen?

第3章 常数测定
Chapter 3 Determination of the Constant

实验5 化学反应速率和活化能

【实验目的】

1．试验浓度、温度及催化剂对化学反应速率的影响。

2．测定过二硫酸铵与碘化钾反应时的速率，并计算反应级数、反应速率常数及反应的活化能。

【实验提要】

在水溶液中过二硫酸铵和碘化钾发生如下反应：

$$S_2O_8^{2-} + 3I^- = 2SO_4^{2-} + I_3^- \tag{3-1}$$

其反应的速率方程可表示为：

$$v = k[S_2O_8^{2-}]^m[I^-]^n$$

式中，v为瞬时反应速率；k为速率常数；m与n的和称为反应级数。

实验所能测量的是在一段时间（Δt）内反应的平均速率（$\bar{v}$），如果在Δt时间内$S_2O_8^{2-}$的浓度变化为$\Delta[S_2O_8^{2-}]$，则平均速率

$$\bar{v} = \Delta[S_2O_8^{2-}]/\Delta t$$

由于本实验在Δt时间内反应物浓度的变化很小，作为近似处理，可用平均速率代替瞬时速率。即

$$\Delta[S_2O_8^{2-}]/\Delta t \approx v = k[S_2O_8^{2-}]^m[I^-]^n$$

为了能够测定出反应在Δt时间内$S_2O_8{}^{2-}$的浓度变化值，本实验做出如下设计：在混合KI和$(NH_4)_2S_2O_8$溶液之前，先加入一定体积已知浓度的$Na_2S_2O_3$溶液和淀粉溶液，这样在反应式（3-1）进行的同时，体系中还进行着如下反应：

$$2S_2O_3^{2-} + I_3^- = S_4O_6^{2-} + 3I^- \tag{3-2}$$

反应式（3-2）进行得非常快，几乎瞬间完成，而反应式（3-1）比反应式（3-2）慢得多，因此由反应式（3-1）生成的I_3^-立即与$S_2O_3^{2-}$反应，生成无色的$S_4O_6^{2-}$和I^-。一旦$Na_2S_2O_3$耗尽，反应（3-1）继续生成的微量I_3^-就与淀粉作用，使溶液显蓝色。

从反应式（3-1）和式（3-2）的关系式可以看出，$S_2O_8^{2-}$与$S_2O_3^{2-}$的变化量有如下关系：

$$\Delta[S_2O_8^{2-}] = \Delta[S_2O_3^{2-}]/2$$

在本实验中，每份混合液中$Na_2S_2O_3$的起始浓度都相同。从反应开始到溶液出现蓝色时$S_2O_3^{2-}$全部耗尽，故从$Na_2S_2O_3$的起始浓度可求出$\Delta[S_2O_3^{2-}]$，进而可计算$\Delta[S_2O_8^{2-}]$和反应的平均速率。

对反应速率方程两边取对数得：

$$\lg v = m\lg[S_2O_8^{2-}] + n\lg[I^-] + \lg k$$

当控制$[I^-]$不变时，以 $\lg v$ 对 $\lg[S_2O_8^{2-}]$作图，可得一直线，斜率为 m。同理，当$[S_2O_8^{2-}]$不变时，以 $\lg v$ 对 $\lg[I^-]$作图，可求出 n。

求出 m 和 n 后，可用一组$[S_2O_8^{2-}]$、$[I^-]$和 v 数据代入速率方程求得反应速率常数 k。

根据阿伦尼乌斯公式，反应速率常数 k 与反应温度 T 之间有以下关系：

$$\lg k = \frac{-E_a}{2.303RT} + A$$

式中，E_a 为反应的活化能，$kJ\cdot mol^{-1}$；R 为气体常数，$R=8.314J\cdot K^{-1}\cdot mol^{-1}$；$T$ 为热力学温度；A 为常数项。

测出不同温度时的 k 值，以 $\lg k$ 对 $\frac{1}{T}$ 作图，可得一直线，由直线斜率（$\frac{-E_a}{2.303R}$）可求得反应的活化能 E_a。

【仪器和药品】

100mL 烧杯 7 只，大试管，量筒，电热恒温水浴槽，秒表，温度计。

$(NH_4)_2S_2O_8$（$0.02mol\cdot L^{-1}$，新配制），KI（$0.20mol\cdot L^{-1}$）❶，KNO_3（$0.20mol\cdot L^{-1}$），$Na_2S_2O_3$（$0.010mol\cdot L^{-1}$），$(NH_4)_2SO_4$（$0.20mol\cdot L^{-1}$），$Cu(NO_3)_2$($0.02mol\cdot L^{-1}$)，淀粉溶液（质量分数为 2%），冰块。

【实验内容】

1．浓度对化学反应速率的影响，求反应级数

在室温下用 3 个量筒（贴有标签）分别量取 KI 溶液 20.0mL、$Na_2S_2O_3$ 溶液 8.0mL 和淀粉溶液 4.0mL，加到 100mL 烧杯中混合均匀，然后用另一量筒量取$(NH_4)_2S_2O_8$ 溶液 20.0mL 迅速倒入烧杯中，同时立即按下秒表，并不断搅拌至溶液刚出现蓝色时，立即停表，记下反应时间，同时记下室温。

用同样方法按照表 3-1 中用量进行另外 4 次实验。为了使每次实验中的溶液离子强度和总体积保持不变，不足的量分别用 KNO_3 溶液和$(NH_4)_2SO_4$ 溶液补足。

2．温度对化学反应速率的影响，求活化能

按表 3-1 中实验Ⅴ的用量，把 KI、$Na_2S_2O_3$、KNO_3 和淀粉溶液加到 100mL 烧杯中，把$(NH_4)_2S_2O_8$ 溶液加在另一个烧杯（或大试管）中，并把它们同时放在冰水中冷却，等烧杯中溶液都冷却到约为 0℃时，把$(NH_4)_2S_2O_8$ 溶液迅速加到混合液中。同时立即按下秒表，并不断搅拌至溶液出现蓝色，停表。记下反应时间Δt，用同样方法在约比室温高 10℃和低 10℃条件下，重复以上实验，这样连同实验 1 中Ⅴ号结果可以得到 4 个不同温度下的反应温度，算出它们的反应速率和速率常数，有关数据和处理结果填入表 3-2。

---

❶ 本实验对试剂有一定要求：

（1）碘化钾固体表面变黄表示变质，有 I_2 析出，不能配制溶液，KI 溶液应贮存于棕色瓶中，因为 KI 溶液遇光会有游离 I_2 析出，使溶液变黄。

（2）$(NH_4)_2S_2O_8$ 易分解，其溶液要新配制。此外，如所配$(NH_4)_2S_2O_8$ 溶液的 pH 值小于 3，则不适合本实验使用，所用试剂中如混有 Cu^{2+}、Fe^{3+}等杂质，对反应会有催化作用，必要时需滴入几滴 EDTA($0.1mol\cdot L^{-1}$)溶液。

表 3-1 浓度对化学反应速率的影响

实验编号		I	II	III	IV	V
试剂用量/mL	$Na_2S_2O_3$ (0.010mol·L^{-1})	8.0	8.0	8.0	8.0	8.0
	淀粉溶液（质量分数为 2%）	4.0	4.0	4.0	4.0	4.0
	KNO_3 (0.20mol·L^{-1})	0	0	0	5.0	10.0
	KI (0.20mol·L^{-1})	20.0	20.0	20.0	15.0	10.0
	$(NH_4)_2SO_4$ (0.20mol·L^{-1})	0	5.0	10.0	0	0
	$(NH_4)_2S_2O_8$ (0.20mol·L^{-1})	20.0	15.0	10.0	20.0	20.0
起始浓度	KI/mol·L^{-1}					
	$(NH_4)_2S_2O_8$/mol·L^{-1}					
$S_2O_8^{2-}$的浓度变化 $\Delta[S_2O_8^{2-}]$/mol·L^{-1}						
反应时间Δt/s						
反应速率 v/mol·L^{-1}·s^{-1}						
lg$v$						
lg $[S_2O_8^{2-}]$						
lg $[I^-]$						
m						
n						
反应速率常数 k						

表 3-2 温度对化学反应速率的影响

实验编号	I	II	III	IV
反应温度 t/℃				
反应时间Δt/s				
反应速率 v/mol·L^{-1}·s^{-1}				
反应速率常数 k				
lg k				
$\frac{1}{T}$ /K^{-1}				
反应活化能 E_a/kJ·mol^{-1}				

3．催化剂对化学反应速率的影响

按表 3-1 实验Ⅴ的用量把 KI、$Na_2S_2O_3$、KNO_3 和淀粉溶液加到 100mL 烧杯中，再加入两滴 $Cu(NO_3)_2$ 溶液，然后迅速加入 $(NH_4)_2S_2O_8$ 溶液，搅拌，计时，把此时实验的反应速率与表 3-1 中实验Ⅴ的反应速率比较。

【问题与讨论】

1．实验中为什么可以由反应溶液出现蓝色的时间长短来计算反应速率？反应溶液出现蓝色后，反应是否就终止了？

2．下列操作情况对实验结果有何影响？

① 取用 6 种试剂的量筒没有分开专用；

② 先加 $(NH_4)_2S_2O_8$ 溶液，最后加 KI 溶液；

③ 慢慢加入 $(NH_4)_2S_2O_8$ 溶液。

3．试分析本实验中 $Na_2S_2O_3$ 的用量过多或过少，对实验结果有什么影响？

Experiment 5 Determination of a Rate Law and Activation Energy

【Objectives】

1. To determine the effects of concentration, temperature and catalyst on the rate of a chemical reaction

2. To determine the rate of the reaction between ammonium persulfate and potassium iodide, and to calculate the reaction order and the activation energy

【Introduction】

The reaction of ammonium persulfate and potassium iodide is

$$S_2O_8^{2-}+3I^- = 2SO_4^{2-}+I_3^- \qquad (3\text{-}1)$$

The rate law for this reaction is

$$v=k[S_2O_8^{2-}]^m[I^-]^n$$

Where v is the instantaneous rate, and k is the reaction rate constant. The superscripts m and n designate the order with respect to each reactant.

This experiment will determine the average rate ($\overline{v}$). If the change of concentration for $S_2O_8^{2-}$ is $\Delta[S_2O_8^{2-}]$ in a period of time (Δt), the average rate can be expressed as

$$\overline{v}=\Delta[S_2O_8^{2-}]/\Delta t$$

In this experiment, the concentration of reactant changes very little in a short period of time (Δt). So the average rate can be regarded as the approximate instantaneous rate.

$$\Delta[S_2O_8^{2-}]/\Delta t \approx v=k[S_2O_8^{2-}]^m[I^-]^n$$

In order to determine the concentration of $S_2O_8^{2-}$, a constant known amount of $Na_2S_2O_3$ and starch solution are added before mixing KI with $(NH_4)_2S_2O_8$. Then two reactions, (3-1) and (3-2), occur at the same time.

$$2S_2O_3^{2-}+I_3^- = S_4O_6^{2-}+3I^- \qquad (3\text{-}2)$$

The rate of reaction (3-2) is very fast, so this reaction completes instantly. But reaction (3-1) is much slower than reaction (3-2). So the I_3^- produced by reaction (3-1) will react with $S_2O_3^{2-}$ instantly to form $S_4O_6^{2-}$ and I^-. Once all of the $S_2O_3^{2-}$ is used up, reaction (3-2) can no longer occur, and the I_3^- still being formed in reaction (3-1) will appear in the solution as deep blue because of the reaction between I_3^- and starch.

From reaction (3-1) and (3-2), we can find that each mole of $S_2O_3^{2-}$ used is equivalent to 1/2 mole of $S_2O_8^{2-}$ used.

$$\Delta[S_2O_8^{2-}]=\Delta[S_2O_3^{2-}]/2$$

In this experiment, the initial concentration of $Na_2S_2O_3$ is the same for each mixture. When the deep blue color appears, the $S_2O_3^{2-}$ is used up. So we would know $\Delta[S_2O_3^{2-}]$ and Δt, and further calculate $\Delta[S_2O_8^{2-}]$ and the average rate.

In logarithmic form, the rate law equation becomes

$$\lg v = m\lg[S_2O_8^{2-}] + n\lg[I^-] + \lg k$$

When $[I^-]$ is kept constant, a plot of $\lg v$ versus $\lg[S_2O_8^{2-}]$ produces a straight line with a slope equal to m. In the same way, a plot of $\lg v$ versus $\lg[I^-]$ produces a straight line with a slope equal to n when $[S_2O_8^{2-}]$ is kept constant.

Then the value of k, the reaction rate constant, can be calculated with the already known m, n, $[S_2O_8^{2-}]$, $[I^-]$ and v.

According to the Arrhenius equation,

$$\lg k = \frac{-E_a}{2.303RT} + A$$

where k——rate constant;

E_a——activation energy, $kJ \cdot mol^{-1}$;

R——gas constant, $R = 8.314 J \cdot K^{-1} \cdot mol^{-1}$;

T——thermodynamic temperature;

A——constant.

Determine the different k values at different temperatures. A plot of $\lg k$ versus $1/T$ produces a straight line with a slope of $(\frac{-E_a}{2.303R})$, so the activation energy can be calculated.

【Apparatus and Chemicals】

7 Beakers (100mL), large test tube, graduated cylinder, thermostatic waterbath, stopwatch, thermometer.

$(NH_4)_2S_2O_8$ ($0.02 mol \cdot L^{-1}$, freshly prepared), KI ($0.20 mol \cdot L^{-1}$), KNO_3 ($0.20 mol \cdot L^{-1}$), $Na_2S_2O_3$ ($0.010 mol \cdot L^{-1}$), $(NH_4)_2SO_4$ ($0.20 mol \cdot L^{-1}$), $Cu(NO_3)_2$ ($0.02 mol \cdot L^{-1}$), starch solution (2%, wt/wt), ice.

【Experimental Procedure】

1. Effect of concentration on the reaction rate

Use different graduated cylinders to measure 20.0mL of KI, 8.0mL of $Na_2S_2O_3$, and 4.0mL of starch solution. Put all of them in a 100mL beaker and mix well. Measure 20.0mL of $(NH_4)_2S_2O_8$ with another graduated cylinder. The reaction begins when the $(NH_4)_2S_2O_8$ is added to the above mixture. Be prepared to start timing the reaction in seconds. Now rapidly add the $(NH_4)_2S_2O_8$ to the mixture in the beaker——START TIME. Swirl the mixture until the deep blue color appears——STOP TIME. Record the time lapse and the temperature of the reaction mixture.

Repeat the remaining kinetic trials in Table 3-1. In order to keep the same total volume and ionic strength for each trial, KNO_3 solution and $(NH_4)_2SO_4$ solution are used.

2. Effect of temperature on the reaction rate

According to the solution volumes of kinetic trial Ⅴ in Table 3-1, mix KI, $Na_2S_2O_3$, KNO_3 and starch solution in a 100mL beaker. Place $(NH_4)_2S_2O_8$ in another beaker. Cool both beakers in ice water until 0℃. Then rapidly add the $(NH_4)_2S_2O_8$ to the mixture in the beaker until the deep blue color appears. Record the time lapse and the temperature of the reaction mixture. Repeat another two trials at approximately 10℃ above and below room temperature, respectively. With trial Ⅴ in Table 3-1, a total of four trials with different temperatures are done. Calculate and record the rate of the reaction and the rate constant in Table 3-2.

Table 3-1　The effect of concentration on the rate of a chemical reaction

	Kinetic trial	Ⅰ	Ⅱ	Ⅲ	Ⅳ	Ⅴ
Solution /mL	$Na_2S_2O_3$ (0.010mol·L^{-1})	8.0	8.0	8.0	8.0	8.0
	Starch solution (2%)	4.0	4.0	4.0	4.0	4.0
	KNO_3 (0.20mol·L^{-1})	0	0	0	5.0	10.0
	KI (0.20mol·L^{-1})	20.0	20.0	20.0	15.0	10.0
	$(NH_4)_2SO_4$ (0.20mol·L^{-1})	0	5.0	10.0	0	0
	$(NH_4)_2S_2O_8$ (0.20mol·L^{-1})	20.0	15.0	10.0	20.0	20.0
Initial concentration	KI/mol·L^{-1}					
	$(NH_4)_2S_2O_8$ /mol·L^{-1}					
	$\Delta[S_2O_8^{2-}]$/mol·L^{-1}					
	Δt/s					
	v/mol·L^{-1}·s^{-1}					
	lg$v$					
	lg $[S_2O_8^{2-}]$					
	lg $[I^-]$					
	m					
	n					
	Rate constant k					

Table 3-2　The effect of temperature on the rate of a chemical reaction

Kinetic trial	Ⅰ	Ⅱ	Ⅲ	Ⅳ
Temperature t/℃				
Δt/s				
v/mol·L^{-1}·s^{-1}				
Rate constant k				
lg k				
$\frac{1}{T}$ /K				
E_a/kJ·mol^{-1}				

3. Effect of catalyst on the reaction rate

According to the solution volumes of kinetic trial Ⅴ in Table 3-1, mix KI, $Na_2S_2O_3$, KNO_3 and starch solution in a 100mL beaker, and then add two drops of $Cu(NO_3)_2$ solution. Rapidly add the $(NH_4)_2S_2O_8$ to the mixture in the beaker until the deep blue color appears. Record the time lapse and the temperature of the reaction mixture. Compare the rate of this reaction with that of trial Ⅴ.

【Questions and Discussion】

1. Why can the reaction rate be calculated by the time lapse from the beginning of the reaction till the deep blue color appears in the experiment? Do the reactions continue or cease when the deep blue color appears?

2. How will the following operations affect the results of the experiment?

① The graduated cylinders for different solutions are mixed up.

② $(NH_4)_2S_2O_8$ is added before KI solution.

③ Slowly adding $(NH_4)_2S_2O_8$ solution to the mixture.

3. How will it affect the results if the amount of $Na_2S_2O_3$ is too much or too little?

实验 6 弱酸电离常数测定（pH 法）

【实验目的】

1．了解弱酸电离常数测定方法。

2．进一步了解电离平衡的基本概念。

3．了解酸度计的使用方法。

【实验提要】

醋酸在水溶液中存在下列平衡：

$$HAc \rightleftharpoons H^+ + Ac^-$$

其电离常数表达式为：

$$K_{HAc}=\frac{[H^+][Ac^-]}{[HAc]} \quad (3\text{-}3)$$

设醋酸的起始浓度为 c，平衡时 $[H^+]=[Ac^-]=x$，代入式（3-3），可以得到：

$$K_{HAc}=\frac{x^2}{c-x} \quad (3\text{-}4)$$

在一定温度下，用酸度计测定一系列已知浓度的醋酸的 pH 值，根据 $pH=-\lg[H^+]$，换算出 $[H^+]$，代入式（3-4）中，可求一系列对应的 K 值，取其平均值，即为该温度下醋酸的电离常数。

【仪器和药品】

酸度计。

滴定管（酸式、碱式）各一只，100mL 烧杯 6 只，HAc（$0.1000mol \cdot L^{-1}$）。

【实验内容】

1．将两支滴定管用蒸馏水洗 3 次，碱式滴定管内装入水至零刻度线，将酸式滴定管用已标定好的 $0.1000mol \cdot L^{-1}$ HAc 溶液洗 3 次后，装入该 HAc 溶液。

2．配制不同浓度的醋酸溶液：将 5 只洗净烘干的烧杯编成 1~5 号，按表 3-3 配制醋酸溶液。

表 3-3 实验数据记录

编 号	HAc 体积（已标定）/mL	H_2O 体积/mL	配制 HAc 浓度/$mol \cdot L^{-1}$	HAc 的 pH 值
1	3.00	45.00		
2	6.00	42.00		
3	12.00	36.00		
4	24.00	24.00		
5	48.00	0.00		

3．数据处理。根据实验数据填写表 3-4。

测定时的温度＝　　℃。

表 3-4 实验数据和计算结果

编 号	HAc 体积/mL	H_2O 体积/mL	配制[HAc]/$mol \cdot L^{-1}$	pH 值	$[H^+]/mol \cdot L^{-1}$	$K=x^2/(c-x)$
1						
2						
3						
4						
5						

最后计算出 K 的平均值为：______________。

【问题与讨论】

1．本实验测定 HAc 电离常数的依据是什么？

2．怎样配好 HAc 溶液？又如何从 pH 值测得其电离常数？

Experiment 6　Determination of Ionization Constant of Acetic Acid

【Objectives】

1. To understand the method of determining the ionization constant of a weak acid.

2. To further understand the concept of ionization equilibrium.

3. To learn how to use the pH meter.

【Introduction】

The ionization equilibrium of acetic acid is

$$HAc \rightleftharpoons H^+ + Ac^-$$

The ionization constant of acetic acid is expressed as

$$K_{HAc}=\frac{[H^+][Ac^-]}{[HAc]} \quad (3\text{-}3)$$

Suppose that the initial concentration of acetic acid is c, and the equilibrium concentration of H^+ equals that of Ac^-, that is $[H^+]=[Ac^-]=x$. So, equation (3-3) can be written as

$$K_{HAc}=\frac{x^2}{c-x} \quad (3\text{-}4)$$

At a certain temperature, if we prepare the acetic acid solution and determine the pH of this equilibrium solution, we can calculate the concentration of H^+ according to the formula, $pH=-\lg[H^+]$. Then, the ionization constant K can be calculated.

【Apparatus and Chemicals】

pH meter.

Acid buret, base buret, 6 beakers (100mL), HAc ($0.1000mol \cdot L^{-1}$).

【Experimental Procedure】

1. Rinse the two burets with distilled water three times, and fill the base buret with distilled water. Then, rinse the acid buret with acetic acid solution with known concentration, and fill the acid buret with acetic acid solution.

2. Prepare acetic acid solutions with different concentrations in five beakers according to Table 3-3.

Table 3-3　Experimental data

Trial	Volume of HAc/mL	Volume of H_2O/mL	Concentration of HAc/$mol \cdot L^{-1}$	pH value of HAc
1	3.00	45.00		
2	6.00	42.00		
3	12.00	36.00		
4	24.00	24.00		
5	48.00	0.00		

3. Calculate and record data in Table 3-4.

Temperature＝ ℃.

Table 3-4 Data and calculation

Trial	Volume of HAc/mL	Volume of H_2O/mL	[HAc]/mol·L^{-1}	pH	[H^+]/mol·L^{-1}	$K=x^2/(c-x)$
1						
2						
3						
4						
5						

The average of K is ________.

【Questions and Discussion】

1. How do you determine the ionization constant of HAc in this experiment?

2. How do you prepare an acetic acid solution? How do you determine its ionization constant from the pH value?

实验 7 硫酸钡溶度积的测定（电导法）

【实验目的】

1．学习硫酸钡沉淀制备的基本方法。

2．学习使用电导仪或电导率仪。

3．测定难溶电解质硫酸钡的溶度积。

【实验提要】

难溶电解质的饱和溶液都是稀溶液，由于在极稀溶液中，电解质分子的解离度接近于 1，离子的活度系数也接近 1，因此处理难溶电解质饱和溶液的平衡问题时，可以只考虑固体与溶液中离子之间的平衡，并用浓度代替活度。

例如，对于难溶电解质 A_mB_n 有：

$$A_mB_n(s) \rightleftharpoons mA^{n+}(aq) + nB^{m-}(aq)$$

$$K_{sp}=[A^{n+}]^m[B^{m-}]^n$$

当不存在同离子效应时，溶液中 A_mB_n 的饱和浓度 $c_{A_mB_n}$ 与离子的平衡浓度 $[A^{n+}]$ 和 $[B^{m-}]$ 有如下关系：

$$c_{A_mB_n}=\frac{1}{m}[A^{n+}]=\frac{1}{n}[B^{m-}]$$

代入

$$K_{sp}=(\frac{n}{m})^n[A^{n+}]^{m+n}=(\frac{m}{n})^m[B^{m-}]^{m+n}$$

$$=m^m n^n c_{A_mB_n}^{m+n}$$

从上述关系式可知，只要已知 $c_{A_mB_n}$、$[A^{n+}]$ 和 $[B^{m-}]$ 三者之中任一个量，都能求出 K_{sp} 的值，因此，溶度积的测定，可转化为对难溶电解质饱和溶液中某一物质的浓度测定。任何能够准确测定稀溶液中电解质或电解质离子浓度（直接或间接）的方法都可以用于测定溶度积常数 K_{sp}。

本实验用电导法（或电导率法）测定硫酸钡的溶度积。具体的方法是测出硫酸钡饱和溶液的电导率（或电导），然后通过电解质溶液的电导率（或电导）与电解质浓度的关系计算出硫酸钡饱和溶液中硫酸钡的浓度，从而求得 K_{sp} 值。

在推导电导率（或电导）与电解质浓度关系时需要先了解电导率（κ）和电解质的摩尔电导率（Λ）的意义及其相互间的关系。

电解质溶液导电能力的大小，通常用电阻 R 或电导 G 来表示。在国际单位制（SI）中，电导的单位是 S，称为西门子（$1S=1A\cdot V^{-1}$）。

（1）电导率 κ 表示放在相距 1m，面积为 $1m^2$ 两平行电极之间溶液的电导。若导体具有均匀截面，则其电导与截面积 A 成正比，与长度 l 成反比，即：

$$G=\kappa\times\frac{A}{l} \tag{3-5}$$

式中，κ 为比例常数，称为电导率，单位为 $S\cdot m^{-1}$。

（2）摩尔电导率指在相距为 1m，面积为 $1m^2$ 的两个平行电极之间，放置含有 1mol 电解质的溶液，此溶液的电导称为摩尔电导率，用λ表示，单位为$S\cdot m^2\cdot mol^{-1}$。摩尔电导率Λ与电导率κ有如下关系：

$$\kappa = \Lambda c \tag{3-6}$$

在使用摩尔这个单位时，必须明确规定基本单元，基本单元可以是分子、原子、离子、电子，或是这些粒子的特定组合，因而表示电解质的摩尔电导率时，亦应标明基本单元。例如若采用$MgCl_2/2$为基本单元，则$\Lambda_{MgCl_2}=2\Lambda_{MgCl_2/2}$。

（3）极限摩尔电导率指当溶液无限稀释时，正、负离子之间的影响趋于零，Λ值达到最大值，用Λ_0表示（Λ_0称为极限摩尔电导率）。实验证明当溶液无限稀释时，每种电解质的极限摩尔电导率Λ_0是两种离子的极限摩尔电导率的简单加和。即：

$$\Lambda_0=\Lambda_{0,+}+\Lambda_{0,-} \tag{3-7}$$

离子的极限摩尔电导率（$\Lambda_{0,+}$或$\Lambda_{0,-}$）可以从物理化学手册上查到。

在硫酸钡（$BaSO_4$）的饱和溶液中，存在如下平衡：

$$BaSO_4(s) \rightleftharpoons Ba^{2+}(aq)+SO_4^{2-}(aq)$$

$$K_{sp,BaSO_4}=c(Ba^{2+})\,c(SO_4^{2-})=c^2(BaSO_4) \tag{3-8}$$

由于$BaSO_4$的溶解度很小，它的饱和溶液可以近似地看成无限稀释的溶液，故有：

$$\Lambda_{0,BaSO_4}=\Lambda_{0,Ba^{2+}}+\Lambda_{0,SO_4^{2-}}$$

25℃时，无限稀释的$Ba^{2+}/2$和$SO_4^{2-}/2$的Λ_0值分别为63.6×10^{-4}和80.0×10^{-4}。

$$\begin{aligned}\Lambda_{0,BaSO_4}&=2\Lambda_{0,BaSO_4/2}=2(\Lambda_{0,Ba^{2+}/2}+\Lambda_{0,SO_4^{2-}/2})\\&=2\times(63.6\times10^{-4}+80.0\times10^{-4})\\&=287.2\times10^{-4}(S\cdot m^2\cdot mol^{-1})\end{aligned}$$

因此，只要测得$BaSO_4$饱和溶液的电导率κ_{BaSO_4}（或电导G_{BaSO_4}），即可由式（3-6）计算出$BaSO_4$饱和溶液的物质的量浓度。

$$c_{BaSO_4}=\frac{\kappa_{BaSO_4}}{1000\Lambda_{0,BaSO_4}}(mol\cdot L^{-1}) \tag{3-9}$$

应注意的是，测定得到的$BaSO_4$饱和溶液的电导率$\kappa_{BaSO_4溶液}$（或电导$G_{BaSO_4溶液}$）都包括了H_2O电离出的H^+和OH^-的电导率κ_{H_2O}（或G_{H_2O}）所以：

$$\kappa_{BaSO_4}=\kappa_{BaSO_4溶液}-\kappa_{H_2O} \tag{3-10}$$

或

$$G_{BaSO_4}=G_{BaSO_4溶液}-G_{H_2O}$$

由式（3-8）～式（3-10）可得：

$$K_{sp,BaSO_4}=\left(\frac{\kappa_{BaSO_4溶液}-\kappa_{H_2O}}{1000\Lambda_{0,BaSO_4}}\right)^2$$

或

$$K_{sp,BaSO_4}=\left[\frac{(G_{BaSO_4溶液}-G_{H_2O})\frac{l}{A}}{1000\Lambda_{0,BaSO_4}}\right]^2$$

式中，$\frac{l}{A}$是电导池常数或电极常数，对某一电极来说，$\frac{l}{A}$为常数。由电极标出。

【仪器和药品】

电导率仪，电动离心机，烧杯（50mL）2 只，量筒（50mL）1 只。

$BaCl_2$ (0.1mol·L^{-1})，Na_2SO_4 (0.1mol·L^{-1})，$AgNO_3$ (0.1mol·L^{-1})，二次蒸馏水。

【实验内容】

1．$BaSO_4$ 沉淀的制备

取 10mL 0.1mol·L^{-1} Na_2SO_4 溶液于干净小烧杯中，另取 10mL 0.1mol·L^{-1} $BaCl_2$ 溶液于另一干净烧杯中，将盛有 Na_2SO_4 溶液的小烧杯加热至近沸时，在搅拌下缓慢滴加 $BaCl_2$ 溶液，直到 $BaCl_2$ 溶液加完后，用 5mL 水洗涤盛有 $BaCl_2$ 的烧杯，并全部加入 Na_2SO_4 溶液中，继续加热煮沸 5min，静置陈化（一般陈化时间需要 15～20min）。用倾析法将 $BaSO_4$ 沉淀上的清液弃去，用近沸的蒸馏水洗涤 $BaSO_4$ 沉淀，至无 Cl^-为止。（可用 $AgNO_3$ 溶液检验）。这样得到了纯净的 $BaSO_4$ 沉淀。

2．$BaSO_4$ 饱和溶液的制备

往 $BaSO_4$ 沉淀中加入 50mL 已测定电导率的蒸馏水（或二次蒸馏水），加热煮沸 3～5min，并不断搅拌，静置，冷却。

3．电导率的测定

（1）取 50mL 纯水[1]，测定其电导率κ_{H_2O}（或 G_{H_2O}）。测定时操作要迅速，测定电导率（或电导）后的纯水立即用于制备 $BaSO_4$ 饱和溶液。

（2）将制得的 $BaSO_4$ 饱和溶液冷却至室温后（上层液是澄清的），用电导率仪尽快测定其饱和溶液的电导率$\kappa_{BaSO_4溶液}$或电导 $G_{BaSO_4溶液}$。并将数据和结果填入表 3-5。

表 3-5　数据记录和计算结果

温度 t/℃	$\kappa_{BaSO_4溶液}$/S·m^{-1}	κ_{H_2O}/S·m^{-1}	$K_{sp,BaSO_4}$

【问题与讨论】

1．为什么纯水也有一定的导电能力？

2．测量纯水和硫酸钡饱和溶液电导时，如果水的纯度不高，或所用的玻璃器皿不够洁净，将对实验结果有何影响？

3．在什么情况下电解质的摩尔电导率是其离子的摩尔电导率的简单加和？

4．一般情况下，难溶电解质溶度积的测定可以转化成什么样的问题加以解决？本实验又是怎样把问题进一步转化的？

Experiment 7　Determination of the Solubility Product Constant of Barium Sulfate

【Objectives】

1. To learn how to prepare the precipitate of barium sulfate.

2. To learn how to use the conductivity meter.

3. To determine the solubility product constant of barium sulfate.

[1] 本实验所用纯水的电导率κ_{H_2O}<5×10^{-4}S·m^{-1}时，可使 $K_{sp,BaSO_4}$测定值接近文献值。

【Introduction】

Strong electrolytes, such as barium sulfate, are hardly soluble in water. The saturated solution of barium sulfate is a dilute solution. In dilute solution, the degree of dissociation of the electrolyte is close to 1, and the activity coefficient of the ion is also close to 1. Therefore, we will use the concentration to replace the activity for the equilibrium of a saturated solution of barium sulfate.

For example, an equilibrium between undissolved A_mB_n solid and its ions is reached when the solution is saturated.

$$A_mB_n(s) \rightleftharpoons mA^{n+}(aq) + nB^{m-}(aq)$$

$$K_{sp}=[A^{n+}]^m[B^{m-}]^n$$

If the saturated concentration of A_mB_n is $c_{A_mB_n}$, and the equilibrium concentrations of A^{n+} and B^{m-} are $[A^{n+}]$ and $[B^{m-}]$, respectively, we can derive the following equation from the above chemical equilibrium:

$$c_{A_mB_n}=\frac{1}{m}[A^{n+}]=\frac{1}{n}[B^{m-}]$$

So, the solubility product constant can be expressed as

$$K_{sp}=\left(\frac{n}{m}\right)^n[A^{n+}]^{m+n}=\left(\frac{m}{n}\right)^m[B^{m-}]^{m+n}$$

$$=m^m n^n c_{A_mB_n}^{m+n}$$

The solubility product constant K_{sp} can be calculated if we know $c_{A_mB_n}$, or $[A^{n+}]$, or $[B^{m-}]$. So, we can determine the solubility product constant K_{sp} by determining the concentration of the solution or ions.

In this experiment, the solubility product constant of barium sulfate will be determined by using a conductivity meter. First we will determine the conductivity of the saturated solution. Then calculate the concentration of the solution by using conductivity. Finally, the solubility product constant K_{sp} will be calculated by the concentration.

In order to understand the relation between the conductivity and the concentration, we need to know the conductivity (κ) and molar conductivity (Λ).

Conductance G (S, Siemens) is defined as reciprocal of the resistance R.

(1) Movement of ions in water can be studied by installing a pair of electrodes into the liquid and by introducing a potential difference between the electrodes. Conductance of a given liquid sample decreases when the distance between the electrodes increases, and increases when the effective area of the electrodes increases. This is shown in the following relation:

$$G=\kappa\times\frac{A}{l} \tag{3-5}$$

Where κ is the conductivity ($S \cdot m^{-1}$), A is the cross-sectional area of the electrodes (m^2), and l is the distance between the electrodes (m).

(2) Molar conductivity Λ ($S \cdot m^2 \cdot mol^{-1}$) is defined as the conductivity of an electrolyte solution divided by the molar concentration of the electrolyte.

$$\kappa=\Lambda c \tag{3-6}$$

Where c is the molar concentration of the electrolyte.

(3) The limiting molar conductivity (Λ_0) is the molar conductivity at infinite dilution. The limiting molar conductivity (Λ_0) can be expressed as a sum of contributions from its individual ions. If the limiting molar conductivity for the cations is $\Lambda_{0,+}$ and for the anions is $\Lambda_{0,-}$, the "law of the independent migration of ions" states:

$$\Lambda_0 = \Lambda_{0,+} + \Lambda_{0,-} \tag{3-7}$$

The limiting molar conductivity for the ions ($\Lambda_{0,+}$ or $\Lambda_{0,-}$) can be found in a handbook of Chemistry and Physics.

The equilibrium for a saturated solution of barium sulfate is:

$$BaSO_4(s) \rightleftharpoons Ba^{2+}(aq) + SO_4^{2-}(aq)$$

$$K_{sp,BaSO_4} = c(Ba^{2+})\, c(SO_4^{2-}) = c^2(BaSO_4) \tag{3-8}$$

The solubility of $BaSO_4$ is very small. So, its saturated solution can be regarded as infinite dilution. Therefore, we can write the following equation:

$$\Lambda_{0,BaSO_4} = \Lambda_{0,Ba^{2+}} + \Lambda_{0,SO_4^{2-}}$$

At 25℃,

$$\Lambda_{0,BaSO_4} = 2\Lambda_{0,\frac{1}{2}BaSO_4} = 2(\Lambda_{0,\frac{1}{2}Ba^{2+}} + \Lambda_{0,\frac{1}{2}SO_4^{2-}})$$
$$= 2\times(63.6\times10^{-4} + 80.0\times10^{-4})$$
$$= 287.2\times10^{-4}(S\cdot m^2\cdot mol^{-1})$$

If we can determine the conductivity of the saturated solution of barium sulfate (κ_{BaSO_4}), its molar concentration can be calculated.

$$c_{BaSO_4} = \frac{\kappa_{BaSO_4}}{1000\Lambda_{0,\,BaSO_4}}\ (mol\cdot L^{-1}) \tag{3-9}$$

When determining the conductivity of $BaSO_4$ saturated solution ($\kappa_{BaSO_4 solution}$), the conductivity of H_2O (κ_{H_2O}) is included, so

$$\kappa_{BaSO_4} = \kappa_{BaSO_4 solution} - \kappa_{H_2O} \tag{3-10}$$

or

$$G_{BaSO_4} = G_{BaSO_4 solution} - G_{H_2O}$$

Thus, a combination of equations (3-8), (3-9) and (3-10) can be expressed as

$$K_{sp,\,BaSO_4} = \left(\frac{\kappa_{BaSO_4\,solution} - \kappa_{H_2O}}{1000\Lambda_{0,\,BaSO_4}}\right)^2$$

or

$$K_{sp,\,BaSO_4} = \left[\frac{\left(G_{BaSO_4\,solution} - G_{H_2O}\right)\frac{l}{A}}{1000\Lambda_{0,\,BaSO_4}}\right]^2$$

where $\frac{l}{A}$ is a constant for an electrode.

【Apparatus and Chemicals】

Conductivity meter, centrifuge, beaker (50mL), graduated cylinder (50mL).

$BaCl_2$ (0.1mol·L^{-1}), Na_2SO_4 (0.1mol·L^{-1}), $AgNO_3$ (0.1mol·L^{-1}), double-distilled water.

【Experimental Procedure】

1. Prepare $BaSO_4$ precipitate

Place 10mL of 0.1mol·L^{-1} Na_2SO_4 solution in a clean beaker, and 10mL of 0.1mol·L^{-1} $BaCl_2$ solution in another clean beaker. Heat the Na_2SO_4 solution till boiling, and then slowly add the $BaCl_2$ solution while stirring. After adding all of the $BaCl_2$ solution, rinse the beaker with 5mL of water, and add all of the water to the Na_2SO_4 solution. Heat and boil for five minutes. Now, let the mixture stand still for about 15 to 20 minutes. Decant the supernatant, and then wash the $BaSO_4$ precipitate with boiled water until there are no chlorine ions (Check with $AgNO_3$ solution if necessary).

2. Prepare $BaSO_4$ saturated solution

Put 50mL of double-distilled water, whose conductivity is already determined, into the $BaSO_4$ precipitate. Heat and boil for three to five minutes while stirring. Let the mixture stand still and cool down.

3. Determine the conductivity

(1) Determine the conductivity of 50mL double-distilled water (κ_{H_2O}). Do the determination quickly. Then use the water to prepare the $BaSO_4$ saturated solution.

(2) After cooling the $BaSO_4$ saturated solution to room temperature (the supernatant should be clear), determine the conductivity of the saturated solution ($\kappa_{BaSO_4 solution}$) quickly. Record the data in Table 3-5.

Table 3-5 Experimental Data

Temperature t /℃	$\kappa_{BaSO_4\ solution}$/S·m^{-1}	κ_{H_2O}/S·m^{-1}	$K_{sp,BaSO_4}$

【Questions and Discussion】

1. Why does pure water have a certain conductivity?

2. When determining the conductivity of the water and saturated solution of barium sulfate, if the water is not pure enough, or if the glassware is not clean enough, how will this affect the result?

3. Under what condition can the molar conductivity be expressed as a sum of contributions from its individual ions?

4. How do we determine the solubility product constant in this experiment?

实验 8 磺基水杨酸铁配合物的组成及稳定常数的测定

【实验目的】

1．了解分光光度法测定溶液中配合物组成及稳定常数的原理和方法。

2．测定 pH≈2 时磺基水杨酸铁的组成及稳定常数。

【实验提要】

磺基水杨酸（简式为 H_3R）与 Fe^{3+}可以形成稳定的配合物。配合物的组成随溶液的 pH 值不同而改变。当 pH<4 时生成 1∶1 紫红色的配合物；pH=4～9 时生成 1∶2 的红色配合物；pH=9～11.5 时生成 1∶3 黄色的配合物。本实验是在 pH≈2 时用分光光度法测定磺基水杨酸配合物的组成及稳定常数。反应式如下：

$$Fe^{3+} + {}^{-}O_3S\text{-}C_6H_3(OH)(COOH) \rightleftharpoons \left[{}^{-}O_3S\text{-}C_6H_3(O)(COO)Fe^{+}\right] + 2H^{+}$$

分光光度法基本原理

当一束具有一定波长的单色光通过一定厚度的有色物质溶液时，有色物质吸收了一部分光能，使透射光的强度（I_t）比入射光的强度（I_0）有所减弱。这种现象称为有色溶液对光的吸收作用。

对光的吸收和透过程度，通常用吸光度（A）和透光率（T）表示。透光率是透过光的强度 I_t 与入射光的强度 I_0 之比。即

$$T=\frac{I_t}{I_0}$$

吸光度是透光率的负对数。

$$A=-\lg T=\lg\frac{I_0}{I_t}$$

A 值越大表示光被有色物质吸收的程度越大。反之，A 值越小，表示有色物质对光的吸收程度越小。按照朗伯-比耳定律，溶液中有色物质对光的吸收程度即吸光度 A 与液层的厚度（b）及有色物质的浓度（c）成正比，即：

$$A=\varepsilon bc$$

ε为摩尔吸光系数，当波长一定时，它是有色物质的特征常数。当液层厚度 b 不变时，吸光度 A 与有色物质的浓度（c）成正比。

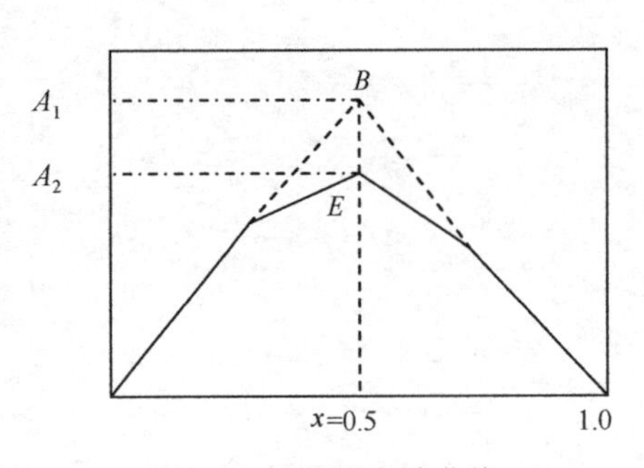

图 3-1 浓比递变法曲线

用分光光度法研究配合物的组成时，常用的一种实验方法是等摩尔系列法（也叫浓比递变法）。即保持溶液中金属离子（M）和配体（L）的总的物质的量不变而 M 和 L 的摩尔分数连续变化，配成一系列的溶液，测定溶液的吸光度。在这一系列溶液中，有一些溶液的金属离子是过量的，而另一些溶液中配体是过量的，这

两部分溶液中配位离子的浓度都不可能达到最大值，只有当溶液中金属离子与配体的摩尔比与配离子的组成一致时，配离子的浓度才能最大，因而吸光度也最大。以吸光度 A 为纵坐标，以体积分数（或摩尔分数）为横坐标绘图（图3-1），得一曲线，将曲线两边的直线部分延长，相交于 B 点。B 点为最大吸收处，对应于吸光度 A_1，由B点的横坐标值 x 可以计算配离子中金属离子与配体的摩尔比，当 x=0.5时，

$$\frac{V_L}{V_L+V_M}=\frac{n_L}{n_L+n_M}=0.5$$

整理可得：

$\frac{n_L}{n_M}=1$ 即金属离子与配体之比是1∶1，该配合物的组成为ML型。

配离子的稳定常数可根据图求得。从图中看出最大吸光度在 B 点，吸光度为 A_1，可认为M与L全部配位。但由于配离子有一部分解离，其浓度要稍小一些，实测得到最大的吸光度在 E 点，其值为 A_2。所以配离子的解离度 $\alpha=\frac{A_1-A_2}{A_1}$。

ML型配位离子的稳定常数 K 与 α 有如下关系：

$$K_{稳}=\frac{[ML]}{[M][L]}=\frac{1-\alpha}{c\alpha^2}$$

式中，c 为 B 点时ML的浓度。

【仪器和药品】

分光光度计，容量瓶（100mL 2只，50mL 11只），吸量管（5mL 3支），移液管（10mL 2支），洗耳球。

$HClO_4$（0. 1000mol·L^{-1}），磺基水杨酸（0.1000mol·L^{-1}），Fe^{3+}(0.1000mol·L^{-1})。

【实验内容】

1．试剂的配制

（1）配制0.0100mol·L^{-1} Fe^{3+}溶液：用移液管精确吸取10mL Fe^{3+}溶液加入100mL容量瓶中，用$HClO_4$溶液稀释至刻度，摇匀备用。

（2）配制 0.0100mol·L^{-1} 磺基水杨酸溶液：用实验室提供的磺基水杨酸溶液配成，方法同上。

2．配制系列溶液

将11个50mL容量瓶洗净编号。

按表3-6中所示的用量用吸量管分别量取$HClO_4$(0.1000mol·L^{-1})、Fe^{3+}(0.0100mol·L^{-1}) 和磺基水杨酸 (0.0100mol·L^{-1}) 各溶液注入已编号的容量瓶中，再用蒸馏水稀释至刻度摇匀。

表3-6 数据记录

编号	V_{HClO_4}/mL	$V_{Fe^{3+}}$/mL	$V_{磺基水杨酸}$/mL	体积分数$\frac{V_L}{V_L+V_M}$	吸光度 A		
					1	2	平均值
1	5.00	5.00	0.00				
2	5.00	4.50	0.50				
3	5.00	4.00	1.00				
4	5.00	3.50	1.50				
5	5.00	3.00	2.00				

续表

编号	V_{HClO_4}/mL	$V_{Fe^{3+}}$ /mL	$V_{磺基水杨酸}$/mL	体积分数$\frac{V_L}{V_L+V_M}$	吸光度 A		
					1	2	平均值
6	5.00	2.50	2.50				
7	5.00	2.00	3.00				
8	5.00	1.50	3.50				
9	5.00	1.00	4.00				
10	5.00	0.50	4.50				
11	5.00	0.00	5.00				

3．测定系列溶液的吸光度

接通分光光度计电源，调整好仪器，在波长 500nm，用（1cm）比色皿，以 1 号或 11 号溶液作参比溶液测定上述 11 个溶液的吸光度。将所测得的吸光度值记录在表 3-6 中（每号溶液测 2 次吸光度，取平均值）。

4．结果处理

作出吸光度 A 对$\frac{V_L}{V_{Fe^{3+}}+V_L}$ 的曲线，从图中找出最大的吸光度 A_1，算出磺基水杨酸铁配离子的组成和稳定常数。

【问题与讨论】

1．实验中每个溶液的 pH 值是否一样？如果不一样对结果是否有影响？

2．用等摩尔系列法测定配合物组成时，为什么说溶液中金属离子的摩尔数与配体的摩尔数之比正好与配合物组成相同时，配合物（配离子）的浓度为最大？试说明 M∶L=1∶2 及 M∶L=1∶3 时吸光度最大值的位置。

Experiment 8 Determination of the Composition and the Stability Constant of an Iron(Ⅲ)-Sulfosalicylate Complex

【Objectives】

1. To learn the spectrophotometric method for determining the composition and the stability constant of a complex.

2. To determine the composition and the stability constant of an Iron(Ⅲ)-sulfosalicylate complex at a pH≈2.

【Introduction】

A stable Iron(Ⅲ)-sulfosalicylate complex can be formed by sulfosalicylic acid and iron ions (Fe^{3+}). The composition of this complex depends upon the pH. The Iron(Ⅲ)-sulfosalicylate complex which contains one ligand is red purple at a pH less than 4. It is red with two ligands at a pH 4~9, and yellow with three ligands at a pH 9～11.5. In this experiment, we will determine the composition and the stability constant of an Iron(Ⅲ)-sulfosalicylate complex at a pH≈2. The reaction is as follows

$$Fe^{3+} + {}^{-}O_3S\text{—}C_6H_3(COOH)\text{—}OH \rightleftharpoons \left[{}^{-}O_3S\text{—}C_6H_3(\text{—}O\text{—})(\text{—}C(=O)\text{—}O\text{—})Fe^{+}\right] + 2H^{+}$$

Spectrophotometry

The Lambert-Beer law relates the absorption of light to the properties of the sample through which the light is travelling. The transmission (or transmissivity) and the absorbance are used to explain the absorption of light.

The transmission (T) is defined as

$$T=\frac{I_t}{I_0}$$

where I_0 and I_t are the intensity of the incident light and the transmitted light, respectively.

The absorbance (A), expressed in terms of the transmission, is defined as

$$A=-\lg T=\lg\frac{I_0}{I_t}$$

If more light is absorbed by the sample, the value of A will be bigger. According to Lambert-Beer law,

$$A=\varepsilon bc$$

the absorbance (A) is related to the distance the light travels through the sample (i.e., the path length, b) and the molar concentration of the sample (c). The molar absorptivity (ε) is a characteristic constant of the sample at a certain wavelength. When the path length (b) is maintained constant, A is in direct proportion to c.

The method of continuous variations is widely used for the spectrophotometric determination of a complex composition. If the total concentration of the metal ion (M) and the ligand (L) is maintained constant and only their ratio is changed, a wavelength of light is selected where the complex absorbs strongly but the metal ion and ligand do not. The maximum absorbance will be obtained when the metal ion and the ligand are in a proper ratio to form the complex. A plot of the mole fraction of ligand versus the absorbance gives the triangular-shaped curve shown in Figure 3-1. The legs of the triangle are extrapolated until they cross at Point B, which corresponds to the maximum absorbance (A_1). The mole fraction at Point B gives the composition of the complex because at this point the metal ion and the ligand are in a proper ratio to form the complex.

As in Figure 3-1, $x = 0.5$, that is

$$\frac{V_L}{V_L+V_M}=\frac{n_L}{n_L+n_M}=0.5$$

therefore,

$$\frac{n_L}{n_M}=1$$

The ratio of the metal ion to the ligand is 1∶1, so the complex is ML.

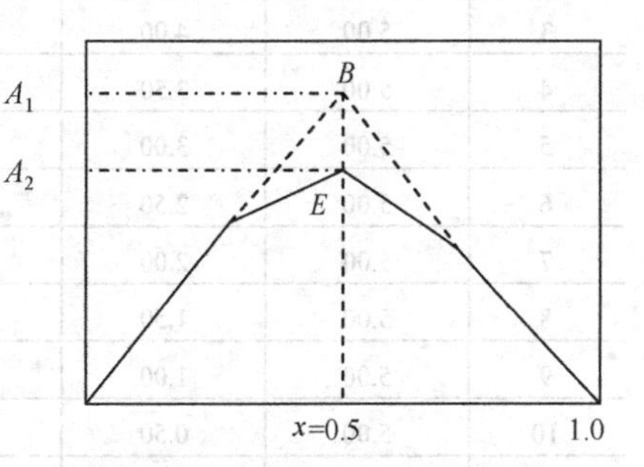

Figure 3-1 Continuous variations plot

The stability constant of the complex can also be calculated according to Figure 3-1. The actual maximum absorbance (Point E, absorbance A_2) observed from the experiment slightly deviates from the extrapolated value (Point *B*, absorbance A_1). The extrapolated value A_1 is the maximum absorbance of the complex if the complex formation is complete. Actually, the complex is slightly dissociated, so the absorbance A_2 is slightly low. The degree of dissociation (α) of the complex can be written as:

$$\alpha = \frac{A_1 - A_2}{A_1}$$

The relation between the stability constant (*K*) and the degree of dissociation (α) of the complex can be expressed as follows:

$$K = \frac{[ML]}{[M][L]} = \frac{1-\alpha}{c\alpha^2}$$

where *c* is the molar concentration of the complex ML at Point *B*.

【Apparatus and Chemicals】

Spectrophotometer, volumetric flask (100mL × 2, 50mL × 11), 3 measuring pipets (5mL), 2 pipets (10mL), rubber pipet bulb.

$HClO_4$ (0.1000mol·L^{-1}), sulfosalicylic acid (0.1000mol·L^{-1}), Fe^{3+} (0.1000mol·L^{-1}).

【Experimental Procedure】

1. Preparation of solutions

(1) Prepare 0.0100mol·L^{-1} Fe^{3+} solution: accurately pipet 10mL of 0.1000mol·L^{-1} Fe^{3+} solution to a 100mL volumetric flask. Dilute the solution to 100mL with $HClO_4$ solution.

(2) Prepare 0.0100mol·L^{-1} sulfosalicylic acid solution: accurately pipet 10mL of 0.1000 mol·L^{-1} sulfosalicylic acid solution to a 100mL volumetric flask. Dilute the solution to 100mL with $HClO_4$ solution.

2. Preparation of a series of solutions

According to Table 3-6, prepare a series of solutions with $HClO_4$ (0.1000mol·L^{-1}), Fe^{3+} solution (0.0100mol·L^{-1}) and sulfosalicylic acid solution (0.0100mol·L^{-1}).

Pipet an appropriate amount of solution into each volumetric flask, and dilute with distilled water.

Table 3-6 Experimental data

Trial	V_{HClO_4}/mL	$V_{Fe^{3+}}$ /mL	V_L/mL	Volume fraction $\frac{V_L}{V_L+V_M}$	Absorbance *A*		
					1	2	average
1	5.00	5.00	0.00				
2	5.00	4.50	0.50				
3	5.00	4.00	1.00				
4	5.00	3.50	1.50				
5	5.00	3.00	2.00				
6	5.00	2.50	2.50				
7	5.00	2.00	3.00				
8	5.00	1.50	3.50				
9	5.00	1.00	4.00				
10	5.00	0.50	4.50				
11	5.00	0.00	5.00				

3. Determination of the absorbance

At the wavelength of 500nm, using Trial 1 or Trial 11 as the blank, determine the absorbance of the solutions. Record your data in Table 3-6. Each solution should be determined twice, and the average absorbance should be calculated.

4. Analysis of data

Plot the absorbance *A* versus the mole fraction of the ligand ($\frac{V_L}{V_{Fe^{3+}}+V_L}$). Then calculate the composition and the stability constant of the Iron(Ⅲ)-sulfosalicylate complex.

【Questions and Discussion】

1. For the series of prepared solutions, do they have the same pH? If not, what effect on the results will be caused by different pH values?

2. The maximum absorbance will be obtained when the metal ion and the ligand are in a proper ratio to form the complex. Why? When the ratio of L to M is 2, at which position should the maximum absorbance be? What about when the ratio of L to M is 3?

实验9　氧化还原反应

【实验目的】

1．理解电池电动势和电极电位的概念。

2．了解并试验浓度、酸度对氧化还原反应的影响。

【实验提要】

反应物之间有电子转移的化学反应，称为氧化还原反应。

原电池是利用氧化还原反应产生电流的装置，如常见的Cu-Zn原电池。在Cu-Zn原电池中，Cu极的电极电位比Zn极高，电流从Cu极流向Zn极。电极电位的大小，不仅取决于电极的性质，还与温度和溶液中离子的浓度、气体的分压有关。

本实验将测定Cu-Zn原电池的电动势，并比较浓度对电动势的影响；比较不同电对电极电位的大小；以及浓度、介质pH对氧化还原反应的影响。

【仪器和药品】

H_2SO_4（$1mol·L^{-1}$），HNO_3（浓，$2mol·L^{-1}$），NaOH（$6mol·L^{-1}$），$NH_3·H_2O$（浓），$CuSO_4$（$0.1mol·L^{-1}$），$ZnSO_4$（$0.1mol·L^{-1}$），KBr（$0.1mol·L^{-1}$），$KMnO_4$（$0.01mol·L^{-1}$），$FeCl_3$（$0.1mol·L^{-1}$），Na_2SO_3（$0.1mol·L^{-1}$），KI（$0.1mol·L^{-1}$），$FeSO_4$（$0.1mol·L^{-1}$），KIO_3（$0.1mol·L^{-1}$），KSCN（$0.1mol·L^{-1}$），溴水，CCl_4，锌粒，Cu电极，Zn电极。

【实验内容】

1. 原电池电动势的测定

在50mL小烧杯中加入15mL $0.1mol·L^{-1}$ $CuSO_4$溶液，在另一小烧杯中加入15mL $0.1mol·L^{-1}$ $ZnSO_4$溶液。然后，在$CuSO_4$溶液中插入已用砂纸擦净的Cu电极，在$ZnSO_4$溶液中插入已用砂纸擦净的Zn电极，两极各连一导线，用盐桥连起来。Cu极导线与伏特计正极相连，Zn极与伏特计的负极相连。测量其电动势。记录下实验数据。

在装有$CuSO_4$的小烧杯中滴加浓氨水，不断搅拌，直至生成的沉淀完全溶解变成深蓝色的$[Cu(NH_3)_4]^{2+}$为止。测量其电动势，记录数据。

再在装有$ZnSO_4$的小烧杯中滴加浓氨水，使沉淀完全溶解变成$[Zn(NH_3)_4]^{2+}$。再测量其电动势，记录数据。

根据上述实验所得的电动势值，利用能斯特方程式说明浓度对原电池电动势的影响，并写出反应式。

2. 比较电极电位的大小

（1）在一支试管中加入1mL $0.1mol·L^{-1}$ KI溶液和5滴$0.1mol·L^{-1}$ $FeCl_3$溶液，振荡后有何现象？再加入0.5mL CCl_4，充分振荡，CCl_4层呈何色？反应的产物是什么？

（2）在另一支试管中加入1mL $0.1mol·L^{-1}$ KBr溶液和5滴$0.1mol·L^{-1}$ $FeCl_3$溶液，振荡后有何现象？再加入0.5mL CCl_4，充分振荡，CCl_4层呈何色？

（3）在一支试管中加入1mL $0.1mol·L^{-1}$ $FeSO_4$溶液，滴加$0.1mol·L^{-1}$KSCN溶液，溶液颜色有无变化？

（4）在另一支试管中加入1mL $0.1mol·L^{-1}$ $FeSO_4$溶液，加数滴溴水，振荡后再滴加0.1

mol·L^{-1} KSCN溶液，溶液呈何色？与上一支试管对照，说明试管中发生何反应？

根据以上实验，定性比较 Br_2/Br^-、I_2/I^-和 Fe^{3+}/Fe^{2+}三个电对的电极电位的高低。何者为强氧化剂？何者为强还原剂？

3. 影响氧化还原反应的因素

（1）浓度对氧化还原反应的影响　在两支盛有一粒锌的试管中，分别加入1mL浓HNO_3和2mol·L^{-1} HNO_3溶液，观察所发生的现象。不同浓度的HNO_3与Zn作用的反应产物和反应速率有何不同？

（2）介质pH值对氧化还原反应的影响

① 介质pH值对氧化还原反应方向的影响　在一支盛有2滴0.1mol·L^{-1} KI溶液的试管中，加入2滴1mol·L^{-1} H_2SO_4酸化，然后逐滴加入0.1mol·L^{-1} KIO_3溶液，振荡并观察现象，写出反应式。然后在该试管中再逐滴加入6 mol·L^{-1} NaOH溶液，振荡后又有何现象产生？写出反应式，说明介质对氧化还原反应方向的影响。

② 介质pH值对氧化还原反应产物的影响　在三支盛有1mL 0.01mol·L^{-1} $KMnO_4$溶液的试管中，分别加入1mol·L^{-1} H_2SO_4溶液、蒸馏水和6mol·L^{-1} NaOH溶液各0.5mL，混合后再逐滴加入0.1mol·L^{-1}Na_2SO_3溶液。观察溶液的颜色变化。写出反应式。

【问题与讨论】

1．CCl_4在反应体系中起何作用？

2．金属铁与HCl和HNO_3作用得到的主要产物是什么？

Experiment 9　Oxidation-Reduction Reactions

【Objectives】

1. To understand the electromotive force of a cell and electrode potential.

2. To understand the effects of concentration and acidity on redox reactions.

【Introduction】

Oxidation-reduction, or redox, reactions involve a transfer of electrons.

The galvanic cell is based on the redox reaction. For the Cu-Zn cell (also called Daniel cell), the electrode potential of the Cu electrode is higher than that of the Zn electrode. So, current flows from the Cu electrode to the Zn electrode. The value of each electrode potential is not only related to its inner substance properties, but also related to the temperature and concentration (or gaseous pressures) of its components.

In this experiment, we will determine the electromotive force (EMF) of a Cu-Zn cell, and compare the effect of concentration on EMF. We will also compare the electrode potentials of redox couples, and compare the effect of concentration and acidity on the redox reaction.

【Apparatus and Chemicals】

H_2SO_4(1mol·L^{-1}),HNO_3(concentrated，2mol·L^{-1}), NaOH(6mol·L^{-1}), $NH_3·H_2O$(concentrated), $CuSO_4$(0.1mol·L^{-1}), $ZnSO_4$(0.1mol·L^{-1}), KBr(0.1mol·L^{-1}), $KMnO_4$(0.01mol·L^{-1}), $FeCl_3$(0.1mol·L^{-1}), Na_2SO_3(0.1mol·L^{-1}), KI(0.1mol·L^{-1}), $FeSO_4$(0.1mol·L^{-1}), KIO_3(0.1mol·L^{-1}), KSCN(0.1mol·L^{-1}), bromine water, CCl_4, Zn, Cu electrode, Zn electrode.

【Experimental Procedure】

1. Determination of the electromotive force of a Cu-Zn cell

Obtain two 50mL beakers. Add 15mL of 0.1mol·L^{-1} $CuSO_4$ solution to one beaker, and 15mL of 0.1 mol·L^{-1} $ZnSO_4$ solution to the other beaker. Then place a clean strip of Cu into the $CuSO_4$ solution, and a clean strip of Zn into the $ZnSO_4$ solution. Connect the two solutions with a salt bridge. Next, connect the Cu strip to the positive terminal of the voltmeter, and connect the Zn strip to the negative terminal of the voltmeter. Read and record the voltage.

Add concentrated $NH_3 \cdot H_2O$ to the beaker with the $CuSO_4$ solution until a dark blue solution remains, which is due to the formation of $[Cu(NH_3)_4]^{2+}$. Read and record the voltage.

Then add concentrated $NH_3 \cdot H_2O$ to the beaker with the $ZnSO_4$ solution until a transparent solution remains, which is due to the formation of $[Zn(NH_3)_4]^{2+}$. Read and record the voltage.

According to the Nernst equation, explain the effect of concentration on the electromotive force of a cell.

2. Comparison of the electrode potential

(1) In a test tube, add 1mL of 0.1mol·L^{-1} KI solution and 5 drops of 0.1mol·L^{-1} $FeCl_3$ solution. Shake well and observe. Then add 0.5mL of CCl_4 and shake well. What is the color of the CCl_4 layer? What is the product?

(2) In the second test tube, add 1mL of 0.1mol·L^{-1} KBr solution and 5 drops of 0.1mol·L^{-1} $FeCl_3$ solution. Shake well and observe. Then add 0.5mL of CCl_4 and shake well. What is the color of the CCl_4 layer?

(3)In the third test tube, add 1mL of 0.1mol·L^{-1} $FeSO_4$ solution and 0.1mol·L^{-1} KSCN solution. What is the color of the solution?

(4) In the fourth test tube, add 1mL of 0.1mol·L^{-1} $FeSO_4$ solution and a few drops of bromine water. Shake well, and add 0.1mol·L^{-1}KSCN solution. What is the color of the solution? Compare it with the third test tube and explain the reason for your observation.

Compare the electrode potentials of Br_2/Br^-, I_2/I^- and Fe^{3+}/Fe^{2+}. Point out the strongest oxidant and strongest reductant among these redox couples.

3. Effects on redox reaction

(1) The effect of concentration on a redox reaction Prepare two test tubes, and put a grain of zinc into each one. Then add 1mL of concentrated HNO_3 to the first test tube, and 1mL of 2 mol·L^{-1} HNO_3 to the second test tube. Compare the products and the rates of the reactions.

(2) The effect of pH on a redox reaction

① The effect of pH on the direction of a redox reaction Add two drops of 0.1mol·L^{-1} KI solution to a test tube, and add two drops of 1mol·L^{-1} H_2SO_4 solution to acidify it. Then add 0.1mol·L^{-1} KIO_3 solution drop by drop. Shake well, observe, and write the chemical equation. Next, add 6mol·L^{-1} NaOH drop by drop to the test tube and shake well. Observe and explain your observations.

② The effect of pH on the product of a redox reaction Add 1mL of 0.01mol·L^{-1} $KMnO_4$

solution to each of three test tubes. Add 0.5mL of $1mol\cdot L^{-1}$ H_2SO_4 solution to the first test tube, add 0.5mL of distilled water to the second test tube, and add 0.5mL of $6mol\cdot L^{-1}$ NaOH solution to the third test tube. Shake well. Then add $0.1mol\cdot L^{-1}$ Na_2SO_3 solution drop by drop to each test tube. Record your observations and write the chemical reactions.

【Questions and Discussion】

1. What is the function of CCl_4?
2. When iron reacts with HCl and HNO_3, respectively, what is the main product?

第4章 化合物的制备与提纯

Chapter 4 Preparation and Purification of Compounds

实验10 硫酸铜的提纯

【实验目的】

1．了解用化学法提纯硫酸铜的方法。

2．练习无机制备的基本操作。

【实验提要】

粗硫酸铜中含有不溶性杂质和可溶性杂质 $FeSO_4$、$Fe_2(SO_4)_3$ 及其它重金属盐等。不溶性杂质可用过滤法除去。Fe^{2+}需由氧化剂 H_2O_2 或 Br_2 氧化成 Fe^{3+}，然后调节溶液的 pH 值，使 Fe^{3+}水解为 $Fe(OH)_3$ 沉淀而除去[1]。反应式如下：

$$2Fe^{2+}+H_2O_2+2H^+ = 2Fe^{3+}+2H_2O$$

$$Fe^{3+}+3H_2O = Fe(OH)_3\downarrow+3H^+ \ (pH\approx4.0)$$

除铁离子后的滤液，即可蒸发结晶。当硫酸铜晶体析出时，其它微量的可溶性杂质仍留在母液中，过滤时可与硫酸铜分离。

各氢氧化物沉淀时的 pH 值及 K_{sp} 值如表 4-1 所示。

表 4-1 氢氧化物沉淀时的 pH 值及其 K_{sp} 值

氢氧化物	开始沉淀的 pH	完全沉淀的 pH	K_{sp}
$Cu(OH)_2$	4.2	6.7	2.2×10^{-20}
$Fe(OH)_3$	1.6～2.0	3.9	4.0×10^{-36}
$Fe(OH)_2$	6.5	9.7	2.0×10^{-14}

【仪器和药品】

天平，研钵，玻璃漏斗，布氏漏斗，抽滤瓶，蒸发皿，25mL 比色管。

粗 $CuSO_4$，$NH_3\cdot H_2O$ (1mol·L^{-1}, 6mol·L^{-1})，H_2SO_4 (1mol·L^{-1})，H_2O_2 (质量分数 3%)，HCl (2mol·L^{-1}，AR)，NaOH（2mol·L^{-1}），$CuSO_4\cdot5H_2O$ 晶种，KSCN (质量分数 25%)，滤纸，pH

---

[1] 使 Fe^{3+}水解析出氢氧化铁沉淀是长期以来一种典型的除铁方法，在冶金和化工生产中得到广泛应用。但这种方法的主要缺点是 $Fe(OH)_3$ 具有胶体性质，沉淀速度慢、过滤困难，且使一些物质被吸附而损失。通常用凝聚剂使氢氧化铁凝聚沉淀或较长时间加热煮沸来破坏胶体。但当 Fe^{3+}浓度较大时，从溶液中分离氢氧化铁仍然有困难。

如果使 Fe(Ⅲ)的硫酸盐在较低的 pH 值下发生水解，这时溶液中存在一些聚合的 $Fe_2(OH)_2^{4+}$、$Fe_2(OH)_4^{2+}$ 而不产生氢氧化铁沉淀。这些离子和 SO_4^{2-} 结合生成一种浅黄色的复盐晶体，其化学式为 $M_2Fe_6(SO_4)_4(OH)_{12}$($M=Na^+$、K^+、NH_4^+)，俗称黄铁矾。黄铁矾在水中的溶解度小，而且颗粒大，沉淀速度快，很容易过滤。黄铁矾法与氢氧化铁法相似，包括：把溶液中的所有铁转化为 Fe^{3+}，常用的氧化剂有 H_2O_2、$NaClO_3$ 等；控制合适的 pH 值(pH＝1.6~1.8)；溶液中生成黄铁矾沉淀，温度控制在 358~368K。有关水解反应方程式可表示为：

$$Fe^{3+}+H_2O \rightleftharpoons Fe(OH)^{2+}+H^+$$

$$2Fe(OH)^{2+}+2H_2O = Fe_2(OH)_4^{2+}+2H^+$$

$$2Fe(OH)^{2+}+2Fe_2(OH)_4^{2+}+4SO_4^{2-}+2Na^++2H_2O = \underset{\text{黄钠铁矾（浅黄色）}}{Na_2Fe_6(SO_4)_4\cdot(OH)_{12}\downarrow}+2H^+$$

试纸。

【实验内容】

1．粗硫酸铜的提纯

（1）称取经研细的粗 $CuSO_4$ 30.0g 于 250mL 烧杯中，加入 120mL 蒸馏水，加热，搅拌，使其完全溶解，放置冷却。

（2）在上述溶液中滴加 4～5mL H_2O_2 溶液（若 Fe^{2+} 的含量高，可适当多加些）。用 $2mol \cdot L^{-1}$ NaOH 溶液调节溶液 pH≈4.0，加热并煮沸数分钟后趁热常压过滤。

（3）将滤液转入蒸发皿中，滴加 $1mol \cdot L^{-1}$ H_2SO_4 溶液调节溶液 pH≈2，水浴加热，蒸发浓缩至液面刚出现结晶膜为止。冷至室温，抽滤至干，取出晶体，用滤纸吸干其表面水分，称量，计算产率。母液回收（每人留 1.0g 产品做质量评定）。

2．晶体的培养

在大小适宜的洁净烧杯中，将产品配成某一温度下的饱和溶液❶。用一头发丝或洁净光滑的丝线系一透明且形状规整的 $CuSO_4 \cdot 5H_2O$ 晶种，在溶液温度高出饱和溶液温度 1~2℃时❷，将晶种悬挂于溶液中，再用一只大烧杯罩上。放置至下次实验时观察单晶的形状。

【选做部分】

按下述步骤对产品进行质量评定。

（1）将 1.0 g 提纯的产品用适量蒸馏水溶解，再用 $1mol \cdot L^{-1}$ H_2SO_4 酸化溶液，并滴加 2mL H_2O_2，充分搅拌后，煮沸片刻。

（2）在搅拌下逐滴加入 $6mol \cdot L^{-1}$ $NH_3 \cdot H_2O$ 至最初生成的浅蓝色沉淀完全溶解并呈现深蓝色溶液。此时 Fe^{3+} 转为 $Fe(OH)_3$ 沉淀，而 Cu^{2+} 成为配离子 $[Cu(NH_3)_4]^{2+}$。

（3）常压过滤，并用滴管将 $1mol \cdot L^{-1}$ $NH_3 \cdot H_2O$ 滴加到滤纸上洗涤，直到蓝色完全洗去为止。滤液回收。

（4）用 3mL 热的 $2mol \cdot L^{-1}$ HCl 滴在滤纸上，使 $Fe(OH)_3$ 溶解，收集溶液。

（5）将上述溶液转入 25mL 的比色管中，加入 1mL KSCN 溶液，用不含氧的蒸馏水稀释至刻度，摇匀，与标准溶液比较，评定级别。

【问题与讨论】

1．粗硫酸铜中的杂质 Fe^{2+} 为什么要氧化成 Fe^{3+} 除去？采用 H_2O_2 作氧化剂比其它氧化剂有什么优点？

2．除 Fe^{3+} 时，要调节 pH≈4.0，若 pH 值太大或太小有什么影响？

3．为什么除 Fe^{3+} 后的滤液还要调节 pH≈2.0 再进行蒸发浓缩？

4．当溶液中 Fe^{3+} 浓度为 $0.1mol \cdot L^{-1}$ 时，$Fe(OH)_3$ 开始沉淀与沉淀完全时的 pH 值（即残留 Fe^{3+} 浓度＜10^{-5} $mol \cdot L^{-1}$）分别是多少？

5．在溶有 15.0g $CuSO_4 \cdot 5H_2O$ 的 50mL 溶液中，若不使 $Cu(OH)_2$ 沉淀析出，必须控制 pH 值为多少？

6．如果粗硫酸铜中含有铅等杂质，它们会在哪一步骤中被除去，可能的存在形式是什么？

---

❶ 为获得较大的 $CuSO_4 \cdot 5H_2O$ 单晶，可控制结晶条件：按 53.2g $CuSO_4 \cdot 5H_2O$/100g H_2O 配制溶液，晶体生长的起始温度为 40℃。

❷ 晶种放入溶液内的理想时间是溶液温度恰好达到饱和温度的时候，加入相同温度的晶种。然而，实际上晶种的温度与室温相同。因此，将晶种加至饱和温度下的溶液时，由于溶液与晶种有温差而常使晶体急剧析出。为此，在溶液温度高出饱和温度 1～2℃时加入晶种为宜。

Experiment 10 Purification of Copper (Ⅱ) Sulfate

【Objectives】

1. To understand the chemical techniques for purification of copper sulfate.

2. To practice the basic operations of preparing inorganic compounds.

【Introduction】

Crude copper sulfate is a mixture of copper (Ⅱ) sulfate with various impurities that may include iron (Ⅱ) sulfate ($FeSO_4$), iron (Ⅲ) sulfate [$Fe_2(SO_4)_3$], and other metal salts. These impurities can be divided into two groups: soluble and insoluble. Insoluble impurities can be removed by dissolution and subsequent filtration. The most common soluble impurities in crude copper sulfate are iron (Ⅱ) sulfate and iron (Ⅲ) sulfate. To remove iron ions, Fe^{2+} is oxidized to Fe^{3+} first by using H_2O_2 or Br_2. Then Fe^{3+} will be removed as a $Fe(OH)_3$ precipitate by adjusting the pH of the solution, as shown in the following reactions:

$$2Fe^{2+}+H_2O_2+2H^+ \Longrightarrow 2Fe^{3+}+2H_2O$$

$$Fe^{3+}+3H_2O \Longrightarrow Fe(OH)_3\downarrow+3H^+ \text{ (pH} \approx 4.0\text{)}$$

The filtrate without Fe ions can be recrystallized. When copper sulfate crystals form in the solution, some other trace amount of soluble impurities, which have not reached saturation, will remain in solution. Therefore pure copper sulfate can be separated by filtration. Table 4-1 shows the pH values of precipitation and K_{sp} of selected hydroxides.

Table 4-1 pH values of precipitation and K_{sp} of selected hydroxides

Hydroxides	pH starting precipitation	pH completing precipitation	K_{sp}
$Cu(OH)_2$	4.2	6.7	2.2×10^{-20}
$Fe(OH)_3$	1.6～2.0	3.9	4.0×10^{-36}
$Fe(OH)_2$	6.5	9.7	2.0×10^{-14}

【Apparatus and Chemicals】

Electronic balance, mortar, funnel, Büchner funnel, filter flask, evaporating dish, 25mL color comparison cylinder.

Crude $CuSO_4$, $NH_3\cdot H_2O$ ($1mol\cdot L^{-1}$, $6mol\cdot L^{-1}$), H_2SO_4 ($1mol\cdot L^{-1}$), H_2O_2 (3%, wt/wt), HCl ($2mol\cdot L^{-1}$, AR), NaOH($2mol\cdot L^{-1}$), $CuSO_4\cdot 5H_2O$ seed crystal, KSCN (25%, wt/wt), filter paper, pH test paper.

【Experimental Procedure】

1. Purification of crude copper sulfate

(1) Weigh 30.0g of a ground powder sample of crude $CuSO_4$, and transfer it into a 250mL beaker. Add 120mL of distilled water to the beaker, heat the mixture, and stir using a stirring rod until it dissolves. Then, cool the solution to room temperature.

(2) Add 4~5mL of H_2O_2 to the above solution. Use $2.0mol\cdot L^{-1}$ NaOH to adjust the pH of the solution to about 4.0. Heat and boil the solution for a few minutes. Then filter the solution.

(3) Transfer the filtrate to an evaporating dish, and add $1.0mol\cdot L^{-1}$ H_2SO_4 to adjust the

solution to a pH 2.0. Use a water bath to evaporate the solution until crystals begin to form a thin layer on the surface. Cool the solution to room temperature, and then use a vacuum filter to collect $CuSO_4$ crystals. Finally, weigh the crystals and calculate the yield.

2. Growth of the copper sulfate crystal

Prepare a copper sulfate saturated solution in a beaker at about 40℃. Tie a transparent and well-shaped $CuSO_4 \cdot 5H_2O$ seed crystal to a piece of hair or a clean silk thread. Hang the seed crystal in the solution at the temperature of 41~42℃. Cover the beaker with another bigger beaker. Observe the crystal until the next laboratory.

【Questions and Discussion】

1. Why should impurity Fe^{2+} be oxidized to Fe^{3+} before removal? What is the advantage of using H_2O_2 as an oxidant?

2. To remove Fe^{3+}, the pH of the solution should be adjusted to about 4.0. What are the effects if the pH is too high or too low?

3. For the filtrate after removing Fe^{3+}, why should the pH of the filtrate be adjusted to about 2.0 before evaporation?

4. For a solution containing $0.1 mol \cdot L^{-1}$ Fe^{3+}, calculate the pH values when $Fe(OH)_3$ starts precipitation and completes precipitation (when the concentration of Fe^{3+} is less than $10^{-5} mol \cdot L^{-1}$ in the solution), respectively.

5. If 15.0g of $CuSO_4 \cdot 5H_2O$ is dissolved to a 50mL solution, and no $Cu(OH)_2$ precipitate forms, what pH value should the solution have been maintained?

6. If the impurities include lead salt, how should the lead salt be removed and in which step of the experiment?

实验 11　氯化钠的提纯

【实验目的】

1．掌握提纯 NaCl 的原理和方法。

2．学习过滤、蒸发、浓缩、结晶等基本操作。

3．了解 SO_4^{2-}、Ca^{2+}、Mg^{2+}等的定性鉴定。

【实验提要】

化学试剂或医药用的 NaCl 都是以粗食盐为原料提纯的，粗食盐中含有 Ca^{2+}、Mg^{2+}、K^+、SO_4^{2-} 等可溶杂质和泥沙等不溶杂质。选择适当的试剂可使 Ca^{2+}、Mg^{2+}、SO_4^{2-} 等生成沉淀而除去。一般是先在食盐溶液中加入 $BaCl_2$ 溶液，除去 SO_4^{2-}

$$Ba^{2+} + SO_4^{2-} = BaSO_4\downarrow$$

然后在溶液中加入 Na_2CO_3 溶液，除去 Ca^{2+}、Mg^{2+}和过量的 Ba^{2+}

$$Ca^{2+} + CO_3^{2-} = CaCO_3\downarrow$$

$$4Mg^{2+} + 5CO_3^{2-} + 2H_2O = Mg(OH)_2 \cdot 3MgCO_3\downarrow + 2HCO_3^-$$

$$Ba^{2+} + CO_3^{2-} = BaCO_3\downarrow$$

过量的 Na_2CO_3 溶液用盐酸中和。粗食盐中的 K^+与这些沉淀剂不起作用，仍留在溶液中，由于 KCl 的溶解度比 NaCl 大，而且在粗食盐中的含量较少，所以在蒸发浓缩食盐溶液时，NaCl 结晶出来，KCl 仍留在溶液中。

【仪器和药品】

HCl ($6mol\cdot L^{-1}$)，HAc ($2mol\cdot L^{-1}$)，NaOH ($6mol\cdot L^{-1}$)，$BaCl_2$ ($0.5mol\cdot L^{-1}$)，Na_2CO_3 (饱和)，$(NH_4)_2C_2O_4$ (饱和)，镁试剂，pH 试纸和粗食盐等。

【实验内容】

1．粗食盐的溶解

称取 5g 粗食盐于 100mL 烧杯中，加 20mL 水，加热搅拌使粗食盐溶解（不溶性杂质沉于底部）。

2．除去 SO_4^{2-}

将溶液加热至沸，边搅拌边滴加 $BaCl_2$ 溶液至 SO_4^{2-} 沉淀完全。继续加热煮沸数分钟，抽滤。

3．除去 Mg^{2+}、Ca^{2+}、Ba^{2+}等阳离子

将所得的滤液加热近沸。边搅拌边滴加饱和 Na_2CO_3 溶液，直至不再产生沉淀为止。再多加 0.5mL Na_2CO_3 溶液，继续加热 5min 后，抽滤，弃去沉淀。

4．除去过量 CO_3^{2-}

往溶液中滴加 $6mol\cdot L^{-1}$ HCl，加热搅拌，中和到溶液的 pH 值为 2～3。

5．浓缩与结晶

把溶液倒入预先称好的蒸发皿中，蒸发浓缩到有大量 NaCl 结晶出现（约为原来体积的 1/4）。冷却，抽滤。将氯化钠晶体转移到蒸发皿中，小火烘干。冷却后称量，计算产率。

6．产品纯度的检验

取产品和原料各 1g，分别溶于 5mL 蒸馏水中，然后进行下列离子的定性检验。

（1）SO_4^{2-}　各取溶液 1mL 于试管中，分别加入 $6mol \cdot L^{-1}$ HCl 溶液 2 滴和 $0.5mol \cdot L^{-1}$ $BaCl_2$ 溶液 2 滴。比较两溶液中沉淀产生的情况。

（2）Ca^{2+}　各取溶液 1mL，加 $2mol \cdot L^{-1}$ HAc 使呈酸性，再分别加入饱和$(NH_4)_2C_2O_4$溶液 3～4 滴，若有白色 CaC_2O_4 沉淀产生，表示有 Ca^{2+}存在❶（该反应可作为 Ca^{2+}的定性鉴定），比较两溶液中沉淀产生的情况。

（3）Mg^{2+}　各取溶液 1mL，加 $6mol \cdot L^{-1}$ NaOH 溶液 5 滴和镁试剂❷2 滴，若有天蓝色沉淀生成，表示有 Mg^{2+}存在（该反应可作为 Mg^{2+}的定性鉴定）。比较两溶液的颜色。

【问题与讨论】

1．在除去 Ca^{2+}、Mg^{2+}、SO_4^{2-} 时，为什么要先加入 $BaCl_2$ 溶液，然后再加入 Na_2CO_3 溶液？

2．为什么用 $BaCl_2$（毒性很大）而不用 $CaCl_2$ 除去 SO_4^{2-}？

3．在除 Ca^{2+}、Mg^{2+}、Ba^{2+}等离子时，能否用其它可溶性碳酸盐代替 Na_2CO_3？

4．加 HCl 除 CO_3^{2-} 时，为什么要把溶液的 pH 值调到 2～3？调至恰为中性好不好？（提示：从溶液中 H_2CO_3，HCO_3^- 和 CO_3^{2-} 浓度的比值与 pH 值的关系去考虑）

Experiment 11　Purification of Sodium Chloride

【Objectives】

1. To learn the principle and method of purifying sodium chloride.
2. To learn the operation of filtration, evaporation, concentration and crystallization.
3. To learn the methods of qualitative test for SO_4^{2-}, Ca^{2+} and Mg^{2+}

【Introduction】

Sodium chloride, which is used as a chemical or medical reagent, is purified from crude salt. There are not only insoluble impurities in the crude salt, such as sediment, but also soluble impurities, such as Ca^{2+}, Mg^{2+}, K^+ and SO_4^{2-}. To remove Ca^{2+}, Mg^{2+} and SO_4^{2-}, add appropriate reagents to produce insoluble precipitates.

First, add $BaCl_2$ to the crude salt solution to remove SO_4^{2-}.

$$Ba^{2+} + SO_4^{2-} = BaSO_4 \downarrow$$

Then add Na_2CO_3 to remove Ca^{2+}, Mg^{2+} and excessive Ba^{2+}.

$$Ca^{2+} + CO_3^{2-} = CaCO_3 \downarrow$$

$$4Mg^{2+} + 5CO_3^{2-} + 2H_2O = Mg(OH)_2 \cdot 3MgCO_3 \downarrow + 2HCO_3^-$$

$$Ba^{2+} + CO_3^{2-} = BaCO_3 \downarrow$$

The excessive Na_2CO_3 can be neutralized with HCl. The low content soluble impurity K^+, having a different solubility from sodium chloride, can be removed by recrystallization. It will be

---

❶ Mg^{2+}对此反应有干扰，也产生草酸盐沉淀。但 MgC_2O_4 溶于 HAc，故加 HAc 可排除 Mg^{2+}干扰。

❷ 对硝基苯偶氮间苯二酚（$O_2N-C_6H_4-N=N-C_6H_3(OH)_2$）俗称镁试剂，在碱性环境下呈红色或红紫色，被 $Mg(OH)_2$ 吸附后呈天蓝色。

retained in the solution when NaCl crystals form.

【Apparatus and Chemicals】

HCl ($6mol \cdot L^{-1}$), HAc ($2mol \cdot L^{-1}$), NaOH ($6mol \cdot L^{-1}$), $BaCl_2$ ($0.5mol \cdot L^{-1}$), Na_2CO_3 (saturated), $(NH_4)_2C_2O_4$ (saturated), magneson, pH test paper, crude salt.

【Experimental Procedure】

1. Dissolving crude salt

Weigh 5.0g of crude salt in a 100mL beaker, add 20mL of water, heat and stir to make it dissolve.

2. Removing SO_4^{2-}

Heat the solution to boiling, and then add $BaCl_2$ solution while stirring until the precipitation is complete. After continuing to boil the mixture for several minutes, vacuum filter the mixture.

3. Removing Mg^{2+}, Ca^{2+} and Ba^{2+}

Heat the above filtrate to boiling. Add saturated Na_2CO_3 solution while stirring until the precipitation is complete. Add an additional 0.5mL of Na_2CO_3 solution, and continue heating for 5 minutes. Vacuum filter the mixture, and discard the precipitates.

4. Removing excessive CO_3^{2-}

Add $6mol \cdot L^{-1}$ HCl to the solution, heat, and stir until the pH is about 2~3.

5. Concentration and crystallization

Transfer the above solution to an evaporating dish which is already weighed. Heat and evaporate until crystals form (The volume of the solution should be about a quarter of the original solution). Cool to room temperature, and vacuum filter the mixture. Transfer the crystals to the evaporating dish, and dry them with low heat. Cool the crystals to room temperature, weigh and calculate the yield.

6. Product purity analysis

Weigh 1.0g of crude salt and purified salt, respectively. Add 5mL of distilled water to dissolve each salt. Then perform the following qualitative analyses.

(1) SO_4^{2-} In two test tubes, transfer 1mL of crude salt solution and 1mL of purified salt solution, respectively. In each test tube, add two drops of $6mol \cdot L^{-1}$ HCl and two drops of $0.5mol \cdot L^{-1}$ $BaCl_2$ solution. Compare the precipitates in the two test tubes.

(2) Ca^{2+} In two test tubes, transfer 1mL of crude salt solution and 1mL of purified salt solution, respectively. In each test tube, add $2mol \cdot L^{-1}$ HAc, and 3~4 drops of saturated $(NH_4)_2C_2O_4$. Compare the white precipitates (CaC_2O_4) in the two test tubes.

(3) Mg^{2+} In two test tubes, transfer 1mL of crude salt solution and 1mL of purified salt solution, respectively. In each test tube, add five drops of $6mol \cdot L^{-1}$ NaOH and 2 drops of magneson. The blue precipitates confirm the presence of Mg^{2+}. Compare the blue precipitates in the two test tubes.

【Questions and Discussion】

1. When removing Ca^{2+}, Mg^{2+} and SO_4^{2-}, why is $BaCl_2$ added first, and then Na_2CO_3 is added?

2．Why is the toxic $BaCl_2$ used to remove SO_4^{2-} instead of $CaCl_2$?

3．Can we use another soluble carbonate to replace Na_2CO_3 in order to remove Ca^{2+}, Mg^{2+} and Ba^{2+}?

4．When using HCl to remove CO_3^{2-}, why should the pH be adjusted to 2~3? Can we adjust the pH to 7?

实验 12　氯化钾的提纯

【实验目的】

1．了解用沉淀法提纯氯化钾的原理和方法。

2．巩固减压过滤、蒸发、结晶等基本操作。

3．学习某些离子的定性检出方法。

【实验提要】

工业氯化钾除了含有不溶于水的泥沙杂质（可用过滤除去）外，还含有Na^+、Al^{3+}、Mg^{2+}、Fe^{3+}和SO_4^{2-} 等可溶性杂质。其中Na^+含量少，当KCl结晶析出时，它未能达到过饱和仍留在母液中与KCl分离除去。其它离子可通过加入合适试剂[1]生成难溶性化合物过滤除去。

本实验通过加入稍过量的$BaCl_2$溶液，则：

$$Ba^{2+} + SO_4^{2-} \longrightarrow BaSO_4\downarrow$$

用KOH溶液调节pH=7～8，则：

$$Al^{3+} + 3OH^- \longrightarrow Al(OH)_3\downarrow$$

$$Fe^{3+} + 3OH^- \longrightarrow Fe(OH)_3\downarrow$$

滤去沉淀，即可除去SO_4^{2-} ，Al^{3+}和Fe^{3+}。滤液中再加入KOH和K_2CO_3溶液，生成沉淀：

$$2Mg^{2+} + 2OH^- + CO_3^{2-} \longrightarrow Mg_2(OH)_2CO_3\downarrow$$

$$Ba^{2+} + CO_3^{2-} \longrightarrow BaCO_3\downarrow$$

滤去沉淀，即可除去Mg^{2+}和外加的过量Ba^{2+}。过量的KOH和K_2CO_3可用盐酸中和除去。

$$H^+ + OH^- \longrightarrow H_2O$$

$$2H^+ + CO_3^{2-} \xlongequal{\triangle} CO_2\uparrow + H_2O$$

【仪器和药品】

天平，烧杯（250mL），量筒，布氏漏斗，抽滤瓶，瓷蒸发皿，石棉网，酒精灯。

氯化钾（工业），HCl ($6mol\cdot L^{-1}$，AR)，KOH ($2mol\cdot L^{-1}$)，K_2CO_3 ($1mol\cdot L^{-1}$)，$BaCl_2$ ($0.5 mol\cdot L^{-1}$)，镁试剂，铝试剂，KSCN ($0.1mol\cdot L^{-1}$)，pH试纸，滤纸。

【实验内容】

1．KCl的提纯

（1）粗KCl溶解　称取30.0g粗氯化钾，加入120mL蒸馏水，加热、搅拌使之溶解。

（2）除SO_4^{2-}　将溶液加热至沸，边搅拌边滴加$BaCl_2$溶液至SO_4^{2-} 沉淀完全[2]。继续加热煮沸数分钟，抽滤。

（3）除Al^{3+}和Fe^{3+}　滤液用KOH溶液调节pH值为7～8，有胶状沉淀，继续加热煮沸数分钟，趁热抽滤，弃去沉淀。

（4）除Mg^{2+}和过量的Ba^{2+}　用KOH-K_2CO_3 (1∶1)混合溶液调节上述滤液的pH值为11

---

[1] 一般除杂质选择沉淀剂的原则是：生成的沉淀溶解度愈小愈好；所用试剂不要引入新的杂质离子，或所引入的杂质离子可在下一步反应除去；尽量使用廉价易得的试剂。

[2] 检验沉淀完全的方法：将溶液停止加热和搅拌，取少量上层溶液过滤（或离心），在滤液中加入几滴沉淀剂，若无浑浊，表示沉淀完全，若有浑浊表示沉淀尚未完全，需继续滴加沉淀剂，直到无浑浊为止。

左右，取液检验Ba^{2+}是否除尽。待除尽后，继续加热煮沸数分钟，抽滤。

（5）除过量的CO_3^{2-} 加热上述滤液，边搅拌边滴加HCl至溶液pH为2～3。

（6）蒸发结晶 加热蒸发浓缩上述溶液至出现一层较厚的晶膜（蒸发过程注意搅拌，防止溅出）。冷至室温，抽滤至干。把晶体转回蒸发皿中，用空气浴小火烘干（石棉网上隔一泥三角，蒸发皿置于泥三角上，加热石棉网），冷至室温，称量，计算产率。保留产品作KNO_3制备的原料。

2．产品纯度的检验

取粗氯化钾和产品各1.0g，分别溶于5mL蒸馏水中，定性检验SO_4^{2-}、Al^{3+}❶、Fe^{3+}和Mg^{2+}❷。列表记录现象，比较检验结果。

【问题与讨论】

1．能否用重结晶的方法提纯氯化钾？

2．除杂质时，能否把步骤颠倒一下，即先除Al^{3+}、Fe^{3+}和Mg^{2+}，再除SO_4^{2-}，其效果有何不同？除Al^{3+}和Mg^{2+}时其沉淀剂都是KOH溶液，能否将两步合并一起过滤？

3．在沉淀完全后，总要加热煮沸数分钟后再进行过滤，这是为什么？

4．能否选用$CaCl_2$代替毒性大的$BaCl_2$来除SO_4^{2-}？

5．用KOH和K_2CO_3两种溶液作为除Mg^{2+}和Ba^{2+}的沉淀剂，有什么优点？如果只用K_2CO_3好吗？可否用其它的碳酸盐如Na_2CO_3或$(NH_4)_2CO_3$代替K_2CO_3？

Experiment 12 Purification of Potassium Chloride

【Objectives】

1. To understand the principle and approach to purifying potassium chloride by the method of precipitation.

2. To consolidate the operation of vacuum filtration, evaporation and crystallization.

3. To study the methods of qualitative test for some ions.

---

❶ Al^{3+}的检出：取5滴试剂，在中性条件下。加入1～2滴铝试剂（玫红三羧酸铵或金精三酸铵盐），搅拌后微热，有鲜红色沉淀产生，示有Al^{3+}存在。反应式如下

[Reaction structure: H₄NOOC, HO, HO, H₄NOOC substituted triphenylmethane dye with C=, =O, COONH₄ $+\frac{1}{3}Al^+ \longrightarrow$ complex with O…$Al_{1/3}$, $+NH_4^+$]

❷ Mg^{2+}的检出：取一滴试剂在点滴板穴中加1滴2mol·L^{-1}NaOH溶液和1滴镁试剂，有天蓝色沉淀，示有Mg^{2+}存在。镁试剂是一种有机染料，称对硝基苯偶氮间苯二酚。其结构式为：

$O_2N-C_6H_4-N{=}N-C_6H_3(OH)_2$ (structure: O_2N, N═N, OH, HO)

它在酸性溶液中为黄色，在碱性溶液中呈红色或紫色，被$Mg(OH)_2$沉淀吸附后为天蓝色。

【Introduction】

There are soluble and insoluble impurities in industrial potassium chloride. The insoluble impurities can be removed by filtration. The soluble impurities, such as Na^+, Al^{3+}, Mg^{2+}, Fe^{3+} and SO_4^{2-}, can be removed by proper methods. Because of the low content of Na^+, it will be retained in the solution while KCl crystals are formed. Other soluble ions can be removed by producing insoluble compounds with the appropriate reagents.

First, excessive $BaCl_2$ is added.

$$Ba^{2+} + SO_4^{2-} \longrightarrow BaSO_4\downarrow$$

Second, KOH is used to adjust the pH of the solution to 7~8.

$$Al^{3+} + 3OH^- \longrightarrow Al(OH)_3\downarrow$$

$$Fe^{3+} + 3OH^- \longrightarrow Fe(OH)_3\downarrow$$

Therefore, SO_4^{2-}, Al^{3+} and Fe^{3+} can be removed by filtration.

Third, add KOH and K_2CO_3 to the filtrate, and precipitates will be formed as follows:

$$2Mg^{2+} + 2OH^- + CO_3^{2-} \longrightarrow Mg_2(OH)_2CO_3\downarrow$$

$$Ba^{2+} + CO_3^{2-} \longrightarrow BaCO_3\downarrow$$

In this step, Mg^{2+} and excessive Ba^{2+} can be removed by filtration. Excessive KOH and K_2CO_3 can be neutralized with HCl.

$$H^+ + OH^- \longrightarrow H_2O$$

$$2H^+ + CO_3^{2-} \xrightarrow{\triangle} CO_2\uparrow + H_2O$$

【Apparatus and Chemicals】

Electronic balance, beaker (250mL), graduated cylinder, Büchner funnel, filter flask, evaporating dish, wire/asbestos gauze, alcohol burner.

Potassium chloride (industrial), HCl ($6mol\cdot L^{-1}$, AR), KOH ($2mol\cdot L^{-1}$), K_2CO_3 ($1mol\cdot L^{-1}$), $BaCl_2$ ($0.5\ mol\cdot L^{-1}$), magneson, aluminon, KSCN ($0.1mol\cdot L^{-1}$), pH test paper, filter paper

【Experimental Procedure】

1. Purification of KCl

(1) Dissolving the crude KCl Weigh 30.0g of crude KCl in a beaker, then add 120mL of distilled water. Heat and stir the solution until the crude KCl dissolves.

(2) Removing SO_4^{2-} Heat the solution to boiling, and then add $BaCl_2$ solution while stirring until the precipitation is complete. After continuing to boil the mixture for several minutes, vacuum filter the mixture.

(3) Removing Al^{3+} and Fe^{3+} Adjust the filtrate pH to 7~8 by using KOH. Gelatinous precipitates will be formed. After continuing to boil the mixture for several minutes, vacuum filter the mixture, and discard the precipitates.

(4) Removing Mg^{2+} and excessive Ba^{2+} Adjust the filtrate to a pH of about 11 by using KOH- K_2CO_3 (1∶1) solution. Test whether or not Ba^{2+} is removed completely. After the complete removal of Ba^{2+}, continue boiling the mixture for several minutes, and vacuum filter the mixture.

(5) Removing excessive CO_3^{2-} Heat the above filtrate, and add HCl into the solution while stirring until the pH is 2~3.

(6) Evaporation and crystallization Evaporate the above solution until crystals start to form a thick layer on the surface (*Caution*: stir the solution to avoid splashing.) Cool to room temperature, and vacuum filter the mixture. Transfer the crystals to the evaporating dish, and dry them with low heat. Cool the crystals to room temperature, weigh and calculate the yield.

2. Product purity analysis

Weigh 1.0g of crude and purified products respectively. Add 5mL of distilled water to dissolve each product. Then determine whether SO_4^{2-}、Al^{3+}、Fe^{3+} and Mg^{2+} are present in the solution by qualitative analysis. Record and compare the results.

【Questions and Discussion】

1. Can KCl be purified by using the method of recrystallization?

2. Can we change the steps in the experimental procedure? What is the difference if we first remove Al^{3+}, Fe^{3+} and Mg^{2+}, and then remove SO_4^{2-}? KOH is used to remove both Al^{3+} and Mg^{2+}, can we put the two steps together and remove the two ions at the same time?

3. Why do we continue boiling the mixture for several minutes after the precipitation is complete?

4. Can we replace the toxic $BaCl_2$ with $CaCl_2$ to remove SO_4^{2-} ?

5. What is the advantage to use both KOH and K_2CO_3 to remove Mg^{2+} and Ba^{2+}? Can we use only K_2CO_3? Can we replace K_2CO_3 with Na_2CO_3 or $(NH_4)_2CO_3$?

实验 13　硝酸钾的制备与提纯

【实验目的】

1．利用温度对物质溶解度影响的差别，通过复分解法制备硝酸钾。

2．继续巩固减压过滤、趁热过滤、蒸发结晶、重结晶等基本操作。

【实验提要】

复分解法是制备无机盐类的常用方法。制备难溶盐比较容易，制备可溶性盐的条件及可能性，则需要根据不同盐类的溶解度差别以及温度对溶解度的影响来确定。

本实验用 KCl 和 $NaNO_3$ 通过复分解反应制备硝酸钾。当 KCl 和 $NaNO_3$ 相混合时，在混合液中同时存在 K^+、Na^+、Cl^-和 NO_3^- 4 种离子。由这 4 种离子组成的 4 种盐（KCl、NaCl、$NaNO_3$、KNO_3）在不同温度时的溶解度数据如表 4-2 所示。

表 4-2　KCl、NaCl、$NaNO_3$、KNO_3 在不同温度时的溶解度　单位：g/100g H_2O

溶质	温度/℃										
	0	10	20	30	40	50	60	70	80	90	100
KCl	27.6	31.0	34.0	37.0	40.0	42.6	45.6	48.7	51.1	54.0	56.7
$NaNO_3$	73	80	88	96	104	114	124	136	148	—	180
NaCl	35.7	35.8	36.0	36.3	36.6	37.0	37.3	37.8	38.4	39.0	39.2
KNO_3	13.3	20.9	31.6	45.8	63.9	85.5	110.0	138	169	202	245

从以上 4 种盐类单独存在时的溶解度数据❶可以看出，随着温度的升高，NaCl 的溶解度几乎没有变化，在较高温度时，它的溶解度最小。KCl 和 $NaNO_3$ 的溶解度也改变不大，而 KNO_3 的溶解度随温度变化大，因此加热 KCl 和 $NaNO_3$ 的混合溶液时，就有 NaCl 晶体析出。用反应式表示为：

$$KCl + NaNO_3 \xlongequal{100℃} NaCl\downarrow + KNO_3$$

趁热过滤除去 NaCl，从而改变了溶液的组成，当滤液冷却时，KNO_3 因溶解度的急剧下降而析出。这时析出的 KNO_3 晶体，一般混有可溶性盐的杂质（哪种盐？），可采取重结晶方法进行提纯。

【仪器和药品】

烧杯（带刻度，100mL、50mL），布氏漏斗，抽滤瓶，量筒，天平，恒温水浴槽，酒精灯。

KCl(s)、$NaNO_3$(s)、$AgNO_3$ (0.1mol·L^{-1})，冰水，滤纸。

【实验内容】

1．硝酸钾的制备

（1）根据自己提纯的 KCl 的质量，称取比计算量多 20%左右的 $NaNO_3$❷，放在带有刻度的 100mL 烧杯中，加入计算量的蒸馏水❸（在杯外壁液面处做记号）。加热，搅拌，使之溶解。

❶ 表 3-2 中的溶解度数据为单组分体系的数据，而混合体系中各种盐的溶解度数据会有差异，但仍可粗略判断制备 KNO_3 的条件。

❷ $NaNO_3$ 用量是按 KCl 的等物质的量计算，多加 20%的量是经验值。

❸ 按 NaCl 在 100℃时的溶解度与生成 NaCl 的理论量之间的关系求出所需的水量。

（2）继续加热（保持沸腾），蒸发至溶液的体积为原来体积的一半时有晶体析出（晶体是什么？）。

（3）趁热迅速抽滤。滤液中立即有晶体析出（是什么？）（抽滤前要预先将布氏漏斗放在水浴锅上用水蒸气加热或置于热水中浸泡加热）。

（4）将抽滤瓶置于热水浴中温热使晶体溶解，将滤液转入小烧杯中，另取 2～3mL 蒸馏水荡洗抽滤瓶并将荡洗液一并转入小烧杯中。

（5）将溶液静置，冷却至 20℃以下（用冰水调节水浴温度），有结晶析出，观察并记录晶体的外观。抽滤，得 KNO_3 粗产品，用滤纸吸干，称量，计算粗产品的理论产量和产率。

2．粗产品的重结晶

（1）称出 0.1g 粗产品供纯度检验外，其余按粗产品：水=2：1（质量比）的比例，将粗产品溶于所需的蒸馏水中，加热使其刚好完全溶解（能否将溶液蒸干？），放置冷却。

（2）冷却至 20℃以下，将析出的晶体抽滤、吸干、称量。

3．产品纯度的检验

称取粗、精产品各 0.1g，分别置于洁净的小烧杯中，用 20mL 蒸馏水溶解，混匀后各取出 1mL 稀释至 10mL，各加 2 滴 $0.1mol\cdot L^{-1}$ $AgNO_3$ 溶液，比较粗、精产品的纯度。

【问题与讨论】

1．本实验用过量的 $NaNO_3$ 与 KCl 反应，其作用何在？

2．实验中为什么要趁热、快速过滤除去 NaCl 晶体？

3．粗产品重结晶时，确定蒸馏水与硝酸钾比例的根据是什么？

4．根据 4 种盐在不同温度时的溶解度数据作溶解度曲线图（以温度作横坐标，溶解度作纵坐标）。试从溶解度曲线图上理解本实验在 20℃以下析出 KNO_3，其作用何在？

Experiment 13　Preparation and Purification of Potassium Nitrate

【Objectives】

1. To prepare potassium nitrate with the theory that different salts have different solubilities at different temperatures, and by the double decomposition method.

2. To further practice the basic operations of vacuum filtration, hot gravity filtration, evaporation and recrystallization.

【Introduction】

The double decomposition method is commonly used to prepare salts. It is easier to prepare insoluble salts than soluble salts. To prepare soluble salts, we need to choose various methods because different salts have different solubilities at different temperatures.

In this experiment, potassium chloride and sodium nitrate will be used to prepare potassium nitrate by double decomposition. When KCl and $NaNO_3$ are mixed, there are four ions in the solution, i. e., K^+, Na^+, Cl^- and NO_3^-. These four ions can form four salts, which are KCl, NaCl, $NaNO_3$ and KNO_3. Table 4-2 lists their solubilities at different temperatures.

Table 4-2 Solubilities of KCl, NaCl, $NaNO_3$ and KNO_3 at different temperatures (g/100gH_2O)

Solute	Temperature/℃										
	0	10	20	30	40	50	60	70	80	90	100
KCl	27.6	31.0	34.0	37.0	40.0	42.6	45.6	48.7	51.1	54.0	56.7
$NaNO_3$	73	80	88	96	104	114	124	136	148	—	180
NaCl	35.7	35.8	36.0	36.3	36.6	37.0	37.3	37.8	38.4	39.0	39.2
KNO_3	13.3	20.9	31.6	45.8	63.9	85.5	110.0	138	169	202	245

It is shown from the Table that the solubility of NaCl does not increase dramatically with the increase of temperature, and it is the least soluble among the four salts at 100℃. However, the solubility of KNO_3 increases greatly with the increase of temperature. So if we heat the mixed solution of KCl and $NaNO_3$, NaCl will form crystals first. The reaction is

$$KCl + NaNO_3 \xlongequal{100℃} NaCl\downarrow + KNO_3$$

Remove NaCl by filtration when the mixture is still hot. Cool the filtrate, and KNO_3 crystals can be separated from the solution. The KNO_3 crystals may have impurities (which salt?). Pure KNO_3 crystals can be obtained by recrystallization.

【Apparatus and Chemicals】

Graduated beakers (100mL and 50mL), Büchner funnel and filter flask, graduated cylinder, electronic balance, thermostatic waterbath, alcohol burner.

KCl (s), $NaNO_3$ (s), $AgNO_3$ (0.1mol·L^{-1}), ice water, filter paper.

【Experimental Procedure】

1. Preparation

(1) Calculate the mass of $NaNO_3$ according to the mass of the purified KCl from the last experiment. The mass of $NaNO_3$ used in this experiment should be 20% more than the calculated mass. Weigh the appropriate amount of $NaNO_3$ in a 100mL beaker, and then add the KCl to the beaker. Next, add distilled water, whose amount should be calculated too. Heat and stir to dissolve the solids.

(2) Keep boiling. Crystals will be formed when the volume of the solution is about half of the original volume. What are the crystals?

(3) Before filtration, preheat the Büchner funnel. Remove the crystals by vacuum filtration when the mixture is hot. Crystals will appear in the filtrate. What are the crystals now?

(4) Put the filter flask in a water bath to dissolve the crystals. Then transfer the filtrate to a clean beaker. Rinse the flask with 2～3mL of distilled water, and transfer the water to the beaker also.

(5) Let the solution stand still and cool down to below 20℃ in a water bath. Adjust the water bath temperature with ice water. Crystals should form. Vacuum filter to get the crystals and calculate the yield.

2. Recrystallization

(1) Weigh 0.1g of crude salt to do the purity analysis. Dissolve the rest of the crude salt in distilled water. The mass ratio of crude salt to water is 2 : 1. Heat until the salt just dissolves (Can

the solution be heated to dry?). Cool the solution down.

(2) Cool down to below 20℃. Vacuum filter to get the crystals and calculate the yield.

3. Product purity analysis

Weigh 0.1g crude and purified products respectively. Add 20mL of distilled water to dissolve each product. Then dilute 1mL of each product to 10mL. Add two drops of 0.1mol·L^{-1} $AgNO_3$ to each product. Record and compare the results.

【Questions and Discussion】

1. Why should excessive $NaNO_3$ be used to react with KCl in this experiment?

2. Why should NaCl crystals be filtered quickly when the mixture is still hot?

3. During the recrystallization procedure, why should the ratio of crude salt to water be 2∶1?

4. Plot the temperature versus the solubility to give the solubility curve. Why can KNO_3 crystals be obtained below 20℃?

实验14　明矾的制备

【实验目的】

1．了解由Al制备明矾的原理及过程，进一步了解Al及$Al(OH)_3$的性质。

2．通过晶体的培养，了解类质同晶的概念。

【实验提要】

矾是指一价碱金属（M^{I}，除Li^+外）或铵离子（NH_4^+）与三价金属（M^{III}=Al^{3+}、Fe^{3+}、Ti^{3+}、Co^{3+}和Cr^{3+}等）硫酸盐的含水复盐的总称。其通式为$M^{I}_2SO_4 \cdot M^{III}_2(SO_4)_3 \cdot 24H_2O$。

明矾是硫酸铝钾的俗称，也叫铝钾矾。它是一种无色晶体。其化学组成为$K_2SO_4 \cdot Al_2(SO_4)_3 \cdot 24H_2O$，可简写成$KAl(SO_4)_2 \cdot 12H_2O$。它的晶体和铁铵矾$(NH_4)_2SO_4 \cdot Fe_2(SO_4)_3 \cdot 24H_2O$、铬钾矾$K_2SO_4 \cdot Cr_2(SO_4)_3 \cdot 24H_2O$的晶体属于类质同晶体。

明矾是工业上十分重要的铝盐，用作净水剂、填料和媒染剂。

本实验是利用金属铝溶于氢氧化钠溶液，先制得四羟基铝酸钠。其反应式为：

$$2Al+2NaOH+6H_2O \longlongequal 2Na[Al(OH)_4]+3H_2\uparrow$$

$$\Delta_r G^{\ominus}_{m,298}=-867.6\ kJ \cdot mol^{-1}$$

再用H_2SO_4调节溶液的pH值为8~9，将四羟基铝酸钠转化为氢氧化铝沉淀，并与其它杂质分离。

$$2Na[Al(OH)_4]+H_2SO_4 == 2Al(OH)_3\downarrow+Na_2SO_4+2H_2O$$

然后用硫酸溶解氢氧化铝生成硫酸铝溶液。

$$2Al(OH)_3+3H_2SO_4 == Al_2(SO_4)_3+6H_2O$$

在硫酸铝溶液中加入等物质的量的硫酸钾即可制得明矾。

$$Al_2(SO_4)_3+K_2SO_4+24H_2O == 2KAl(SO_4)_2 \cdot 12H_2O$$

室温下，由于明矾在水中的溶解度比组成它的任何一组分的溶解度都要小，由此可以从含有$Al_2(SO_4)_3$和K_2SO_4的混合溶液中析出明矾晶体。它们的溶解度如表4-3所示。

表4-3　三种化合物在不同温度下的溶解度　　单位：g/100g H_2O

化合物	温度/℃					
	0	20	30	40	60	100
K_2SO_4	7.4	11.1	13.0	14.8	18.2	24.1
$Al_2(SO_4)_3$	31.2	36.4	40.4	45.8	59.2	89.0
$KAl(SO_4)_2$	3.0	5.90	8.39	11.7	24.8	109.0（90℃）

【仪器和药品】

锥形瓶（250mL），烧杯（250mL、100mL），天平，抽滤装置。

铝屑，K_2SO_4(s)，明矾微晶，NaOH(s)，H_2SO_4（3 $mol \cdot L^{-1}$，6$mol \cdot L^{-1}$），$Cr_2(SO_4)_3$（1$mol \cdot L^{-1}$）。

【实验内容】

1．四羟基铝酸钠的制备

迅速称取2.0g NaOH固体，置于250mL锥形瓶中，加入25mL蒸馏水溶解，分几次加入1.0g铝屑（反应剧烈，防止溅入眼中!），待反应完毕再加入蒸馏水，使总体积约40mL，趁热抽滤。

2．氢氧化铝的生成和洗涤

将上述滤液转入 250mL 烧杯中加热至沸，并保持沸腾，边搅拌边以细流状加入一定量的 $3mol \cdot L^{-1}$ H_2SO_4，使溶液的 pH=8~9。继续加热并搅拌数分钟，然后抽滤，并用热水洗涤沉淀，当滤液的 pH=7~8 时，停止洗涤。

3．明矾的制备

将沉淀转移到 100mL 蒸发皿中，加入 12mL $6mol \cdot L^{-1}$ H_2SO_4 溶液，小心加热并搅拌使 $Al(OH)_3$ 完全溶解。另将 3.3g K_2SO_4 配成热的饱和溶液，然后将两溶液相混合，并搅拌均匀，自然冷却溶液，待结晶完全后，抽滤，称量，计算产率。

【选做部分】

明矾单晶的培养：按 100g 蒸馏水溶解 29.5g 明矾的比例配成溶液，将溶液分成两份，分别置于 100mL 烧杯中，其中一份溶液加入 1mL $Cr_2(SO_4)_3$ 溶液，待溶液冷至 46~47℃，将一根头发丝系牢的明矾微晶悬挂在溶液中，使晶体生成。用一只大烧杯罩住小烧杯，下次实验观察晶形。

【问题与讨论】

1．试说明本实验用碱溶解 Al 而不用酸溶解的道理。

2．铝屑中铁杂质在哪一步操作中除去？

3．在生成 $Al(OH)_3$ 的过程中，为什么要加热煮沸并不断搅拌？用热水洗涤 $Al(OH)_3$ 沉淀是除去什么离子？

4．计算用 1.0 g 金属铝能生成多少克硫酸铝？若将硫酸铝全部转变成明矾，需要多少克硫酸钾？

5．由 $Na[Al(OH)_4]$ 转为 $Al(OH)_3$ 沉淀时，除了用 H_2SO_4 调节溶液 pH 值外，还可以用饱和 NH_4HCO_3 或 $(NH_4)_2CO_3$ 来调节 pH 值。你认为用哪一种试剂较好？

Experiment 14 Synthesis of Potassium Alum

【Objectives】

1. To learn how to prepare potassium alum from aluminum, and further study the properties of Al and $Al(OH)_3$.

2. To understand the isomorphism.

【Introduction】

An alum is a hydrated double sulfate salt with the general formula $M_2^{I}SO_4 \cdot M_2^{III}(SO_4)_3 \cdot 24H_2O$. M^{I} is a univalent cation. Some common univalent cations are Na^+, K^+, Tl^+, NH_4^+ or Ag^+. M^{III} is a trivalent cation, commonly Al^{3+}, Fe^{3+}, Cr^{3+}, Ti^{3+}, or Co^{3+}.

Potassium alum (potassium aluminum sulfate dodecahydrate) is a colorless crystal, and its formula is $K_2SO_4 \cdot Al_2(SO_4)_3 \cdot 24H_2O$ or $KAl(SO_4)_2 \cdot 12H_2O$. Potassium alum crystal, ammonium alum crystal $(NH_4)_2SO_4 \cdot Fe_2(SO_4)_3 \cdot 24H_2O$, and chrome alum crystal $K_2SO_4 \cdot Cr_2(SO_4)_3 \cdot 24H_2O$ are isomorphous crystals.

Potassium alum is very important in industry, and is commonly used in water purification, sewage treatment, leather tanning, dyeing, baking powder, fire textiles and fire extinguishers.

In this experiment, aluminum metal reacts first with aqueous NaOH producing a soluble sodium aluminate salt solution.

$$2Al+2NaOH+6H_2O \xlongequal{} 2Na[Al(OH)_4]+3H_2\uparrow$$

$$\Delta_r G^{\ominus}_{m,298} = -867.6\ kJ\cdot mol^{-1}$$

Second, H_2SO_4 is added to adjust the pH to 8~9, and the aluminate ion, $Al(OH)_4^-$, will precipitate as aluminum hydroxide.

$$2Na[Al(OH)_4]+H_2SO_4 \xlongequal{} 2Al(OH)_3\downarrow+Na_2SO_4+2H_2O$$

Then, H_2SO_4 is used to dissolve the aluminum hydroxide.

$$2Al(OH)_3+3H_2SO_4 \xlongequal{} Al_2(SO_4)_3+6H_2O$$

Finally, K_2SO_4, which has the same number of moles as $Al_2(SO_4)_3$, is added to produce the potassium alum.

$$Al_2(SO_4)_3+K_2SO_4+24H_2O \xlongequal{} 2KAl(SO_4)_2\cdot 12H_2O$$

Under room temperature, the solubility of the potassium alum is lower than that of $Al_2(SO_4)_3$ or K_2SO_4. Therefore, the potassium alum forms octahedral-shaped crystals when the nearly saturated solution cools. Table 4-3 shows their solubilities.

Table 4-3 Solubilities of three compounds at different temperatures (g/100g H_2O)

Compound	Temperature /℃					
	0	20	30	40	60	100
K_2SO_4	7.4	11.1	13.0	14.8	18.2	24.1
$Al_2(SO_4)_3$	31.2	36.4	40.4	45.8	59.2	89.0
$KAl(SO_4)_2$	3.0	5.90	8.39	11.7	24.8	109.0（90℃）

【Apparatus and Chemicals】

Erlenmeyer flask (250mL), beaker (250mL, 100mL), electronic balance, filter flask and Büchner funnel.

Al, K_2SO_4 (s), potassium alum microcrystal, NaOH (s), H_2SO_4 ($3mol\cdot L^{-1}$, $6mol\cdot L^{-1}$), $Cr_2(SO_4)_3$ ($1mol\cdot L^{-1}$).

【Experimental Procedure】

1. Producing the sodium aluminate salt solution

Weigh 2.0g of solid NaOH, and put it into a 250mL Erlenmeyer flask. Add 25mL of distilled water to dissolve the NaOH. Add 1.0g of Al pieces slowly. (*Caution*: Move the flask to a well-ventilated area, such as a fume hood. Do not splatter the solution. NaOH is caustic.) When no further reaction is evident, add distilled water to a total volume of 40mL. Vacuum filter the warm solution.

2. Formation of aluminum hydroxide

Boil the above filtrate in a 250mL beaker. While stirring, slowly add $3mol\cdot L^{-1}$ H_2SO_4 until the pH is 8~9. Continue heating and stirring for several minutes, and then vacuum filter the mixture. Wash the $Al(OH)_3$ precipitates with hot water until the pH of the filtrate is 7~8.

3. Formation of potassium alum

Transfer the precipitates to a 100mL evaporating dish, and add 12mL of $6mol\cdot L^{-1}$ H_2SO_4. Heat carefully while stirring until the $Al(OH)_3$ dissolves. Add a hot saturated K_2SO_4 solution, which is prepared with 3.3g K_2SO_4, into the above solution. Stir and cool the solution. After the crystals are formed completely, vacuum filter the mixture. Finally, weigh and calculate the yield.

【Questions and Discussion】

1. Why do we use the NaOH instead of an acid to dissolve Al?

2. In which step will the iron impurity in the Al pieces be removed?

3. Why do we boil and stir the solution when producing $Al(OH)_3$? Which ion do we want to remove when we wash the $Al(OH)_3$ precipitates with hot water?

4. How many grams of aluminum sulfate can be produced from 1.0 g of Al? If we want to get potassium alum with the above aluminum sulfate, how many grams of potassium sulfate do we need?

5. In the step of $Al(OH)_3$ formation, the H_2SO_4 is used to adjust the pH. Saturated NH_4HCO_3 or $(NH_4)_2CO_3$ can also be used to adjust the pH. Which is better and why?

实验15 莫尔盐（硫酸亚铁铵）的制备

【实验目的】

1．练习无机制备的基本操作。

2．了解复盐的一般特征和制备。

3．掌握某些铁化合物的基本性质。

【实验提要】

莫尔盐即硫酸亚铁和硫酸铵的复盐。它比一般的亚铁盐稳定，在空气中不易被氧化，溶于水而不溶于乙醇。

像所有复盐一样，莫尔盐在水中的溶解度比组成它的任何组分的溶解度都要小（见表4-4），因此从含有 $FeSO_4$ 和 $(NH_4)_2SO_4$ 的混合溶液中很容易得到结晶的莫尔盐 $(NH_4)_2SO_4 \cdot FeSO_4 \cdot 6H_2O$。

本实验是先将金属铁溶于稀硫酸得到亚铁盐溶液，然后往亚铁盐溶液中加入硫酸铵，使之全部溶解，经加热浓缩，放置冷却则析出复盐。

$$Fe+H_2SO_4(稀) = FeSO_4+H_2\uparrow$$

$$FeSO_4+(NH_4)_2SO_4+6H_2O = (NH_4)_2SO_4 \cdot FeSO_4 \cdot 6H_2O$$

铁屑表面常含有油污，反应前须经净化处理。铁屑与酸反应过程中有 H_2S、PH_3 等有毒气体放出，反应须在良好的通风条件下进行。加热可促进铁屑与酸的反应，但较长时间的加热容易析出 $FeSO_4 \cdot H_2O$ 化合物。

产品主要杂质是Fe(Ⅲ)，利用 Fe^{3+} 与 SCN^- 形成血红色配离子 $[Fe(SCN)_n]^{(3-n)}$ 的颜色深浅，通过目视比色可判断产品中 Fe^{3+} 的含量（评定产品级别）。评定办法是：称取1.0g产品置于25mL比色管中，用15mL不含氧的蒸馏水（指不溶解有氧气，下同）溶解，再加入2mL稀 $HCl(3mol \cdot L^{-1})$ 和1mL质量分数为25%的KSCN溶液，再用不含氧的蒸馏水稀释至刻度线，摇匀，并与 Fe^{3+} 标准液比较，确定产品级别。标准溶液由实验室提供。各级标准液含 Fe^{3+} 量为：一级（0.05mg），二级（0.10mg），三级（0.20mg）。

表4-4 三种盐的溶解度 单位：g/100g H_2O

温度/℃	$FeSO_4 \cdot 7H_2O$	$(NH_4)_2SO_4$	$(NH_4)_2SO_4 \cdot FeSO_4 \cdot 6H_2O$
10	20.0	73	17.2
20	26.5	75.4	21.6
30	32.9	78	28.1

【仪器和药品】

25mL比色管，蒸发皿，电热恒温水浴锅，抽滤瓶，布氏漏斗，锥形瓶（250mL），天平。

铁屑，稀 $H_2SO_4\,(2mol \cdot L^{-1})$，$HCl\,(3mol \cdot L^{-1}, AR)$，$(NH_4)_2SO_4(s)$，$Na_2CO_3$（质量分数为10%），KSCN（质量分数为25%），Fe^{3+} 标准溶液，滤纸。

【实验内容】

1．铁屑的净化

称取4.0g铁屑，放入250mL锥形瓶中，加入20mL Na_2CO_3 溶液，小火加热10min，倾去碱液，再用水将铁屑洗净。

2．$FeSO_4$溶液的制备

在上述锥形瓶中，加入稀H_2SO_4（体积计算值×110%，mL），于通风橱中在水浴上加热，至仅有很少气泡放出时趁热抽滤，用 5mL 水洗涤锥形瓶及漏斗上残渣，抽干，滤液转入蒸发皿中，残渣连滤纸烘干后称重（过滤前滤纸先称重并记录质量）。

3．硫酸亚铁铵的制备

用实际反应的铁屑量确定$(NH_4)_2SO_4$固体所用的量。称取所需的$(NH_4)_2SO_4$固体，加到$FeSO_4$溶液中，此时溶液的 pH 值应小于 2。如 pH 值偏高，可用H_2SO_4溶液调节。待固体全部溶解后，水浴蒸发浓缩至表面刚出现结晶膜为止（注意浓缩过程不宜过多搅拌）。放置冷却，得硫酸亚铁铵晶体。抽滤，把晶体转移至表面皿上晾干片刻，观察晶体的颜色和形状，称量、计算产率。

【选做部分】

质量评定：按提要中介绍的方法进行，但要求做空白实验加以比较，或者先于比色管中加入 15mL 不含氧的蒸馏水，加入 2mL HCl 和 1mL KSCN 溶液，观察溶液颜色后再加入产品 1.0g，溶解后继续加不含氧的水至 25mL 刻度线，摇匀，与标准色阶比色，确定产品级别。

【问题与讨论】

1．溶解铁屑时，为什么H_2SO_4要略过量？若H_2SO_4量不足，可能会有什么后果？

2．怎样制备不含氧的蒸馏水？

3．浓缩硫酸亚铁铵溶液时，能否把溶液蒸干？若把溶液蒸干，可能会有什么结果？

4．如果产品质量在三级以下，可用什么方法来处理产品以提高质量（写出具体措施）？

Experiment 15 Synthesis of Mohr's Salt (Ammonium Ferrous Sulfate)

【Objectives】

1．To practice the laboratory operations of synthesizing compounds.

2．To understand the characterization and synthesis of double salts.

3．To learn the characterization of some iron compounds.

【Introduction】

Ammonium iron(Ⅱ) sulfate, or Mohr's salt, is the inorganic compound with the formula $(NH_4)_2SO_4 \cdot FeSO_4 \cdot 6H_2O$. It is a double salt of ferrous sulfate and ammonium sulfate. Mohr's Salt is more stable than common ferrous salt, and will not be oxidized in the air.

The solubility of double salt is lower than its component (Table 4-4). Therefore, if $FeSO_4$ solution and $(NH_4)_2SO_4$ solution are mixed, crystals of Mohr's salt can be obtained after concentration and crystallization.

In this experiment, metal iron is first dissolved in dilute sulfuric acid to produce $FeSO_4$. Then $(NH_4)_2SO_4$ is added until it is dissolved completely. Mohr's salt can be formed after concentration and cooling.

$$Fe+H_2SO_4\,(dilute) = FeSO_4+H_2\uparrow$$

$$FeSO_4+(NH_4)_2SO_4+6H_2O = (NH_4)_2SO_4 \cdot FeSO_4 \cdot 6H_2O$$

The scrap iron, usually having grease stain on the surface, needs to be cleaned before the experiment. The poisonous gas, H_2S or PH_3, will be produced during the experiment. So this reaction need to be done at a well-ventilated area, such as a fume hood. Heating can increase the reaction rate, but compound $FeSO_4 \cdot H_2O$ will precipitate after heating for a long time. The major impurity in the product is Fe(Ⅲ). The content of Fe(Ⅲ) can be determined by a colorimetric method. Fe^{3+} reacts with SCN^- to form the blood-red complex $[Fe(SCN)_n]^{(3-n)}$. The color of the solution indicates the concentration of Fe^{3+}. The procedure is as follows: Weigh 1.0 g product and transfer it into a 25mL color comparison cylinder. Dissolve the product with distilled water without dissolved oxygen. Add 2mL of dilute HCl ($3mol \cdot L^{-1}$) and 1mL of KSCN (25%, wt/wt), and continue adding distilled water to dilute the solution to the volume of 25mL. Shake well, compare the color with standard solutions, and determine the quality grade of the product. The limit of amount of Fe^{3+} per gram of ammonium ferrous sulfate in 1st grade, 2nd grade and 3rd grade is 0.05mg, 0.10mg and 0.20mg, respectively.

Table 4-4 Solubilities of three salts（g/100g H_2O）

Temperature/℃	$FeSO_4 \cdot 7H_2O$	$(NH_4)_2SO_4$	$(NH_4)_2SO_4 \cdot FeSO_4 \cdot 6H_2O$
10	20.0	73	17.2
20	26.5	75.4	21.6
30	32.9	78	28.1

【Apparatus and Chemicals】

25mL Color comparison cylinder, evaporating dish, electric-heated thermostatic water bath, filter flask and Büchner funnel, Erlenmeyer flask (250mL), electronic balance.

Scrap iron, dilute H_2SO_4 ($2mol \cdot L^{-1}$), HCl ($3mol \cdot L^{-1}$, AR), $(NH_4)_2SO_4$ (s), Na_2CO_3 (10%, wt/wt), KSCN (25%, wt/wt), Fe^{3+} standard solutions, filter paper.

【Experimental Procedure】

1. Cleaning of the scrap iron

Weigh 4.0g of scrap iron, and place it in a 250mL Erlenmeyer flask. Add 20mL of Na_2CO_3 solution, and heat with low heat for 10min. Decant the alkaline solution, and wash the scrap iron with water.

2. Preparation of $FeSO_4$

Add dilute H_2SO_4 (the volume equals the calculated value ×110%) to the Erlenmeyer flask with iron. In a fume hood, heat the mixture with a water bath. When no bubbles emerge, vacuum filter the mixture while it is still hot. Wash the precipitate in the Erlenmeyer flask and Büchner funnel with 5mL of water. Transfer the filtrate to an evaporating dish. Weigh the dried precipitate and filter paper together. (*Caution*: the filter paper should also be weighed before vacuum filtering.)

3. Synthesis of ammonium ferrous sulfate

According to the amount of iron in solution, calculate and add the proper amount of $(NH_4)_2SO_4$ to the $FeSO_4$ solution. The pH of the solution should be less than 2. Use H_2SO_4 to adjust the pH if it is too high. Heat and stir the solution in the water bath until all the reagents dissolve completely. Continue heating the solution until there is a thin layer of solid on the surface of the solution. Do not stir during the concentration. Cool the mixture to room temperature. Vacuum filter to get the crystals. Observe the obtained crystals on a watch glass. Weigh and calculate the yield.

【Questions and Discussion】

1. Why should the excessive H_2SO_4 be used to dissolve the scrap iron? What will happen if the amount of H_2SO_4 is not enough?

2. How do you properly prepare distilled water without dissolved oxygen?

3. Can we evaporate the mixture to dryness when doing the concentration of ammonium ferrous sulfate? What will happen if we evaporate the mixture to dryness?

4. If the product quality is lower than 3rd grade, what can you do to improve the quality? Please write the detailed procedures.

实验16　草酸配合物的合成

【实验目的】

1．学习两种草酸配合物 $K_3[M(C_2O_4)_3]\cdot 3H_2O$ (M=Cr，Fe)的合成方法，加深对配合物性质的了解。

2．进一步巩固无机化学实验操作。

【实验提要】

（1）$K_3[Cr(C_2O_4)_3]\cdot 3H_2O$ 可由草酸钾和草酸与重铬酸钾反应制得：

$$Cr_2O_7^{2-}+6H_2C_2O_4+2H^+ = Cr_2(C_2O_4)_3+6CO_2(g)+7H_2O$$

$$Cr_2(C_2O_4)_3+3K_2C_2O_4+6H_2O = 2K_3[Cr(C_2O_4)_3]\cdot 3H_2O \text{（深绿色）}$$

总反应式为：

$$K_2Cr_2O_4+7H_2C_2O_4+2K_2C_2O_4 = 2K_3[Cr(C_2O_4)_3]\cdot 3H_2O+6CO_2(g)+H_2O$$

（2）$K_3[Fe(C_2O_4)_3]\cdot 3H_2O$ 可由草酸亚铁 $FeC_2O_4\cdot 2H_2O$ 在草酸钾和草酸存在下用过氧化氢氧化而成：

$$2FeC_2O_4\cdot 2H_2O+H_2O_2+3K_2C_2O_4+H_2C_2O_4 = 2K_3[Fe(C_2O_4)_3]\cdot 3H_2O \text{（绿色）}$$

加入适量乙醇，其晶体便从溶液析出。

本实验中，草酸亚铁由莫尔盐和草酸反应而得：

$$(NH_4)_2Fe(SO_4)_2\cdot 6H_2O+H_2C_2O_4 = FeC_2O_4\cdot 2H_2O\downarrow+(NH_4)_2SO_4+H_2SO_4+4H_2O$$

【仪器和药品】

温度计（0~100℃），量筒（10mL，100mL），漏斗，漏斗架，布氏漏斗，抽滤瓶，表面皿，烧杯，天平。

H_2SO_4(1mol·L^{-1})，$H_2C_2O_4$(1mol·L^{-1})，$K_2C_2O_4$（饱和），H_2O_2(质量分数为3%)，$K_2Cr_2O_7$（s），$(NH_4)_2Fe(SO_4)_2\cdot 6H_2O$(s)，乙醇（体积分数95%），丙酮。

【实验内容】

（1）在200mL烧杯中加入10.0mL $K_2C_2O_4$（饱和）和8.0mL $H_2C_2O_4$溶液，再慢慢加入3.0 g磨细了的$K_2Cr_2O_7$(s)，并不断搅拌待反应完毕，蒸发至溶液近干，使晶体析出，冷却，过滤。用丙酮洗涤得深绿色 $K_3[Cr(C_2O_4)_3]\cdot 3H_2O$ 晶体，在110℃下烘干。

（2）把5.0 g $(NH_4)_2Fe(SO_4)_2\cdot 6H_2O$ (s)放入200mL烧杯中，加入1mL H_2SO_4和15mL蒸馏水，加热使其溶解。继续加热至沸腾，同时滴入$H_2C_2O_4$溶液并不断搅拌（$H_2C_2O_4$实际用量20~25mL）。静置，待黄色晶体沉降后倾去上层清液，往沉淀中加入20mL蒸馏水，搅拌并加热至沸，静置，倾去清液。

于上述沉淀中加入$K_2C_2O_4$饱和溶液（约10mL）。水浴加热至40℃，缓慢加入H_2O_2溶液（20mL），不断搅拌并保持温度在40℃左右（此时含有氢氧化铁沉淀）。将溶液煮沸，再加入$H_2C_2O_4$（总量8mL，分两次加入，第一次一次性加入5mL，第二次缓慢滴加3mL），待溶液近沸腾时趁热过滤（用100mL烧杯承接滤液），于滤液中加入95%的乙醇（约10mL），如有晶体析出，应温热使之溶解。用表面皿盖住烧杯，冷却放置过夜，晶体完全析出后，抽滤，称量并计算产率。

【问题与讨论】

1．在制备 $K_3[Cr(C_2O_4)_3]\cdot 3H_2O$ 时，能否把溶液完全蒸干使晶体析出？用丙酮洗涤 $K_3[Cr(C_2O_4)_3]\cdot 3H_2O$ 结晶的作用是什么？

2．在制备 $K_3[Cr(C_2O_4)_3]\cdot 3H_2O$ 时，为什么反应中 $H_2C_2O_4$、$K_2C_2O_4$、H_2O_2 均要求过量？

3．于 $K_3[Fe(C_2O_4)_3]\cdot 3H_2O$ 溶液中加入乙醇的作用是什么？

4．注意第二步反应（即氧化-配位反应）中的条件控制和操作程序，试分析说明下述操作的意义：

① 加入 H_2O_2 时强调缓慢滴加，且温度不得高于 40℃；

② 加入 H_2O_2 后要加热至沸腾；

③ $K_2C_2O_4$ 需要在加入 H_2O_2 前加入，而 $H_2C_2O_4$ 则需在加入 H_2O_2 后加入；

④ 加入 $H_2C_2O_4$ 时分两次进行，且第二次要缓慢滴加。

5．本实验为什么不直接用三价铁盐与草酸钾反应来制备三草酸根合铁（Ⅲ）酸钾？

Experiment 16 Synthesis of Oxalate Complexes

【Objectives】

1. To learn how to synthesize the oxalate complexes $K_3[M(C_2O_4)_3]\cdot 3H_2O$ (M=Cr, Fe), and to further understand the properties of the coordination compounds.

2. To develop laboratory techniques.

【Introduction】

（1）$K_3[Cr(C_2O_4)_3]\cdot 3H_2O$ is obtained by the following reactions:

$$Cr_2O_7^{2-}+6H_2C_2O_4+2H^+ = Cr_2(C_2O_4)_3+6CO_2(g)+7H_2O$$

$$Cr_2(C_2O_4)_3+3K_2C_2O_4+6H_2O = 2K_3[Cr(C_2O_4)_3]\cdot 3H_2O \text{ (dark green)}$$

The overall reaction is:

$$K_2Cr_2O_4+7H_2C_2O_4+2K_2C_2O_4 = 2K_3[Cr(C_2O_4)_3]\cdot 3H_2O+6CO_2(g)+H_2O$$

（2）Together with $K_2C_2O_4$ and $H_2C_2O_4$, $FeC_2O_4\cdot 2H_2O$ will be oxidized to $K_3[Fe(C_2O_4)_3]\cdot 3H_2O$ by H_2O_2. Potassium trioxalatoferrate (Ⅲ) is a green crystal.

$$2FeC_2O_4\cdot 2H_2O+H_2O_2+3K_2C_2O_4+H_2C_2O_4 = 2K_3[Fe(C_2O_4)_3]\cdot 3H_2O \text{(green)}$$

Adding ethanol to the above solution will form green crystals.

$FeC_2O_4\cdot 2H_2O$ is obtained by the reaction of $(NH_4)_2Fe(SO_4)_2\cdot 6H_2O$ with oxalic acid.

$$(NH_4)_2Fe(SO_4)_2\cdot 6H_2O+H_2C_2O_4 = FeC_2O_4\cdot 2H_2O\downarrow+(NH_4)_2SO_4+H_2SO_4+4H_2O$$

【Apparatus and Chemicals】

Thermometer (0~100℃), graduated cylinder (10mL, 100mL), funnel, filter flask and Büchner funnel, watch glass, beaker, electronic balance.

H_2SO_4 (1mol·L^{-1}), $H_2C_2O_4$ (1mol·L^{-1}), $K_2C_2O_4$ (saturated), H_2O_2 (3%, wt/wt), $K_2Cr_2O_7$(s), $(NH_4)_2Fe(SO_4)_2\cdot 6H_2O$ (s), ethanol (95%, vol/vol), acetone.

【Experimental Procedure】

(1) In a 200mL beaker, add 10.0mL of saturated $K_2C_2O_4$ and 8.0mL of $H_2C_2O_4$ solution. Then slowly add 3.0g of ground $K_2Cr_2O_7$ while stirring until the reaction is finished. Evaporate the

solution to get the crystals. Cool the mixture and filter it. Use acetone to wash the dark green crystals $K_3[Cr(C_2O_4)_3]\cdot 3H_2O$, and dry the crystals at 110℃.

(2) Weigh 5.0g of solid $(NH_4)_2Fe(SO_4)_2\cdot 6H_2O$, place it in a 200mL beaker, and add 1mL of H_2SO_4 and 15mL of distilled water. Dissolve the solid while heating. Continue heating the solution to boiling, and at the same time add $H_2C_2O_4$ (about 20~25mL) while stirring. Allow the yellow crystals to settle to the bottom of the beaker, and decant the supernatant. Next, add 20mL of distilled water to the crystals, stir, and heat till boiling. Allow the yellow crystals to settle to the bottom of the beaker, and decant the supernatant again.

Add saturated $K_2C_2O_4$ solution (about 10mL) to the above crystals. Heat the mixture in a water bath to 40℃, and then slowly add 20mL of H_2O_2. Stir continuously and keep the temperature at 40℃. Then, heat the mixture to boiling, and add $H_2C_2O_4$ solution (The first time add 5mL directly into the mixture, and the second time add 3mL by dripping slowly, for a total of 8mL). When the solution is almost boiling, filter the solution while it is hot. Finally, add 10mL of 95% ethanol into the filtrate. If crystals appear, warm the solution till the crystals dissolve. Cover the beaker with the watch glass. Cool down overnight and crystals will form. Vacuum filter and weigh the dry crystals. Calculate the yield.

【Questions and Discussion】

1. When obtaining $K_3[Cr(C_2O_4)_3]\cdot 3H_2O$ crystals, can we evaporate the solution to dryness to get the crystals? Why do we use acetone to wash $K_3[Cr(C_2O_4)_3]\cdot 3H_2O$ crystals?
2. Why should excessive $H_2C_2O_4$, $K_2C_2O_4$ and H_2O_2 be used to prepare $K_3[Cr(C_2O_4)_3]\cdot 3H_2O$?
3. What is the purpose of adding ethanol to the $K_3[Fe(C_2O_4)_3]\cdot 3H_2O$ solution?
4. Refer to the reaction condition and operation sequence in step two.

① The H_2O_2 has to be added slowly, and the temperature should not be above 40℃, why?

② The solution should be heated to boiling after adding H_2O_2, why?

③ $K_2C_2O_4$ is added before H_2O_2, while $H_2C_2O_4$ is added after H_2O_2, why?

④ $H_2C_2O_4$ has to be added in two portions, and the second portion should be added slowly, why?

5. Why don't we use ferric (Ⅲ) salt and potassium oxalate to react directly to get potassium trioxalatoferrate (Ⅲ)?

实验 17　硫代硫酸钠的制备

【实验目的】

1. 熟悉以 Na_2S 和 Na_2SO_3 为原料制备 $Na_2S_2O_3\cdot 5H_2O$ 的基本原理，反应条件及实验操作，了解复相反应的基本特点。

2. 制备 $Na_2S_2O_3\cdot 5H_2O$，进一步巩固 $S_2O_3^{2-}$ 的定性鉴定反应。

【实验提要】

工业上制备硫代硫酸钠通常用亚硫酸钠溶液在沸腾温度下直接与硫粉化合：

$$Na_2SO_3+S \xlongequal{\triangle} Na_2S_2O_3$$

从溶液中结晶得到的硫代硫酸钠一般为 $Na_2S_2O_3\cdot 5H_2O$，俗称大苏打或海波。

本实验用 Na_2SO_3 和 Na_2S 为原料来制备 $Na_2S_2O_3$。在酸性溶液中，Na_2S 首先被 Na_2SO_3 氧化成单质 S：

$$2S^{2-}(aq)+SO_3^{2-}(aq)+6H^+(aq) = 3S(s)+3H_2O(l) \qquad \Delta_r G^{\ominus}_{m,298}=-310.4\ kJ\cdot mol^{-1}$$

反应生成的 S 活性很高，很容易与溶液中的 Na_2SO_3 进一步反应生成 $Na_2S_2O_3$。由于 $Na_2S_2O_3$ 对酸不稳定，故控制好体系酸度是实验成功的关键。本实验的设计是先用浓盐酸与固体亚硫酸钠反应生成 SO_2 气体：

$$Na_2SO_3(s)+2HCl(浓) = SO_2(g)+2NaCl+H_2O$$

然后把反应生成的 SO_2 气体通入 Na_2S 和 Na_2SO_3 的混合溶液中，SO_2 被吸收

$$S^{2-}(aq)+H_2O(l)+SO_2(g) = SO_3^{2-}(aq)+H_2S(aq) \qquad \Delta_r G^{\ominus}_{m,298}=-62.63\ kJ\cdot mol^{-1}$$

$$2H_2S(aq)+SO_2(g) = 3S(s)+2H_2O(l) \qquad \Delta_r G^{\ominus}_{m,298}=-118.6\ kJ\cdot mol^{-1}$$

$$S(s)+SO_3^{2-}(aq) \xlongequal{333\sim353K} S_2O_3^{2-}(aq)$$

总反应为：

$$2Na_2S+Na_2SO_3+3SO_2 \xlongequal{333\sim353K} 3Na_2S_2O_3$$

实际上，也可以把 SO_2 气体通入 Na_2S 和 Na_2CO_3 的混合溶液中，这时部分 SO_2 气体与 Na_2CO_3 反应转化为 Na_2SO_3：

$$CO_3^{2-}(aq)+SO_2(g) = SO_3^{2-}(aq)+CO_2(g) \qquad \Delta_r G^{\ominus}_{m,298}=-52.87\ kJ\cdot mol^{-1}$$

总反应为：

$$2Na_2S(aq)+Na_2CO_3(aq)+4SO_2(g) \xlongequal{333\sim353K} 3Na_2S_2O_3(aq)+CO_2(g)$$

反应是定量进行的，反应过程中溶液的 pH 值逐渐下降，至接近中性（pH≥7.0）时即可停止通入 SO_2 气体。溶液经浓缩、冷却结晶，抽滤即可得到 $Na_2S_2O_3\cdot 5H_2O$。

【仪器和药品】

硫代硫酸钠制备装置如图 4-1 所示。

原料及辅助药品用量与配制。

A 瓶：浓 HCl 50mL；

B 瓶：Na_2SO_3(s)约 25g；

C 瓶：$Na_2S\cdot 6H_2O$ 25g（或 $Na_2S\cdot 9H_2O$，33g），Na_2SO_3 8.5g（或 Na_2CO_3，7.1g），蒸馏

水 150mL；

D 瓶：$3mol \cdot L^{-1}$ NaOH 约 100mL。

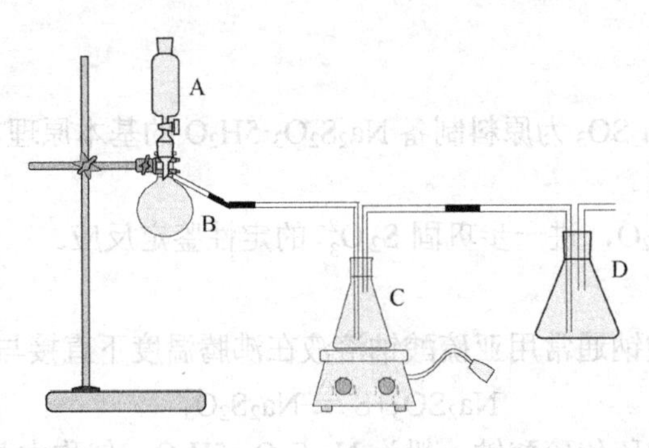

图 4-1　硫代硫酸钠制备装置

【实验内容】

按图 4-1 所示装置，根据制备规模要求定量配制好原料和辅助药品，检查并确保装置密闭后，开启电磁搅拌器并加热 C 瓶溶液，温度控制在 333～353K 范围。小心开启分液漏斗 A 的活塞，让 HCl 缓慢滴下，控制 HCl 的滴加速度使 B 瓶反应生成 SO_2，SO_2 缓慢而均匀地通入反应瓶 C 中。C 瓶开始出现浑浊，随着反应的进行，C 瓶溶液又逐渐变清，这预示 C 瓶中反应接近终点。反应后期可适当加热 B 瓶。当 C 瓶溶液刚好完全变清晰透明时停止通 SO_2。检查 C 瓶溶液 pH 值（若 pH＜7.0，用适量 Na_2CO_3 溶液调节，确保溶液 pH≥7.0），提高加热温度使 C 瓶溶液近沸腾，趁热抽滤。滤液转入蒸发皿蒸发浓缩至溶液表面出现晶膜（注意浓缩过程要搅拌），静置冷却，抽滤得结晶 $Na_2S_2O_3 \cdot 5H_2O$，在 315K 以下烘干 40～60min，称重，计算产率，并作产品定性鉴定。

【注意事项】

1．安装实验装置时，在各连接处（尤其是磨砂接口）涂上适量凡士林，可增强系统气密性和易于拆装。

2．C 瓶中气体导管应尽量插入溶液底部。

3．如果反应一开始 C 瓶就有大量 S 析出，表明 SO_2 生成的反应速率过快，应适当减慢 HCl 滴入速度。

4．注意避免倒吸。

【问题与讨论】

1．为什么投料时，浓盐酸和 B 瓶中 Na_2SO_3 固体用量要比理论量多？

2．若反应结束后 $Na_2S_2O_3$ 溶液的 pH＜7.0，如果没有调节 pH 值便直接蒸发浓缩，会有什么不良后果？

3．本实验为什么不把盐酸直接滴加到 Na_2SO_3 混合溶液中来制备 $Na_2S_2O_3$？

【安全知识】

二氧化硫具有刺激性气味，给人体及环境带来毒害与污染。主要对人体造成黏膜及呼吸道损害，引起流泪、流涕、咽干、咽痛等症状及呼吸系统炎症，大量吸入会导致窒息死亡。因此凡涉及产生二氧化硫的反应都要采取相应措施，减少二氧化硫逸出并在通风橱内进行。

Experiment 17 Preparation of Sodium Thiosulfate

【Objectives】

1. To learn the principle, the reaction conditions, and the operation of how to prepare $Na_2S_2O_3 \cdot 5H_2O$ with Na_2S and Na_2SO_3

2. To prepare $Na_2S_2O_3 \cdot 5H_2O$, and to further learn the identification of $S_2O_3^{2-}$.

【Introduction】

Sodium thiosulfate ($Na_2S_2O_3$) is a colorless crystalline compound that is more familiar as the pentahydrate, $Na_2S_2O_3 \cdot 5H_2O$, an efflorescent, monoclinic crystalline substance also called sodium hyposulfite or hypo.

On an industrial scale, sodium thiosulfate is prepared by directly heating an aqueous solution of sodium sulfite with sulfur.

$$Na_2SO_3 + S \xlongequal{\triangle} Na_2S_2O_3$$

In this experiment, Na_2SO_3 and Na_2S will be used to prepare $Na_2S_2O_3$. In acid solution, Na_2S will be oxidized by Na_2SO_3 to produce elemental S.

$$2S^{2-}(aq) + SO_3^{2-}(aq) + 6H^+(aq) = 3S(s) + 3H_2O(l) \qquad \Delta_r G^{\ominus}_{m,298} = -310.4\ kJ \cdot mol^{-1}$$

The S that is produced is highly reactive, and easily reacts with Na_2SO_3 to produce $Na_2S_2O_3$. Because $Na_2S_2O_3$ is unstable in acid, it is important to control the acidity in this experiment. Therefore, SO_2 gas is produced first by using concentrated HCl and solid sodium sulfite.

$$Na_2SO_3(s) + 2HCl(conc.) = SO_2(g) + 2NaCl + H_2O$$

Then, put SO_2 gas through the mixed solution of Na_2S and Na_2SO_3, and SO_2 gas will be absorbed.

$$S^{2-}(aq) + H_2O(l) + SO_2(g) = SO_3^{2-}(aq) + H_2S(aq) \qquad \Delta_r G^{\ominus}_{m,298} = -62.63\ kJ \cdot mol^{-1}$$

$$2H_2S(aq) + SO_2(g) = 3S(s) + 2H_2O(l) \qquad \Delta_r G^{\ominus}_{m,298} = -118.6\ kJ \cdot mol^{-1}$$

$$S(s) + SO_3^{2-}(aq) \xlongequal{333 \sim 353K} S_2O_3^{2-}(aq)$$

The overall reaction is

$$2Na_2S + Na_2SO_3 + 3SO_2 \xlongequal{333 \sim 353K} 3Na_2S_2O_3$$

The mixed solution of Na_2S and Na_2CO_3 can also be used to absorb SO_2 gas, because Na_2CO_3 reacts with SO_2 gas to produce Na_2SO_3.

$$CO_3^{2-}(aq) + SO_2(g) = SO_3^{2-}(aq) + CO_2(g) \qquad \Delta_r G^{\ominus}_{m,298} = -52.87\ kJ \cdot mol^{-1}$$

The overall reaction is

$$2Na_2S(aq) + Na_2CO_3(aq) + 4SO_2(g) \xlongequal{333 \sim 353K} 3Na_2S_2O_3(aq) + CO_2(g)$$

The pH of the solution will be reduced during the reaction. Stop the reaction when pH≥7.0. Concentrate and cool the solution down to form the crystals, $Na_2S_2O_3 \cdot 5H_2O$.

【Apparatus and Chemicals】

Figure 4-1 is the setup of the apparatus for the preparation of sodium thiosulfate.

Preparation of chemicals in the apparatus.

A: 50mL of concentrated HCl;

B: 25g of Na_2SO_3 (s);

C: 25g of $Na_2S{\cdot}6H_2O$ (or 33g of $Na_2S{\cdot}9H_2O$), 8.5g of Na_2SO_3 (or 7.1g of Na_2CO_3), 150mL of distilled water;

D: 100mL of $3mol{\cdot}L^{-1}$ NaOH.

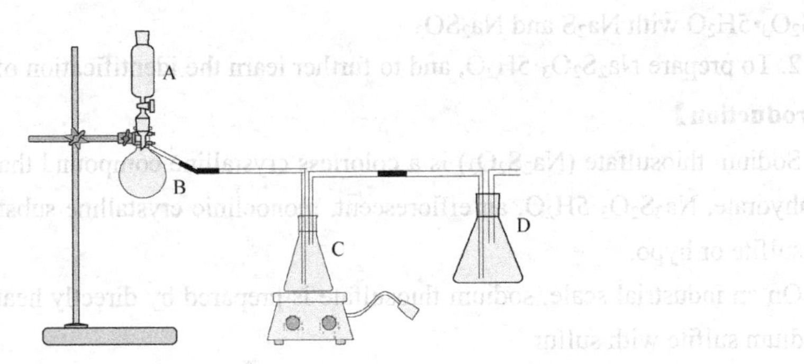

Figure 4-1 Setup of the apparatus for the preparation of sodium thiosulfate

【Experimental Procedure】

Assemble the apparatus as shown in Figure 4-1, and add the appropriate amount of chemicals into the apparatus. Make sure the connections are tight. Start to heat flask C, and control the temperature in the range of 333～353K. Turn on the stopcock of separating funnel A, so HCl can be added to Flask B to produce SO_2 gas. Because of the reaction with SO_2 gas, Flask C should produce precipitate first, and then the precipitate disappears and the solution starts to clear again. This means the reaction is close to the end. Flask B can be heated at a later time in the experiment if necessary. Stop SO_2 gas when the solution in Flask C just turns totally clear. Test the pH of the solution, and make sure the pH$\geqslant$7.0. Use Na_2CO_3 solution to adjust the pH if it is less than 7.0. Increase the temperate and heat the solution to boiling. Vacuum filter while the solution is hot. Transfer the filtrate to an evaporating dish and concentrate the solution, while stirring, until a film of crystals forms. Finally cool the solution down, and vacuum filter to obtain $Na_2S_2O_3{\cdot}5H_2O$ crystals. Dry the crystals at 315K for 40 to 60 minutes. Weigh and calculate the yield. Qualitatively determine the product.

【Questions and Discussion】

1. Why is the amount of concentrated HCl and Na_2SO_3 solid used for the experiment more than the theoretical calculation?

2. If the pH of $Na_2S_2O_3$ solution is less than 7.0 when the reaction is over, what will happen when we evaporate and concentrate the solution without adjusting pH?

3. Why don't we directly add HCl to Na_2SO_3 mixed solution to prepare $Na_2S_2O_3$?

第5章 元素及其化合物的性质
Chapter 5 Properties of Elements and Compounds

实验18 s区元素及其化合物

【实验目的】

1. 了解金属钠和镁的活泼性。
2. 试验并了解少数锂、钠和钾盐的微溶性。
3. 试验并掌握镁、钙和钡的氢氧化物、硫酸盐、碳酸盐、草酸盐以及铬酸盐的溶解性。
4. 练习用焰色反应鉴定某些碱金属和碱土金属离子的方法。

【实验提要】

s区元素的单质都是较活泼的金属，特别是碱金属，其熔点、沸点和硬度都较低，密度也较小，是典型的轻金属。例如：钾和钠在空气中易被氧化，用小刀切割金属钾和钠，切割后的新鲜表面可以看到银白色的光泽，但接触空气后，由于生成一层氧化物而颜色变暗。钠在空气中燃烧直接得到过氧化钠，它们与水又能剧烈作用，因此一般储存于煤油中。碱土金属的活泼性比碱金属略差，如镁与冷水反应很慢，在加热时反应加快。

碱金属的氢氧化物易溶于水，固体碱吸湿性强，易潮解，因此固体NaOH是常用的干燥剂。碱土金属的氢氧化物在水中的溶解度一般都不大，同族元素氢氧化物的溶解度从上到下逐渐增大，这是由于随着离子半径的增大，阳离子和阴离子之间的吸引力逐渐减小，容易被水分子拆开的缘故。

碱金属的盐类一般都易溶于水，但也有若干锂盐和少数具有较大阴离子的盐，如$Na[Sb(OH)_6]$、$KHC_4H_4O_6$、$KClO_4$等较难溶于水。碱土金属盐的重要特征是其微溶性，除氯化物、硝酸盐、硫酸镁、铬酸镁易溶于水外，其余的碳酸盐、硫酸盐、草酸盐和铬酸盐等皆难溶于水。

碳酸铵溶液与镁盐溶液只有在煮沸或持久放置时才生成白色碱式碳酸镁沉淀。如果有强酸的铵盐存在，则无沉淀形成。因为当碳酸铵加入时，高浓度的铵离子将减少溶液中的CO_3^{2-}的浓度，以致不能达到碳酸镁的溶度积。

碱金属钙、锶、钡等挥发性盐在无色的火焰中灼烧时，能使火焰呈现出一定的颜色，这叫"焰色反应"。 利用焰色反应，可以根据火焰的颜色定性鉴别这些元素的存在。碱金属和几种碱土金属的焰色如表5-1所示。

表5-1 碱金属和几种碱土金属的焰色

离子	Li^+	Na^+	K^+	Rb^+	Cs^+	Ca^{2+}	Sr^{2+}	Ba^{2+}
焰色	红	黄	紫	紫红	紫红	橙红	洋红	黄绿

【仪器和药品】

酒精喷灯，漏斗，镍丝，坩埚钳，钴玻璃，酒精灯，试管，试管夹，滤纸，pH试纸。

（除特别注明外，试剂浓度单位为 $mol \cdot L^{-1}$）HCl (2)，HAc (2)，HNO_3 (浓)，$NH_3 \cdot H_2O$ (2, 1)，NaOH (2，新配制)，NaOH (s，0.5)，NaF (0.5)，LiCl (0.5)，NaCl (0.5)，KCl (0.5)，$MgCl_2$ (0.5)，$CaCl_2$ (0.5)，$BaCl_2$ (0.5)，$SrCl_2$ (0.5)，NH_4Cl (3)，Na_2CO_3 (0.5)，$(NH_4)_2CO_3$ (0.5)，K_2CrO_4 (0.5)，Na_2SO_4 (0.5)，$(NH_4)_2C_2O_4$ (饱和)，Na_2HPO_4 (0.5)，$K[Sb(OH)_6]$ (饱和)，$NaHC_4H_4O_6$ (饱和)，Na(s)，镁条，未知液（可能含有 K^+、Mg^{2+}、Ca^{2+}、Ba^{2+}）。

【实验内容】

1．碱金属与碱土金属活泼性的比较

（1）钠、镁与水的作用

① 用镊子取出一小块存放在煤油中的金属钠，置于滤纸上，并用滤纸吸干表面的煤油，然后用小刀切下米粒大小的金属钠，迅速观察新鲜表面的颜色和变化。再将它投入盛有半杯水的烧杯中，立即用大小合适的漏斗盖好，观察反应的情况，并检验溶液的酸碱性。

② 取擦净的镁条，放入盛有冷水的试管中，观察反应的情况，检验溶液的酸碱性。加热试管，情况又如何？

（2）钠、镁与氧气的反应

① 在蒸发皿上铺上用水润湿的滤纸，滤纸上放一小粒金属钠，不久冒出白烟并立即着火燃烧，产生黄色火焰和白烟。设法检验产物的酸碱性和氧化还原性。

② 用坩埚钳夹住 2cm 长的镁条一端，点燃后立即离开灯火，观察燃烧的情况及产物的颜色。设法检验产物的溶解性和酸碱性。

通过以上实验，写出实验现象及有关反应式，并比较ⅠA 族与ⅡA 族单质的活泼性。

2．氢氧化物的性质

（1）在玻璃片上放少量固体 NaOH，并置于空气中，过几分钟后，观察其变化，解释现象。

（2）于试管中加入 2mL $MgCl_2$ 溶液，再逐滴加入 $2mol \cdot L^{-1}$ $NH_3 \cdot H_2O$，观察生成沉淀的颜色。然后将沉淀分别与 $2mol \cdot L^{-1}$ HCl、$2mol \cdot L^{-1}$ NaOH 和 $3mol \cdot L^{-1}$ NH_4Cl 溶液作用，并观察现象。

（3）分别以 $MgCl_2$、$CaCl_2$ 和 $BaCl_2$ 3 种溶液与新配制的 $2mol \cdot L^{-1}$ NaOH 溶液为试剂，设计系列试管实验，说明碱土金属氢氧化物溶解度的大小顺序。

3．碱金属和碱土金属的难溶盐

（1）碱金属微溶盐

① 锂盐　取少量 LiCl 溶液分别与 NaF、Na_2CO_3 和 Na_2HPO_4 溶液反应。观察现象并写出反应式（必要时可加热）。

② 钠盐　取 1mL $0.5mol \cdot L^{-1}$ NaOH 溶液与饱和的 $K[Sb(OH)_6]$溶液作用，放置数分钟后，如无晶体析出，可用玻璃棒摩擦试管的内壁，观察晶体沉淀 $Na[Sb(OH)_6]$的生成。

③ 钾盐　取 1mL KCl 溶液与饱和的酒石酸氢钠（$NaHC_4H_4O_6$）溶液作用，放置数分钟后，如无结晶析出，可用玻璃棒摩擦试管的内壁，观察 $KHC_4H_4O_6$ 晶体的生成。

通过以上实验现象，可以对碱金属盐的溶解性能得出什么结论？

（2）碱土金属难溶盐

① 草酸盐　分别向 $MgCl_2$、$CaCl_2$、$BaCl_2$ 溶液中滴加饱和的$(NH_4)_2C_2O_4$ 溶液，观察生成沉淀的颜色，经离心分离后，分别试验沉淀与 $2mol \cdot L^{-1}$ HAc 和 $2mol \cdot L^{-1}$ HCl 的作用。

② 铬酸盐　用 K_2CrO_4 溶液代替饱和$(NH_4)_2C_2O_4$ 溶液，重复操作①，观察现象。

③ 碳酸盐　用 Na_2CO_3 溶液代替饱和$(NH_4)_2C_2O_4$ 溶液，重复操作①，观察现象。

用 5 滴 $1mol \cdot L^{-1}$ $NH_3 \cdot H_2O$+5 滴 $0.5mol \cdot L^{-1}$ $(NH_4)_2CO_3$+2 滴 $3mol \cdot L^{-1}$ NH_4Cl 的溶液代替

Na_2CO_3溶液重做上述实验，观察是否有沉淀生成？为什么？

④ 硫酸盐　分别向$MgCl_2$、$CaCl_2$和$BaCl_2$溶液加Na_2SO_4溶液，观察沉淀的颜色，并试验沉淀是否溶于浓HNO_3。

根据上述实验现象：

a．将沉淀及颜色填入表5-2。

b．讨论各难溶盐溶于酸的情况，并写出有关反应的离子方程式。

表5-2　离子的沉淀及颜色

试剂	离子的沉淀及颜色		
	Mg^{2+}	Ca^{2+}	Ba^{2+}
$(NH_4)_2C_2O_4$溶液			
K_2CrO_4溶液			
Na_2CO_3溶液			
$NH_3·H_2O+NH_4Cl+(NH_4)_2CO_3$溶液			
Na_2SO_4溶液			

⑤ 磷酸铵镁的生成　于0.5mL $MgCl_2$溶液中加入数滴2mol·L^{-1} HCl溶液和2滴Na_2HPO_4溶液，最后滴加2mol·L^{-1} $NH_3·H_2O$，振荡试管，观察白色$Mg(NH_4)PO_4$晶体的生成并写出反应式。

【选做部分】

（1）焰色反应　用一端弯成小圈的镍丝浸蘸浓HCl，在灯的氧化焰中灼烧，反复几次直至火焰接近无色，然后分别浸蘸0.5 mol·L^{-1} LiCl、NaCl、KCl、$MgCl_2$、$CaCl_2$、$SrCl_2$和$BaCl_2$溶液，并在灯的氧化焰中灼烧，观察颜色（观察K^+的焰色应通过钴玻璃）。

（2）未知液的分析　混合液中可能含有K^+、Mg^{2+}、Ca^{2+}和Ba^{2+}，试设计出分离和鉴定有关离子的方案，并试验之。

提示：

（1）可根据有关盐的溶解性设计分离步骤，并用焰色反应或其它特征反应鉴定各离子是否存在。

（2）可取1mL未知液于离心试管中进行试验，每当沉淀生成，要离心分离，将清液倒入另一离心试管中，然后分别继续下面的实验。

【问题与讨论】

1．为什么在试验$Mg(OH)_2$、$Ca(OH)_2$和$Ba(OH)_2$的溶解度时，所用的NaOH溶液必须是新配的？如何配制不含CO_3^{2-}的NaOH溶液？

2．$MgCl_2$溶液和$NH_3·H_2O$反应生成$Mg(OH)_2$和NH_4Cl，但$Mg(OH)_2$沉淀又能溶于3 mol·L^{-1} NH_4Cl溶液，试从平衡移动的原理加以解释。

3．如何从$CaCO_3$和$BaCO_3$沉淀中分离Ca^{2+}与Ba^{2+}？根据是什么？

4．在分析未知液时，若要沉淀碳酸盐，为什么采用$NH_3·H_2O+(NH_4)_2CO_3$溶液+NH_4Cl溶液（适量）？Mg^{2+}能被沉淀吗？为什么？

Experiment 18　Elements in s-block and Their Compounds

【Objectives】

1. To understand the reactivity of sodium and magnesium.

2. To test and understand the solubility of lithium, sodium and potassium salts.

3. To test and learn the solubility of alkali earth metals salts.

4. To learn the flame test.

【Introduction】

Metals in the s-block are highly reactive, especially alkali metals. The melting points, boiling points, hardness and densities of alkali metals are low. Potassium and sodium are easily oxidized in air. They can be cut easily with a knife, due to their softness, exposing a shiny surface that will tarnish rapidly in air due to oxidation. Sodium burns in air to form sodium peroxide. Because of their high reactivity, sodium and potassium must be stored in kerosene. The reactivity of alkali earth metals is less than that of alkali metals. For example, sodium reacts vigorously with cold water, but magnesium reacts slowly with cold water. The reaction of magnesium with water will be accelerated if heated.

Hydroxides of the alkali metals are highly soluble in water. Because solid sodium hydroxide readily absorbs moisture from the air, it is deliquescent and a frequently-used desiccant. The solubilities of hydroxides of the alkali earth metals are usually low, and the solubilities become higher going down the group in the periodic table.

Most salts of alkali metals are soluble in water. But, some salts, such as $Na[Sb(OH)_6]$, $KHC_4H_4O_6$ and $KClO_4$, are insoluble in water. Salts of alkali earth metals are slightly soluble. Their chlorides and nitrates, magnesium sulfate and magnesium chromate are soluble in water. But other salts, such as their carbonates, sulfates, oxalates and chromates are insoluble in water.

The most common magnesium carbonate forms are magnesite ($MgCO_3$), barringtonite ($MgCO_3 \cdot 2H_2O$), nesquehonite ($MgCO_3 \cdot 3H_2O$), and lansfordite ($MgCO_3 \cdot 5H_2O$). Some basic forms are artinite [$MgCO_3 \cdot Mg(OH)_2 \cdot 3H_2O$], hydromagnesite [$4MgCO_3 \cdot Mg(OH)_2 \cdot 4H_2O$], and dypingite [$4MgCO_3 \cdot Mg(OH)_2 \cdot 5H_2O$].

A flame test is an analytic procedure to detect the presence of certain elements, primarily metal ions. The test involves introducing a sample of the element to a hot, non-luminous flame, and observing the color that results. Table 5-1 lists the flame colors of alkali elements and alkali earth elements.

Table 5-1 Flame colors of alkali elements and alkali earth elements

Ion	Li^+	Na^+	K^+	Rb^+	Cs^+	Ca^{2+}	Sr^{2+}	Ba^{2+}
Color	red	yellow	lilac	red-violet	red-violet	brick red	crimson	yellowish green

【Apparatus and Chemicals】

Bunsen burner, funnel, nickel wire, crucible tongs, cobalt glass, alcohol burner, test tubes, test tube clamp, filter paper, pH test strips.

(The following reagents utilize the concentration unit $mol \cdot L^{-1}$)

HCl (2), HAc (2), HNO_3 (concentrated), $NH_3 \cdot H_2O$ (2, 1), NaOH (2, freshly prepared), NaOH (s,0.5), NaF (0.5), LiCl (0.5), NaCl (0.5), KCl (0.5), $MgCl_2$ (0.5), $CaCl_2$ (0.5), $BaCl_2$ (0.5), $SrCl_2$ (0.5), NH_4Cl (3), KI (0.1), Na_2CO_3 (0.5), $(NH_4)_2CO_3$ (0.5), K_2CrO_4 (0.5), Na_2SO_4 (0.5), $(NH_4)_2C_2O_4$ (saturated), Na_2HPO_4 (0.5), $K[Sb(OH)_6]$ (saturated), $NaHC_4H_4O_6$ (saturated), Na（s),

Mg (s), unknown solution (may contain K^+, Mg^{2+}, Ca^{2+} and Ba^{2+}).

【Experimental Procedure】

1. Reactivity of alkali metals and alkali earth metals

(1) Reaction of sodium or potassium with water

① Use tweezers to remove a small piece of sodium stored in kerosene. Use filter paper to absorb the kerosene. Cut a pea-size piece of sodium with a knife, and quickly observe the cut surface. Then put it into a beaker with water. Cover the beaker with a proper funnel. Test if the solution is acidic or alkaline.

② Put polished magnesium ribbon into a test tube with cold water. Observe and test if the solution is acidic or alkaline. Heat the test tube, what will happen?

(2) Reaction of sodium or magnesium with oxygen

① Put a piece of filter paper moistened with water in an evaporating dish. Then put a small piece of sodium on the filter paper. Observe the white smoke and the yellow flame. Test the redox property of the product, and if the product is acidic or alkaline.

② Observe the combustion of magnesium ribbon (about 2cm in length) by using a pair of crucible tongs. Test the solubility, and if the product is acidic or alkaline.

Record your observations, write the chemical equations, and compare the reactivity of elements in Group ⅠA and ⅡA.

2. Hydroxides

(1) Put some solid NaOH on a watch glass. Observe and explain the change.

(2) Add 2mL of $MgCl_2$ to a test tube, and then add $2mol \cdot L^{-1}$ $NH_3 \cdot H_2O$ drop by drop to the test tube. Observe the color of the produced precipitates. Test the reaction of the precipitate with $2mol \cdot L^{-1}$ HCl, $2mol \cdot L^{-1}$ NaOH, and $3mol \cdot L^{-1}$ NH_4Cl, respectively.

(3) Test the solubility sequence of hydroxides of alkali earth metals by the reaction of $2mol \cdot L^{-1}$ NaOH solution with $MgCl_2$, $CaCl_2$, and $BaCl_2$, respectively.

3. Insoluble salts of alkali metals and alkali earth metals

(1) Slightly soluble salts of alkali metals

① Lithium salt. Use LiCl solution to react with NaF, Na_2CO_3, and Na_2HPO_4 respectively (Heat if necessary). Record your observations and write the chemical equations.

② Sodium salt. Use 1mL of $0.5mol \cdot L^{-1}$ NaOH solution to react with saturated $K[Sb(OH)_6]$ solution. If there is no precipitate, use a glass rod to rub the inner wall of the test tube. Observe the $Na[Sb(OH)_6]$ precipitate.

③ Potassium salt. Use 1mL of KCl solution to react with saturated $NaHC_4H_4O_6$ solution. If there is no precipitate, use a glass rod to rub the inner wall of the test tube. Observe the $KHC_4H_4O_6$ crystals.

Summarize the solubility of salts of alkali metals.

(2) Insoluble salts of alkali earth metals

① Oxalate. Put $MgCl_2$ solution in a test tube, and then add saturated $(NH_4)_2C_2O_4$ solution. Observe the precipitate. After centrifugation, test the reaction of the precipitate with $2mol \cdot L^{-1}$ HAc or $2mol \cdot L^{-1}$ HCl. Use $CaCl_2$ and $BaCl_2$ to replace $MgCl_2$ to replicate the experiment.

② Chromate. Use K_2CrO_4 solution to replace saturated $(NH_4)_2C_2O_4$ solution to replicate Experiment ①.

③ Carbonate. Use Na_2CO_3 solution to replace saturated $(NH_4)_2C_2O_4$ solution to replicate Experiment ①.

Use the mixed solution (5 drops of 1mol·L^{-1} $NH_3·H_2O$+5 drops of 0.5mol·L^{-1} $(NH_4)_2CO_3$+2 drops of 3mol·L^{-1} NH_4Cl) to replace Na_2CO_3 solution. Observe if a precipitate is produced. Why?

④ Sulfate. Add Na_2SO_4 solution to $MgCl_2$, $CaCl_2$ and $BaCl_2$, respectively. Observe the precipitates. Test if the precipitates can be dissolved in concentrated HNO_3.

Complete the following:

a. Fill in Table 5-2 with your observations.

b. Discuss the solubility of the insoluble salts in acid. Write the ionic equations.

Table 5-2 Precipitates and colors

Reagent	Precipitates and colors		
	Mg^{2+}	Ca^{2+}	Ba^{2+}
$(NH_4)_2C_2O_4$ solution			
K_2CrO_4 solution			
Na_2CO_3 solution			
$NH_3·H_2O$+NH_4Cl+$(NH_4)_2CO_3$ solution			
Na_2SO_4 solution			

⑤ Magnesium ammonium phosphate. Put a few drops of 2mol·L^{-1} HCl and two drops of Na_2HPO_4 solution into 0.5mL of $MgCl_2$ solution. Observe the formation of white crystals $Mg(NH_4)PO_4$. Write the chemical equation.

【Questions and Discussion】

1. Why is freshly prepared NaOH solution used to test the solubility of $Mg(OH)_2$, $Ca(OH)_2$ and $Ba(OH)_2$? How do we prepare a NaOH solution without CO_3^{2-} ?

2. $MgCl_2$ solution reacts with $NH_3·H_2O$ to produce $Mg(OH)_2$ and NH_4Cl. But $Mg(OH)_2$ precipitate can also be dissolved in 3mol·L^{-1} NH_4Cl. Use the principle of equilibrium shift to explain the reason.

3. How do we separate Ca^{2+}, Ba^{2+} from $CaCO_3$, $BaCO_3$ precipitates?

实验19 卤素及其化合物

【实验目的】

1．理解次氯酸盐和氯酸盐的制备原理，掌握卤素的歧化反应和卤素含氧酸盐的氧化性。

2．学习和掌握 Cl^-、Br^-和 I^-混合离子的分离与鉴定方法。

【实验提要】

关于卤素化合物的酸碱性质、配位性质等内容已分散在其它有关实验中练习了，本实验着重于掌握卤素（除氟外）及其化合物的氧化还原性质。

除了 X^-外，卤素的任何形态均有较强的氧化性，在酸性溶液中其含氧酸表现得更突出。常见的还原剂如：H_2S、H_2SO_3、Sn^{2+}和 Fe^{2+}（对 Fe^{2+}来说，I_2除外）等均能将其还原成 X^-。例如：

$$H_2S+I_2 = 2I^- + S\downarrow + 2H^+$$

$$2ClO_3^- + 5SO_3^{2-} + 2H^+ = Cl_2\uparrow + 5SO_4^{2-} + H_2O$$

卤素及其化合物之间的氧化还原反应主要有置换和倒置换反应，歧化和逆歧化反应及分解反应等。

常见的卤素置换和倒置换反应有：

$$X_2+2(X')^- = 2X^- + X'_2 \qquad (\text{X 比 X}'\text{的电负性大})$$

$$Cl_2+2BrO_3^- = 2ClO_3^- + Br_2$$

$$I_2+2XO_3^- = 2IO_3^- + X_2 \qquad (X=Cl,\ Br)$$

常见的卤素歧化和逆歧化反应有：

$$X_2+H_2O \rightleftharpoons H^+ + X^- + HXO \qquad (X=Cl,\ Br,\ I)$$

$$3HXO \overset{\triangle}{\rightleftharpoons} 3H^+ + 2X^- + XO_3^- \qquad (X=Cl,\ Br,\ I)$$

$$4KXO_3(s) \overset{\triangle}{=} KX(s) + 3KXO_4(s) \qquad (X=Cl)$$

常见的卤素含氧酸及其盐的热分解反应有：

$$2HXO \xrightarrow{\text{光}} 2HX + O_2\uparrow \qquad (X=Cl,\ Br,\ I)$$

$$26HClO_3 \overset{\triangle}{=} 10HClO_4 + 8Cl_2\uparrow + 15O_2\uparrow + 8H_2O$$

$$2KClO_3(s) \xrightarrow[\triangle]{\text{催化剂}} 2KCl(s) + 3O_2(g)$$

$$KClO_4(s) \xrightarrow{610^\circ C} KCl(s) + 2O_2(g)$$

卤素单质间常见的氧化还原反应有：

$$5Cl_2+X_2+6H_2O = 10Cl^- + 2XO_3^- + 12H^+ \qquad (X=I)$$

综上所述，卤素的含氧化物均不太稳定，常发生歧化分解，或分解放出氧气。现以氯为例，概括其化合物的基本性质及转化关系：

氧化能力减弱 / 热稳定性增强 ↓　　热稳定性增强，氧化能力减弱，酸性增强 →			
HClO	$HClO_2$	$HClO_3$	$HClO_4$
KClO	$KClO_2$	$KClO_3$	$KClO_4$

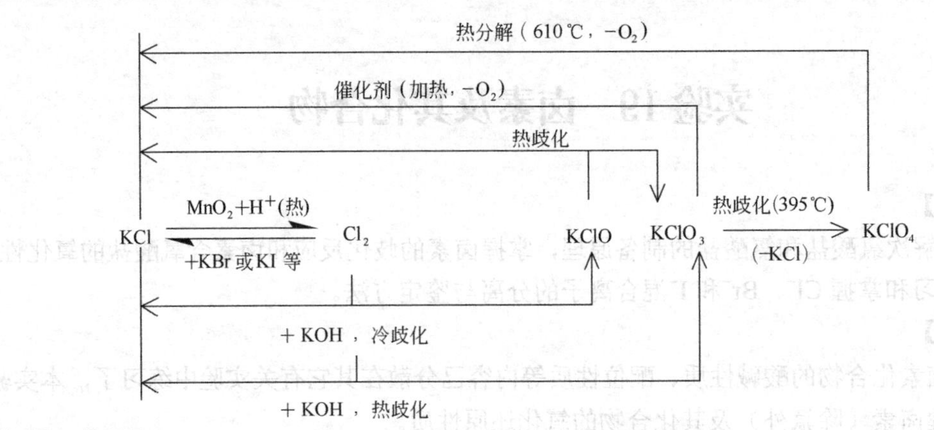

【仪器和药品】

蒸馏瓶、分液漏斗（恒压），烧杯（250mL），抽滤瓶，玻璃洗瓶。

（除特别注明外，试剂浓度单位为 $mol \cdot L^{-1}$）H_2SO_4 (2，1，浓)，HNO_3 (6)，HCl(浓)，KOH (6)，NaOH (3)，NaCl (0.1，饱和)，Na_2CO_3 (0.1)，Na_2SO_4 (0.1)，KBr (0.1)，KI (0.1)，$MnSO_4$ (0.1)，$AgNO_3$ (0.1)，$(NH_4)_2CO_3$ (质量分数为 12%)，$KMnO_4$ (s)，KI (s)，KBr (s)，NaCl (s)，I_2 (s)，Cl_2 水(新鲜配制)，淀粉溶液，品红溶液，Zn (粉)，$KClO_3$(s)，CCl_4，KI-淀粉试纸，$Pb(Ac)_2$ 试纸，玻璃丝，冰。

【实验内容】

用表格报告下列实验结果，解释实验现象时，若涉及化学反应，均要求写出反应方程式。

1．卤离子 X^- 的还原性

往盛有少量（绿豆大小，下同）KI 固体的试管中加入数滴浓 H_2SO_4，观察反应产物的颜色和状态。把湿的 $Pb(Ac)_2$ 试纸放在试管口，检验气体产物（H_2SO_4 的还原产物）。

分别用 KBr 和 NaCl 固体代替 KI 做上述试验，用 KI-淀粉试纸代替醋酸铅试纸检验气体产物。判断 H_2SO_4 是否被还原并比较其还原程度。

2．碘的溶解性

取少量碘晶体放在洁净的试管中，加入 2mL 蒸馏水，摇荡试管，观察溶液的颜色。再加入几滴 KI 溶液，摇匀，观察溶液颜色的变化。把溶液分成两份，其中一份加入数滴 CCl_4；另一份加 2 滴淀粉溶液，摇荡试管，观察溶液颜色的变化。溶液留做实验 3。

3．碘的歧化、逆歧化反应

往上述两支试管中加入数滴 NaOH 溶液，摇匀观察变化，然后用稀 H_2SO_4 酸化，又有什么现象。

4．氯酸盐的氧化性

（1）取少量 $KClO_3$ 晶体，加入约 1mL 浓 HCl 溶液，观察产生的气体颜色。反应式：

$$8KClO_3+24HCl = 9Cl_2+8KCl+6ClO_2\text{(黄)}+12H_2O$$

（2）取少量 $KClO_3$ 晶体，加入约 1mL 水使之溶解，再加入几滴 KI 溶液和 0.5 mL CCl_4，摇荡，观察现象。再加入 1mL 稀 H_2SO_4，摇荡并再观察有何变化。

（3）取少量 $KClO_3$ 晶体和少量 I_2 (s)，加入适量水使之溶解，并滴加数滴浓 H_2SO_4，微热，检验 Cl_2 的生成。

5．卤化物的配位与沉淀平衡

取几滴 NaCl 溶液，滴加 2～4 滴 $AgNO_3$ 溶液，待有沉淀生成，再滴加饱和 NaCl 溶液至沉淀溶解为止。将溶液分成两份，一份加水稀释，另一份滴加KBr溶液，观察有何变化。

6．从混合液中检出Cl^-、Br^-、I^-

取1mL混合液进行试验（可参考图5-1所示方案），记录混合液编号和试验结果[1]。

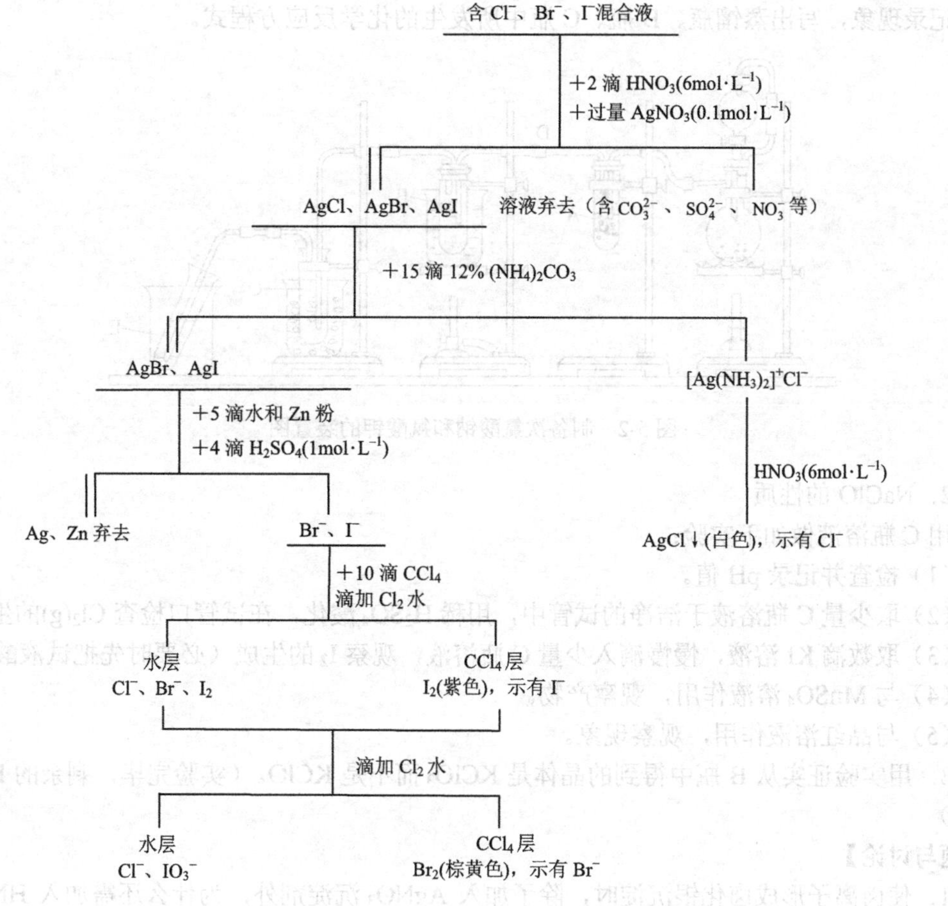

图5-1 从混合液中检出Cl^-、Br^-、I^-示意图

【选做部分】

1．次氯酸盐、氯酸盐的制备

按图5-2所示，在蒸馏瓶内放入20g $KMnO_4$，分液漏斗中加入浓HCl 35mL；A瓶装入玻璃丝；B瓶加入$6mol \cdot L^{-1}$ KOH溶液20mL，B瓶置于水浴中；C瓶加入20mL $3mol \cdot L^{-1}$ NaOH，并置于冰浴中；尾气经稀碱液吸收，接抽气系统。连接各部分仪器，检查装置，确保系统严密。加热B瓶的水浴并保持近沸。开启分液漏斗活塞，使浓HCl缓慢而均匀地滴入蒸馏瓶中，反应生成的Cl_2气均匀地通入反应瓶B和C（注意：防止B瓶溶液倒吸入A瓶，在反应后阶段，可适当加热蒸馏瓶）。当C瓶碱液上部呈现浅黄色时停止滴加HCl，将A瓶与B瓶之间的导管拆开，同时迅速将A瓶放出的尾气通入预先准备好的另一稀碱液中，待整个实验完毕

---

[1] 混合液由实验室提供，共4组，每组溶液含有3～4种阴离子，其中卤素离子X^-不少于2种。混合液由下列单一盐溶液混合而成：NaCl，Na_2CO_3，Na_2SO_4，KBr，KI。

学生报告试验结果时只需列出所含卤素离子，但应指出CO_3^{2-}和SO_4^{2-}的干扰是怎样排除的。

后处理。继续加热B瓶片刻，拆除B瓶、C瓶和尾气吸收装置。

将B瓶溶液冷却，搅拌，待晶体析出后过滤，用少量冰水洗涤晶体一次，抽干，检查滤液中 Cl^- 的存在。所得的晶体及C瓶溶液分别留作下面有关实验用。

记录现象，写出蒸馏瓶、B瓶、C瓶中所发生的化学反应方程式。

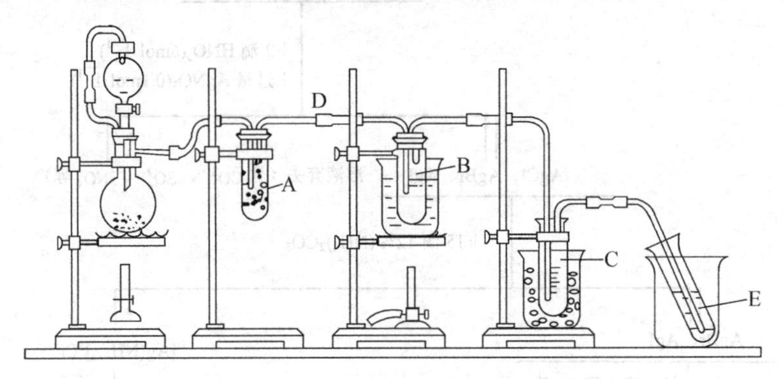

图 5-2　制备次氯酸钠和氯酸钾的装置图

2．NaClO 的性质

用C瓶溶液做如下实验：

（1）检查并记录 pH 值。

（2）取少量C瓶溶液于洁净的试管中，用稀 H_2SO_4 酸化，在试管口检查 Cl_2(g)的生成。

（3）取数滴 KI 溶液，慢慢滴入少量C瓶溶液，观察 I_2 的生成（必要时先把试液酸化）。

（4）与 $MnSO_4$ 溶液作用，观察产物。

（5）与品红溶液作用，观察现象。

3．用实验证实从B瓶中得到的晶体是 $KClO_3$ 而不是 KClO。（实验完毕，剩余的 $KClO_3$ 回收）

【问题与讨论】

1．使卤离子形成卤化银沉淀时，除了加入 $AgNO_3$ 沉淀剂外，为什么还需加入 HNO_3？在什么情况下可以不必加入 HNO_3？在什么情况下必须加入 HNO_3？若向一未知溶液中加入 $AgNO_3$ 时不产生沉淀，能否认为溶液中不存在卤素离子？

2．将 AgCl 与 AgBr 和 AgI 分离时为何不用氨水？试说明本实验采用 $(NH_4)_2CO_3$ 的依据和优点。写出有关反应方程式。

3．甲和乙两人用同样的C瓶溶液与 KI 溶液反应，瞬间同样观察到有 I_2 产生，但当他们用稀 H_2SO_4 酸化各自的反应混合物时却观察到不同现象，甲：I_2 消失，并有 Cl_2 产生；乙：I_2 不消失，没有 Cl_2 生成。试分析产生不同现象的可能原因，并写出有关的反应方程式。

4．在什么情况下，B瓶的溶液容易倒吸进A瓶？应怎样防止这种情况发生。

5．反应结束后怎样处理蒸馏瓶中的残留物？

【安全知识】

1．氯气有毒和刺激性，吸入少量会刺激鼻咽部，引起咳嗽和喘息。大量吸入会导致严重损害，甚至死亡。因此，进行有关氯气实验时必须在通风橱内进行。

2．溴蒸气对气管、肺部、眼鼻喉都有强烈的刺激作用。进行有关溴的试验，应在通风橱内进行，不慎吸入溴蒸气时，可吸入少量氨气和新鲜空气解毒。液态溴具有很强的腐蚀性，能灼烧皮肤，严重时会使皮肤溃烂。移取液态溴时，需戴橡皮手套。溴水的腐蚀性虽比液溴弱些，但在使用时，也不允许直接由瓶内倒出，而应用滴管移取，以防溴水接触皮肤。如果不慎把溴水溅在手上，应及时用水冲洗，再用稀硫代硫酸钠溶液充分浸透的绷带包扎处理。

3．氯酸钾是强氧化剂，保存不当时容易引起爆炸，它与硫、磷的混合物是炸药，因此，绝对不允许将它们混在一起。氯酸钾容易分解，不宜大力研磨、烘干或烤干。在进行有关氯酸钾的实验时，与其它强氧化性物质实验一样，应将剩下的试剂倒入回收瓶内回收处理。

Experiment 19 Halogens and Their Compounds

【Objectives】

1．To learn the principles in the preparation of hypochlorites and chlorates. To learn the disproportionation reactions of halogens, and the oxidizing strength of their oxysalts.

2．To learn the separation and identification of Cl^-, Br^- and I^- in an aqueous solution.

【Introduction】

In this experiment, we will learn the redox properties of halogens and their compounds (except fluorine).

All chemical forms of halogens (except X^-) are strong oxidizing agents, especially their oxyacids in acidic solution. Reducing agents, such as H_2S, H_2SO_3, Sn^{2+} and Fe^{2+}, can reduce them to X^-. I_2 can not be reduced by Fe^{2+}. For example:

$$H_2S+I_2 = 2I^-+S\downarrow+2H^+$$

$$2ClO_3^-+5SO_3^{2-}+2H^+ = Cl_2\uparrow+5SO_4^{2-}+H_2O$$

The oxidation-reduction (redox) reactions for halogens and their compounds are displacement reactions, disproportionation reactions, and decomposition reactions, etc.

Some displacement reactions are,

$$X_2+2(X')^- = 2X^-+X'_2 \quad \text{(The electronegativity of X is greater than X')}$$

$$Cl_2+2BrO_3^- = 2ClO_3^-+Br_2$$

$$I_2+2XO_3^- = 2IO_3^-+X_2 \quad (X=Cl, Br)$$

Some disproportionation reactions are,

$$X_2+H_2O \rightleftharpoons H^++X^-+HXO \quad (X=Cl, Br, I)$$

$$3HXO \xrightleftharpoons{\triangle} 3H^++2X^-+XO_3^- \quad (X=Cl, Br, I)$$

$$4KXO_3(s) \xlongequal{\triangle} KX(s)+3KXO_4(s) \quad (X=Cl)$$

Here are some decomposition reactions.

$$2HXO \xrightarrow{光} 2HX+O_2\uparrow \quad (X=Cl, Br, I)$$

$$26HClO_3 \xlongequal{\triangle} 10HClO_4+8Cl_2\uparrow+15O_2\uparrow+8H_2O$$

$$2KClO_3(s) \xrightarrow[\triangle]{催化剂} 2KCl(s)+3O_2(g)$$

$$KClO_4(s) \xrightarrow{610℃} KCl(s)+2O_2(g)$$

The redox reaction between diatomic halogen molecules can be written as:

$$5Cl_2+X_2+6H_2O = 10Cl^-+2XO_3^-+12H^+ \quad (X=I)$$

In conclusion, oxyacids and oxysalts of halogens are not stable. The following is a summary of properties for Chlorine and its compounds.

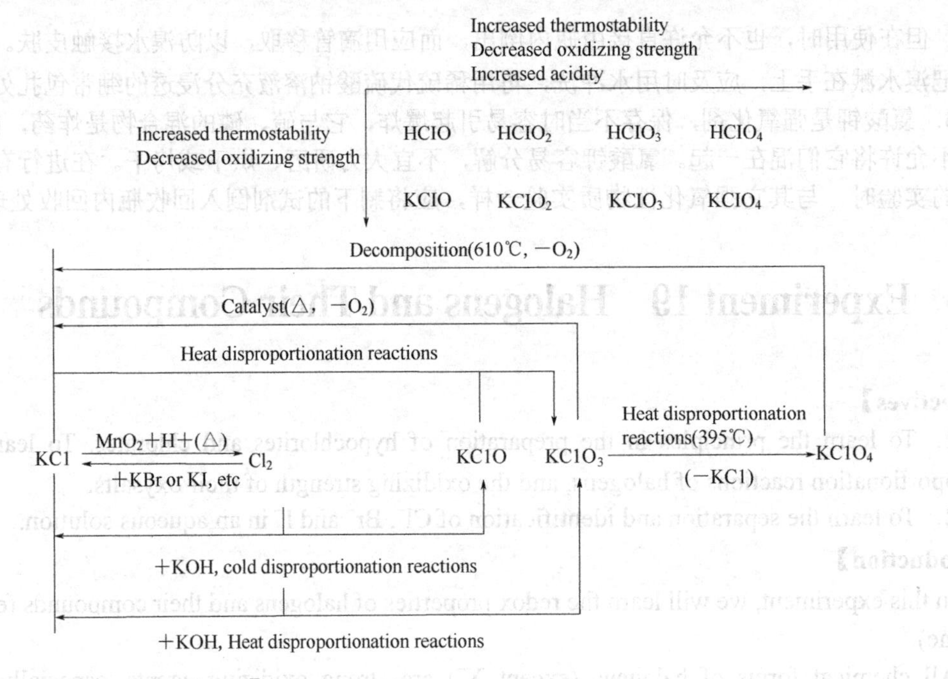

【Apparatus and Chemicals】

Distilling flask, separating funnel, beaker (250 mL), filter flask, glass wash bottle.

(The following reagents utilize the concentration unit $mol\cdot L^{-1}$)

H_2SO_4 (2, 1, concentrated), HNO_3 (6), HCl (concentrated), KOH (6), NaOH (3), NaCl (0.1, saturated), Na_2CO_3 (0.1), Na_2SO_4 (0.1), KBr (0.1), KI (0.1), $MnSO_4$ (0.1), $AgNO_3$ (0.1), $(NH_4)_2CO_3$ (12%, wt/wt), $KMnO_4$ (s), KI (s), KBr (s), NaCl (s), I_2 (s), Cl_2 water (freshly prepared), starch solution, fuchsin solution, zinc powder, $KClO_3$(s), CCl_4, KI-starch test paper, $Pb(Ac)_2$ test paper, glass silk, ice.

【Experimental Procedure】

Fill in the appropriate table and write the chemical equations in your report.

1. Reducibility of halogen ions X^-

Add a few drops of concentrated H_2SO_4 to a test tube containing a pea-size solid of KI. Observe the reaction. Test the produced gas with lead acetate test paper.

Replicate the experiment with KBr and then NaCl. Test the produced gas with KI-starch test paper. Compare the reducing strength of halogen ions.

2. Solubility of Iodine

Put a small amount of Iodine crystal in a test tube, and add 2mL of distilled water. Shake the test tube and observe the color of the solution. Add a few drops of KI solution. Shake well and observe the color of the solution. Then equally divide the solution into two test tubes. Add a few drops of CCl_4 in one test tube, and add two drops of starch solution in the other test tube. Observe the color of the solutions. Now continue doing the following experiments with these two solutions.

3. Disproportionation reactions of Iodine

Add a few drops of NaOH solution to the above two solutions. Shake the test tubes and observe. Then, add dilute H_2SO_4 to the two test tubes. What are your observations?

4. Oxidizability of chlorate

(1) Add 1mL of concentrated HCl to a small amount of $KClO_3$ crystal, and observe the color of the gas. The reaction is:

$$8KClO_3+24HCl = 9Cl_2+8KCl+6ClO_2\,(\text{yellow})+12H_2O$$

(2) Dissolve a small amount of $KClO_3$ crystal with 1mL of distilled water, and add a few drops of KI solution and 0.5mL of CCl_4 solution. Shake well and observe. Then add 1mL of dilute H_2SO_4, and observe.

(3) Dissolve a small amount of $KClO_3$ crystal and solid I_2. Add a few drops of concentrated H_2SO_4, and heat slightly. Test if Cl_2 is produced.

5. Coordination equilibrium and precipitation equilibrium of halides

Add a few drops of NaCl solution to a test tube, and add 2 to 4 drops of $AgNO_3$ solution to it. Precipitates should be formed. Then add saturated NaCl solution until the precipitates disappear. Equally divide the solution into two test tubes. Add water to one test tube, then add KBr solution to the other. Observe the changes in the two test tubes.

6. Separation and identification of a mixed solution of Cl^-, Br^- and I^-

Use 1mL of mixed solution to do the following experiments according to Figure 5-1. Observe and write down your results.

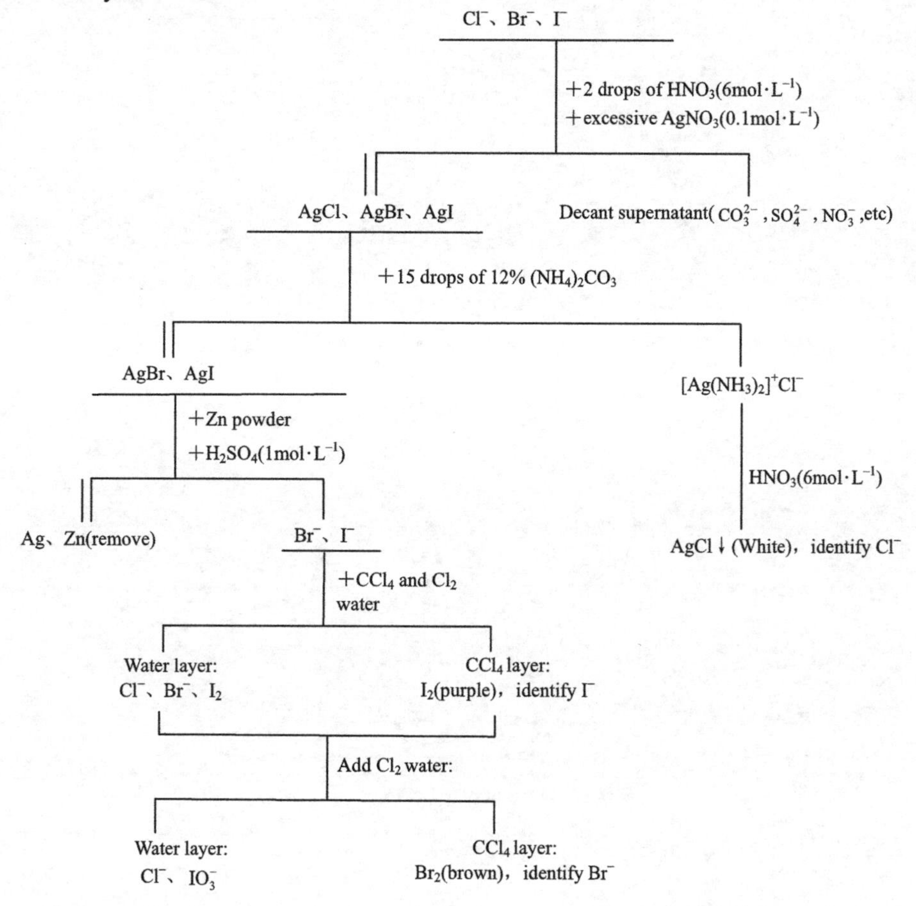

Figure 5-1 Flow diagram of separation and identification of a mixed solution of Cl^-, Br^- and I^-

【Questions and Discussion】

1. To precipitate halogen ions, $AgNO_3$ is added to form the precipitates. Why is HNO_3 also needed? When is or isn't HNO_3 needed? If no precipitates form after adding $AgNO_3$ to an unknown solution, can we say there are no halogen ions in this solution?

2. When separating AgCl from AgBr and AgI, why can't we use ammonia water? Why do we use $(NH_4)_2CO_3$ in this experiment? Record the chemical equations.

实验 20　过氧化氢和硫的化合物

【实验目的】

1．试验并了解 H_2O_2 的某些重要性质。

2．试验并了解 S 的某些化合物的性质。

3．掌握 S^{2-}、SO_3^{2-}、$S_2O_3^{2-}$ 和 SO_4^{2-} 的分离和鉴定方法。

【实验提要】

1．H_2O_2

纯的 H_2O_2 是近于无色的黏稠液体。但通常所用的是质量分数为 3%或 30%的 H_2O_2 水溶液。

H_2O_2 分子中含有过氧基（—O—O—），由于过氧基的键能较小，所以 H_2O_2 分子不稳定，特别是光照、加热和增大溶液的碱度，可加快其分解，某些重金属离子对 H_2O_2 的分解也有加速作用。在 H_2O_2 中，氧的氧化值为−1，处于中间氧化态，因此它既有氧化性又有还原性。但在酸性介质中，其氧化性表现尤为突出。

在铬酸盐的酸性溶液中，加入 H_2O_2，则生成深蓝色的过氧化铬 $CrO(O_2)_2$：

$$HCrO_4^- + 2H_2O_2 + H^+ = CrO(O_2)_2 + 3H_2O$$

（深蓝色）

$CrO(O_2)_2$ 不稳定，立即分解放出氧气并生成 Cr^{3+}，蓝色消失。

$$4CrO(O_2)_2 + 12H^+ = 4Cr^{3+} + 7O_2\uparrow + 6H_2O$$

由于 $CrO(O_2)_2$ 能与某些有机溶剂如乙醚、戊醇等形成比较稳定的配合物。故此反应常用来鉴定 H_2O_2 和 Cr(Ⅵ)。

2．硫的某些重要化合物及其性质

硫的标准电位图 $E_A^\ominus$/V（方括号内为碱性溶液中的形态和 $E_B^\ominus$/V）如下：

$$S_2O_8^{2-} \xrightarrow{2.01} SO_4^{2-} \frac{+0.17}{[-0.93]} \underset{[SO_3^{2-}]}{H_2SO_3} \frac{+0.4}{[-0.57]} S_2O_3^{2-} \frac{+0.50}{[-0.74]} S \frac{+0.14}{[-0.508]} \underset{[S^{2-}]}{H_2S}$$

硫的氢化物有 H_2S 和 H_2S_x（当 x=2 时为过硫化氢），其相应的盐是硫化物和多硫化物。由于 S^{2-} 的半径比较大，其变形性大，与重金属离子结合为硫化物时，其化学键显共价性，难溶于水，且各硫化物有明显不同的颜色和难溶程度，故常用硫化物的生成和溶解来分离和鉴定离子。在可溶性的硫化物浓溶液中加入硫粉时，硫溶解而生成相应的多硫化物：

$$Na_2S + (x-1)S = Na_2S_x$$

$$(NH_4)_2S + (x-1)S = (NH_4)_2S_x$$

多硫化物在酸性溶液中很不稳定，易歧化分解为 H_2S 和 S：

$$S_2^{2-} + 2H^+ = H_2S\uparrow + S\downarrow$$

硫在许多化合物中，可取代氧而生成硫代化合物。常见和重要的硫代化合物有硫代硫酸钠（$Na_2S_2O_3$）和硫代乙酰胺（CH_3CSNH_2，简称为 TAA）等。

由于 $S_2O_3^{2-}$ 中，两个 S 原子的平均氧化值为+2，故 $S_2O_3^{2-}$ 具有还原性，是一个中等强度的还原剂。例如：

$$2S_2O_3^{2-}+I_2 = S_4O_6^{2-}+2I^-$$
$$S_2O_3^{2-}+4Cl_2+5H_2O = 2SO_4^{2-}+8Cl^-+10H^+$$

$S_2O_3^{2-}$还具有很强的配位能力，例如：

$$AgBr+2S_2O_3^{2-} = [Ag(S_2O_3)_2]^{3-}+Br^-$$

另外，$S_2O_3^{2-}$在强酸性溶液中（pH＜4.6），立即分解：

$$S_2O_3^{2-}+2H^+ = SO_2\uparrow+S\downarrow+H_2O$$

TAA在加热时与酸或碱反应，能缓慢地释放出H_2S或S^{2-}。故在分析化学中，常用来代替H_2S溶液或Na_2S溶液作为金属离子的沉淀剂，具有安全、沉淀均匀等优点。

硫还能形成过硫酸，常见的是过二硫酸（$H_2S_2O_8$）及其盐$K_2S_2O_8$或$(NH_4)_2S_2O_8$等。过二硫酸盐的氧化性很强。例如：

$$5S_2O_8^{2-}+2Mn^{2+}+8H_2O \xrightarrow{Ag^+} 2MnO_4^-+10SO_4^{2-}+16H^+$$

3．S^{2-}、SO_3^{2-}、$S_2O_3^{2-}$和SO_4^{2-}的分离与检出

（1）SO_4^{2-}的检出　试液用HCl酸化，在所得清液中加入$BaCl_2$溶液，生成白色$BaSO_4$沉淀，示有SO_4^{2-}存在。

（2）S^{2-}的检出　试液中S^{2-}含量多时，可酸化试液，用$Pb(Ac)_2$试纸检查H_2S。S^{2-}含量少时，可在碱性溶液中加入$Na_2[Fe(CN)_5NO]$溶液检验。

$$S^{2-}+[Fe(CN)_5NO]^{2-} = [Fe(CN)_5(NOS)]^{4-}\text{（紫红色）}$$

由于S^{2-}干扰SO_3^{2-}和$S_2O_3^{2-}$的检出，因此在检出SO_3^{2-}和$S_2O_3^{2-}$前必须除去S^{2-}。其方法是加入固体$CdCO_3$，借助沉淀的转化，将S^{2-}变成CdS沉淀而除去。SO_3^{2-}和$S_2O_3^{2-}$则留在溶液中。

（3）$S_2O_3^{2-}$的检出　在除去S^{2-}的溶液中加入稀HCl并加热，溶液变浑浊，示有$S_2O_3^{2-}$存在。

$$S_2O_3^{2-}+2H^+ = S\downarrow+SO_2\uparrow+H_2O$$

也可在试液中加入过量的$AgNO_3$溶液，则生成白色$Ag_2S_2O_3$沉淀。此沉淀不稳定，立即水解，水解过程伴有颜色变化（白色→黄色→棕色→黑色）。

$$Ag_2S_2O_3+H_2O = Ag_2S\downarrow+H_2SO_4$$

（4）SO_3^{2-}的检出　$S_2O_3^{2-}$干扰SO_3^{2-}的检出，故在检出SO_3^{2-}之前应把它除去。方法是在除去S^{2-}的溶液中加入$SrCl_2$溶液，因$SrSO_3$和$SrSO_4$溶解度很小而生成沉淀，而$S_2O_3^{2-}$留在溶液中，过滤后弃去。在沉淀中加HCl，并滴入I_2-淀粉溶液，若溶液褪色，则示有SO_3^{2-}存在。

S^{2-}、SO_3^{2-}、$S_2O_3^{2-}$和SO_4^{2-}混合离子的分离与检出步骤如图5-3所示。

【仪器和药品】

离心机，刻度离心试管，酒精灯。

（除特别注明外，试剂浓度单位均为$mol\cdot L^{-1}$）$MnSO_4$(0.1)，$K_2Cr_2O_7$(0.1)，Na_2SO_4(0.1)，Na_2S(0.1)，$(NH_4)_2S$(饱和，新配制)，$AgNO_3$(0.1)，H_2O_2(质量分数为3%)，$Na_2S_2O_3$(0.1)，Na_2SO_3(0.1)，$ZnSO_4$(0.1)，$CdSO_4$(0.1)，$FeCl_3$(0.1)，TAA(质量分数为5%)，$Na_2[Fe(CN)_5NO]$(质量分数为1%)，HCl(6，2，1)，H_2SO_4(3，2，1)，NaOH(2)，$NH_3\cdot H_2O$(2)，汞，乙醚，硫

粉，碘水，$K_2S_2O_8$(s)，MnO_2(s)，pH 试纸，$Pb(Ac)_2$试纸。

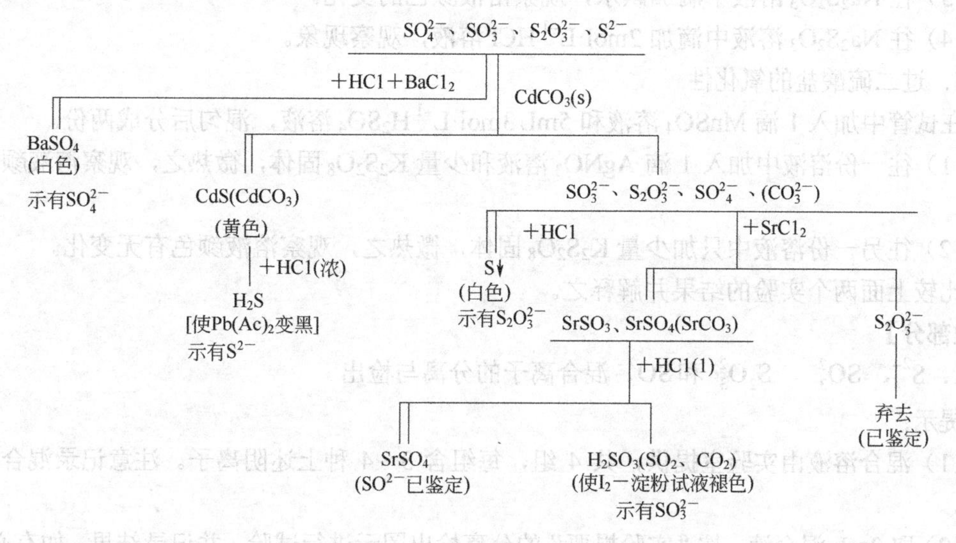

图 5-3　S^{2-}、SO_3^{2-}、$S_2O_3^{2-}$ 和 SO_4^{2-} 混合离子的分离与检出

【实验内容】

在解释下列实验现象时，若涉及化学反应，均要求写出方程式。

1．H_2O_2 的性质

（1）H_2O_2 的分解　4 支试管各加入质量分数为 3%的 H_2O_2 溶液 2mL，然后在第 1 支试管中加入数滴 2 mol·L^{-1} NaOH 溶液；在第 2 支试管中加入数滴 $FeCl_3$ 溶液；在第 3 支试管中加入几粒 MnO_2；第 4 支试管留作比较。将 4 支试管同时放在水浴中加热，观察并比较各试管中 H_2O_2 分解的情况。

（2）H_2O_2 的氧化还原性

① 在 1mL 质量分数为 3%的 H_2O_2 溶液中，加入数滴 $MnSO_4$ 溶液，然后滴加 2mol·L^{-1} NaOH 溶液，观察现象。最后再滴加 1mol·L^{-1} H_2SO_4 溶液酸化，又发生什么变化？解释现象。

② 在一只 50mL 小烧杯中，加入约 20mL 质量分数为 3%的 H_2O_2 溶液，再放入一小颗汞（黄豆般大小），观察 H_2O_2 在汞表面分解和汞表面的变化（由教师做演示实验）。解释现象。

（3）H_2O_2 的鉴定　在一试管中加入 1mL 质量分数为 3%的 H_2O_2 溶液，加数滴乙醚和数滴 2mol·L^{-1} H_2SO_4，再滴加 $K_2Cr_2O_7$ 溶液，振荡试管，观察溶液和乙醚层颜色的变化。

2．H_2S 的还原性和 S^{2-}的鉴定

（1）在试管中加入数滴 Na_2SO_3 溶液和 Na_2S 溶液，然后用稀硫酸酸化。

（2）取少量 Na_2S 溶液，试验 S^{2-}与 $Na_2[Fe(CN)_5NO]$的作用。

3．多硫化物的制备和性质

将少量的硫粉加入饱和$(NH_4)_2S$ 溶液，加热至沸，观察硫粉的溶解及溶液颜色的变化。将上述溶液离心分离，取少量清液，逐滴加入 6mol·L^{-1} HCl 溶液，观察现象，并检验 H_2S 气体的生成。

4．硫代硫酸盐的性质

（1）取 1mL $Na_2S_2O_3$ 溶液，滴加 1～2 滴 $AgNO_3$，振荡试管，观察现象。

（2）取2～3滴$AgNO_3$溶液，逐滴加入$Na_2S_2O_3$溶液，观察变化。

（3）往$Na_2S_2O_3$溶液中滴加碘水，观察溶液颜色的变化。

（4）往$Na_2S_2O_3$溶液中滴加$2mol \cdot L^{-1}$ HCl溶液，观察现象。

5．过二硫酸盐的氧化性

在试管中加入1滴$MnSO_4$溶液和5mL $3mol \cdot L^{-1}$ H_2SO_4溶液，混匀后分成两份。

（1）往一份溶液中加入1滴$AgNO_3$溶液和少量$K_2S_2O_8$固体，微热之，观察溶液颜色的变化。

（2）往另一份溶液中只加少量$K_2S_2O_8$固体，微热之，观察溶液颜色有无变化。

比较上面两个实验的结果并解释之。

【选做部分】

1．S^{2-}、SO_3^{2-}、$S_2O_3^{2-}$和SO_4^{2-}混合离子的分离与检出

提示：

（1）混合溶液由实验室提供，共4组，每组含3～4种上述阴离子。注意记录混合液的编号。

（2）取2mL混合液，按“实验提要”的分离检出图示进行试验，并记录结果。如有必要，可与单一离子做对照试验。

2．CdS和ZnS的分离

在离心试管中，加入$ZnSO_4$溶液和$CdSO_4$溶液各4滴，用$2mol \cdot L^{-1}$ $NH_3 \cdot H_2O$调节pH＝4，并加蒸馏水稀释到2.5mL，然后再加入1mL $1mol \cdot L^{-1}$ HCl溶液和1.5mL TAA溶液。在沸水浴中加热10～15min，观察黄色沉淀的生成。溶液冷却至室温，离心分离。溶液转移至另一离心试管，用$NH_3 \cdot H_2O$调节pH＝9，再次在沸水浴中加热5～10min，观察白色沉淀的生成。离心分离，弃去清液，沉淀用含NH_4Cl的水溶液洗涤1次，向沉淀逐滴加入稀HCl，观察沉淀的溶解，并检验是否有H_2S逸出。

【问题与讨论】

1．为什么H_2O_2既可作氧化剂又可作还原剂？在什么条件下，H_2O_2可将Mn^{2+}氧化为MnO_2？在什么条件下，MnO_2又能将H_2O_2氧化而产生O_2？

2．长久放置的H_2S、Na_2S和Na_2SO_3溶液会发生什么变化？为什么？

3．试验室用酸分解FeS以制备H_2S时，应选用什么酸（HCl，HNO_3或浓H_2SO_4）？为什么？

4．$Na_2S_2O_3$溶液与$AgNO_3$溶液反应时，为什么有时生成Ag_2S沉淀，而有时却生成$[Ag(S_2O_3)_2]^{3-}$配离子？

5．在鉴定$S_2O_3^{2-}$和SO_3^{2-}之前，如何除去S^{2-}的干扰？

6．在混合离子的分离和检出实验中，怎样使SO_3^{2-}和$S_2O_3^{2-}$分离？

7．通过计算说明，用H_2S饱和水溶液分离Cd^{2+}和Zn^{2+}时，溶液的酸度应控制在什么范围才能使两种离子顺利地分离？（有关数据查附录）

【安全知识】

硫化氢具有强烈的臭鸡蛋气味，是毒性较大的气体。主要引起中枢神经系统中毒，与呼吸酶中的铁质结合使酶活性减弱，造成黏膜损害及呼吸系统损害。轻度产生头晕、头痛、呕

吐，严重时可引起昏迷、意识丧失、窒息而致死亡。因此，凡涉及硫化氢参与的反应都应在通风橱内进行。

Experiment 20　Hydrogen Peroxide and Compounds of Sulfur

【Objectives】

1．To evaluate the important properties of hydrogen peroxide.

2．To evaluate the important properties of some compounds of sulfur.

3．To separate and identify S^{2-}、SO_3^{2-} 、$S_2O_3^{2-}$ and SO_4^{2-} .

【Introduction】

1．H_2O_2

Hydrogen peroxide is a colorless, syrupy liquid, slightly more viscous than water. Dilute hydrogen peroxide solutions (3% or 30% by mass) are usually used in the laboratory.

The bond energy of —O—O— in H_2O_2 is small, so H_2O_2 is not stable. Hydrogen peroxide readily decomposes when exposed to sunlight, heated or in the presence of certain metal ions. The oxidation number of oxygen in H_2O_2 is −1, so H_2O_2 can act as an oxidizing agent or a reducing agent. It is a strong oxidizing agent in acidic solution.

Dark blue $CrO(O_2)_2$ can be formed by adding H_2O_2 to chromate in an acidic solution.

$$HCrO_4^- + 2H_2O_2 + H^+ = CrO(O_2)_2 + 3H_2O$$

(dark blue)

$CrO(O_2)_2$ is not stable, and decomposes immediately to form Cr^{3+}.

$$4CrO(O_2)_2 + 12H^+ = 4Cr^{3+} + 7O_2\uparrow + 6H_2O$$

$CrO(O_2)_2$ can form stable complexes with organic solvents, such as diethyl ether and pentanol, which is used to identify H_2O_2 and Cr(Ⅵ).

2．Compounds of sulfur and their properties

The standard potentials $E_A^\ominus$/V of sulfur compounds are as follows (data or compounds in square brackets are the standard potentials $E_B^\ominus$/V or compounds in basic solution):

$$S_2O_8^{2-} \xrightarrow{2.01} SO_4^{2-} \xrightarrow[{[-0.93]}]{+0.17} H_2SO_3\,[SO_3^{2-}] \xrightarrow[{[-0.57]}]{+0.4} S_2O_3^{2-} \xrightarrow[{[-0.74]}]{+0.50} S \xrightarrow[{[-0.508]}]{+0.14} H_2S\,[S^{2-}]$$

Hydrogen sulfide (H_2S) and H_2S_x (it is hydrogen persulfide when x=2) are the hydrides of sulfur. Their corresponding salts are sulfide and polysulfide. The radius of S^{2-} is large, so its deformability is strong, and it forms covalent bonds with metal ions to produce insoluble sulfides. Different sulfides have different colors and insolubilities which can be used to separate and identify ions. Add sulfur powder to concentrated soluble sulfide, the sulfur will be dissolved and the polysulfide will be formed.

$$Na_2S + (x-1)S = Na_2S_x$$

$$(NH_4)_2S + (x-1)S = (NH_4)_2S_x$$

Polysulfide is not stable in an acidic solution, and it will decompose into H_2S and S.

$$S_2^{2-}+2H^+ \longrightarrow H_2S\uparrow+S\downarrow$$

Sulfur can replace oxygen in many compounds. Some important compounds are sodium thiosulfate ($Na_2S_2O_3$) and thioacetamide (CH_3CSNH_2, TAA for short), etc.

The oxidation number of sulfur in $S_2O_3^{2-}$ is +2, so it is a moderate reducing agent. For example:

$$2S_2O_3^{2-}+I_2 \longrightarrow S_4O_6^{2-}+2I^-$$

$$S_2O_3^{2-}+4Cl_2+5H_2O \longrightarrow 2SO_4^{2-}+8Cl^-+10H^+$$

$S_2O_3^{2-}$ is a strong ligand, too. For example:

$$AgBr+2S_2O_3^{2-} \longrightarrow [Ag(S_2O_3)_2]^{3-}+Br^-$$

Moreover, $S_2O_3^{2-}$ decomposes immediately in strong acidic solutions (pH<4.6).

$$S_2O_3^{2-}+2H^+ \longrightarrow SO_2\uparrow+S\downarrow+H_2O$$

Heating TAA, or reacting TAA with an acid or base, can slowly produce H_2S or S^{2-}. So, TAA is usually used to replace H_2S or Na_2S solution to precipitate metal ions. TAA is also safe to use.

There are also peroxosulfuric acids, such as peroxydisulfuric acid ($H_2S_2O_8$). Their salts, $K_2S_2O_8$ or $(NH_4)_2S_2O_8$, are strong oxidizing agents. For example:

$$5S_2O_8^{2-}+2Mn^{2+}+8H_2O \xrightarrow{Ag^+} 2MnO_4^-+10SO_4^{2-}+16H^+$$

3. Separation and identification of S^{2-}, SO_3^{2-}, $S_2O_3^{2-}$ and SO_4^{2-}

(1) Identification of SO_4^{2-} Acidify the test solution with HCl, and add $BaCl_2$ to the solution. The white precipitate $BaSO_4$ confirms the presence of SO_4^{2-} in the test solution.

(2) Identification of S^{2-} If the test solution contains a large amount of S^{2-}, acidify the test solution first, and then test the produced H_2S with $Pb(Ac)_2$ test paper. If the test solution contains a small amount of S^{2-}, add $Na_2[Fe(CN)_5NO]$ solution to the basic test solution to identify S^{2-}.

$$S^{2-}+[Fe(CN)_5NO]^{2-} \longrightarrow [Fe(CN)_5(NOS)]^{4-} \text{ (red purple)}$$

The presence of S^{2-} will interfere the identification of SO_3^{2-} and $S_2O_3^{2-}$. So, it needs to be removed by adding solid $CdCO_3$ to produce CdS precipitate.

(3) Identification of $S_2O_3^{2-}$ After removal of S^{2-}, add dilute HCl to the test solution and heat the solution. The precipitate confirms the presence of $S_2O_3^{2-}$ in the test solution.

$$S_2O_3^{2-}+2H^+ \longrightarrow S\downarrow+SO_2\uparrow+H_2O$$

Excessive $AgNO_3$ solution can also be used to identify $S_2O_3^{2-}$. A white precipitate ($Ag_2S_2O_3$) will be formed. But it is not stable, and will be hydrolysed immediately. The color will change from white to yellow, then brown, and finally black.

$$Ag_2S_2O_3+H_2O \longrightarrow Ag_2S\downarrow+H_2SO_4$$

(4) Identification of SO_3^{2-} Remove $S_2O_3^{2-}$ before identifying SO_3^{2-}. After removal of S^{2-}, add $SrCl_2$ solution to the test solution and precipitates ($SrSO_3$ and $SrSO_4$) will be formed. Decant the supernatant containing $S_2O_3^{2-}$ after centrifugation. Add HCl to the precipitates, and add I_2-starch solution. The diminished color of the solution confirms the presence of SO_3^{2-} in the test solution.

Figure 5-3 shows the procedure for separation and identification of S^{2-}, SO_3^{2-}, $S_2O_3^{2-}$ and SO_4^{2-} mixed solution.

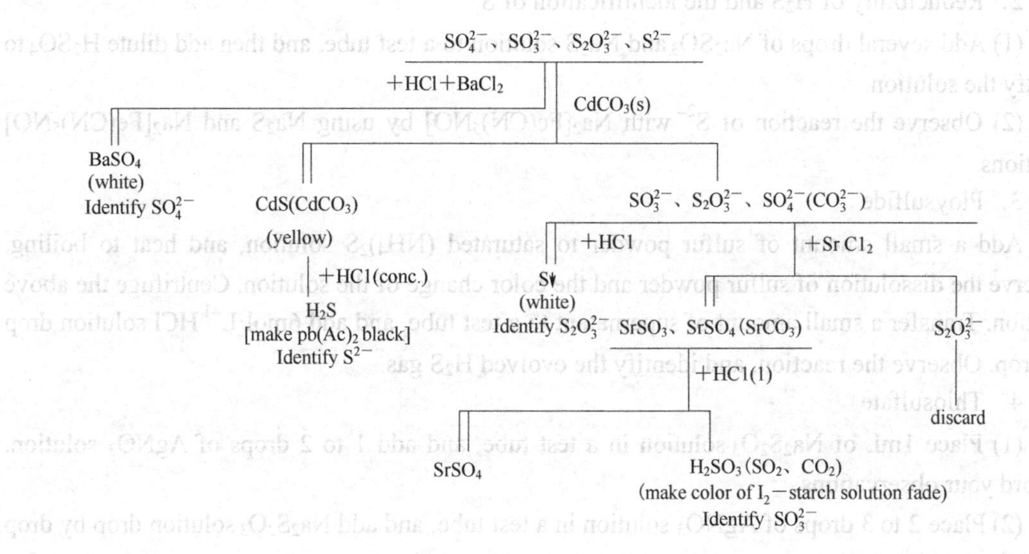

Figure 5-3　Separation and identification of S^{2-}, SO_3^{2-}, $S_2O_3^{2-}$ and SO_4^{2-} mixed solution

【Apparatus and Chemicals】

Centrifuge, graduated centrifuge tubes, alcohol burner.

(The following reagents utilize the concentration unit $mol \cdot L^{-1}$)

$MnSO_4$ (0.1), $K_2Cr_2O_7$ (0.1), Na_2SO_4 (0.1), Na_2S (0.1), $(NH_4)_2S$ (saturated, freshly prepared), $AgNO_3$ (0.1), H_2O_2 (3%, wt/wt), $Na_2S_2O_3$ (0.1), Na_2SO_3 (0.1), $ZnSO_4$ (0.1), $CdSO_4$ (0.1), $FeCl_3$ (0.1), TAA(5%, wt/wt), $Na_2[Fe(CN)_5NO]$ (1%, wt/wt), HCl (6, 2, 1), H_2SO_4 (1, 2, 3), NaOH (2), $NH_3 \cdot H_2O$ (2), mercury, diethyl ether, sulfur powder, Iodine water, $K_2S_2O_8$ (s), MnO_2 (s), pH test paper, $Pb(Ac)_2$ test paper.

【Experimental Procedure】

Fill in the appropriate table and write the chemical equations in your report.

1. H_2O_2

(1) Decomposition of H_2O_2　Prepare four test tubes. Add 2mL of 3% (wt/wt) H_2O_2 solution to each test tube. Then add several drops of $2mol \cdot L^{-1}$ NaOH solution to the first test tube; add several drops of $FeCl_3$ solution to the second test tube; add a small amount of MnO_2 to the third test tube; use the fourth test tube as the control. Put all four test tubes in a water bath and compare the decomposition of H_2O_2.

(2) Redox properties of H_2O_2

① Place 1mL of 3% (wt/wt) H_2O_2 solution in a test tube, and add several drops of $MnSO_4$ solution. Then add $2mol \cdot L^{-1}$ NaOH solution, and observe the reaction. Finally, add $1mol \cdot L^{-1}$ H_2SO_4 to acidify the solution. Observe and explain the reactions in your report.

② Instructor demonstration only. Place 20mL of 3% (wt/wt) H_2O_2 solution in a 50mL beaker, and add a pea-size sample of mercury into the beaker. Observe and explain the reaction.

(3) Identification of H_2O_2　Place 1mL of 3% (wt/wt) H_2O_2 solution in a test tube, and add

several drops of diethyl ether and $2mol \cdot L^{-1}$ H_2SO_4. Then add $K_2Cr_2O_7$ solution. Observe the color changes of the solution and the diethyl ether layer.

2. Reducibility of H_2S and the identification of S^{2-}

(1) Add several drops of Na_2SO_3 and Na_2S solution to a test tube, and then add dilute H_2SO_4 to acidify the solution.

(2) Observe the reaction of S^{2-} with $Na_2[Fe(CN)_5NO]$ by using Na_2S and $Na_2[Fe(CN)_5NO]$ solutions.

3. Ploysulfide

Add a small amount of sulfur powder to saturated $(NH_4)_2S$ solution, and heat to boiling. Observe the dissolution of sulfur powder and the color change of the solution. Centrifuge the above solution. Transfer a small amount of supernatant to a test tube, and add $6mol \cdot L^{-1}$ HCl solution drop by drop. Observe the reaction, and identify the evolved H_2S gas.

4. Thiosulfate

(1) Place 1mL of $Na_2S_2O_3$ solution in a test tube, and add 1 to 2 drops of $AgNO_3$ solution. Record your observations.

(2) Place 2 to 3 drops of $AgNO_3$ solution in a test tube, and add $Na_2S_2O_3$ solution drop by drop. Record your observations.

(3) Add iodine water to $Na_2S_2O_3$ solution. Record your observations.

(4) Add $2mol \cdot L^{-1}$ HCl to $Na_2S_2O_3$ solution. Record your observations.

5. Peroxydisulphate

Mix one drop of $MnSO_4$ solution and 5mL of $3mol \cdot L^{-1}$ H_2SO_4 solution, and equally divide the mixture into two test tubes.

(1) Add one drop of $AgNO_3$ solution and a small amount of solid $K_2S_2O_8$ to the first test tube, and gently heat the test tube. Observe and record the color change of the solution.

(2) Only add a small amount of solid $K_2S_2O_8$ to the second test tube, and gently heat the test tube. Observe and record if there is a color change.

Compare the results and explain.

【Questions and Discussion】

1. Why dose H_2O_2 act as both oxidizing agent and reducing agent? Under what condition can Mn^{2+} be oxidized to MnO_2 by H_2O_2? Under what condition can H_2O_2 be oxidized to O_2 by MnO_2?

2. If H_2S, Na_2S and Na_2SO_3 solutions are stored for a long time, what will happen? Why?

3. We use FeS and an acid to prepare H_2S in the laboratory. Which acid should be used, HCl, HNO_3 or concentrated H_2SO_4? Why?

4. When $Na_2S_2O_3$ solution reacts with $AgNO_3$ solution, sometimes an precipitate will be formed, but sometimes $[Ag(S_2O_3)_2]^{3-}$ complex will be formed. Why?

5. Why does S^{2-} need to be removed before identifying $S_2O_3^{2-}$ and SO_3^{2-}?

6. How do we separate SO_3^{2-} and $S_2O_3^{2-}$?

7. When we use H_2S saturated solution to separate Cd^{2+} and Zn^{2+}, we have to maintain the pH value within a suitable range. Calculate the pH range.

实验21 氮 和 磷

【实验目的】

1．试验并了解氮、磷的某些重要化合物的性质。

2．掌握铵离子、硝酸根、亚硝酸根、磷酸根、焦磷酸根和偏磷酸根的鉴定方法和原理。

【实验提要】

在掌握氮、磷及其化合物性质时，应注意与氧、硫相比较，掌握其异同点，并从它们所在周期表中的位置及原子结构上的相似性和不同点加以理解。

1．NH_3、N_2H_4与H_2O、H_2O_2

NH_3、N_2H_4（联氨）与H_2O、H_2O_2一样，存在分子间氢键，它们与同族其它元素的氧化物相比，具有特别高的沸点。NH_3和N_2H_4是弱碱，H_2O_2是弱酸，H_2O作为溶剂，显中性。NH_3和H_2O一样，对金属离子有很强的配位能力和溶剂化能力。联氨和过氧化氢有相似的氧化还原性质，例如N_2H_4和H_2O_2既可作氧化剂，又可作还原剂，不论在酸性还是碱性介质中，它们都易发生歧化分解，尤其在受热的情况下：

$$3N_2H_4 \xlongequal{250^{\circ}C} 4NH_3+N_2\uparrow$$

$$2H_2O_2 = 2H_2O+O_2\uparrow$$

2．铵盐

（1）NH_4^+与K^+和Rb^+的电荷相同，离子半径相近（分别为143 pm、133 pm、148pm），故铵盐与钾盐和铷盐的晶型相同，溶解度也相近。

（2）所有的铵盐都发生一定程度的水解：

$$2NH_4^+ + SO_4^{2-} + H_2O \rightleftharpoons 2NH_3\cdot H_2O + HSO_4^- + H^+$$

$$2NH_4^+ + CO_3^{2-} + H_2O \rightleftharpoons 2NH_3\cdot H_2O + H_2CO_3$$

$$NH_4^+ + AlO_2^- + H_2O \rightleftharpoons NH_3\uparrow + Al(OH)_3\downarrow$$

（3）铵盐的热稳定性都较差。铵盐的热分解按组成铵盐的酸的性质及分解条件的不同，其分解产物不同。总的来说，若酸是无氧化性的挥发酸（HCl），则分解产物氨和酸一起挥发，遇冷又结合成盐；若酸的氧化性很弱且难挥发（H_3PO_4，H_2SO_4），则氨挥发，酸或酸式盐留下来；若相应的酸（如HNO_3、HNO_2、$H_2Cr_2O_7$）有氧化性，则分解产生的NH_3会立即被氧化，产物视反应条件而异，一般是氨被氧化为氮气。

3．氮的含氧酸及其盐

HNO_2和HNO_3都是较强的氧化剂。硝酸作为氧化剂时，其还原产物是多种多样的。通常稀HNO_3被还原为NO（活泼金属把稀HNO_3还原成NH_4^+），浓HNO_3被还原为NO_2。冷的浓HNO_3会使某些金属（Ti、V、Cr、Al、Fe、Co、Ni等）钝化。亚硝酸HNO_2很不稳定，很容易发生歧化分解：

$$2HNO_2 = NO_2\uparrow + NO\uparrow + H_2O$$

当遇到强氧化剂时，亚硝酸或亚硝酸盐表现出还原性：

$$2MnO_4^- + 5HNO_2 + H^+ = 2Mn^{2+} + 5NO_3^- + 3H_2O$$

$$2MnO_4^- + NO_2^- + 2OH^- = 2MnO_4^{2-} + NO_3^- + H_2O$$

亚硝酸盐的热稳定性一般比硝酸盐强。硝酸盐热分解产物随金属元素活泼性不同而不同。

4．磷的含氧酸及其盐

磷酸容易通过分子间缩水而形成环状或链状的多磷酸。链状的多磷酸通式为 $H_{n+2}P_nO_{3n+1}$，随着链增长（n=16～90），就成为高分子多磷酸，常称为高聚磷酸。

磷酸根有较强的配位（螯合）能力。例如，借助可溶的无色配合物 $H_3[Fe(PO_4)_2]$或$H[Fe(HPO_4)_2]$的形成，可以掩蔽 Fe(Ⅲ)。在锅炉水中加入可溶性的磷酸二氢钠盐（加热后生成格雷恩盐），它可与水中所含的 Ca^{2+}、Mg^{2+}等离子形成可溶性的螯合物，从而防止了锅炉水垢的产生。焦磷酸盐作为重要的电镀液用于无氰电镀，主要是因为它对金属有很强的配位能力。

与多元酸分级解离相对应，易溶的多元酸盐发生分级水解。相应的盐有正盐和酸式盐。在难溶盐中，正盐的溶解度最小。

所有磷酸盐的热稳定性都较强。

$$\left[-O-\overset{\overset{O}{\uparrow}}{P}-O-\overset{\overset{O}{\uparrow}}{P}-O-\right]_n \quad M=Ca、Mg、Fe、Cu \quad n=15\sim45$$

（两个P原子下方的O原子与M配位）

5．有关离子的鉴定

（1）NH_4^+的鉴定（气室法） 于一表面皿上盛试液，并加入浓 NaOH。在另一表面皿上贴上一条润湿的红色石蕊试纸或浸有奈斯勒试剂（$K_2[HgI_4]$ +KOH）的滤纸，立即盖在第一块表面皿上形成气室。若试纸变蓝或滤纸变成棕黄色，示有NH_4^+存在。反应式为：

$$NH_4^+ + OH^- = NH_3\uparrow + H_2O$$

$$NH_3 + 2[HgI_4]^{2-} + 3OH^- = \left[O\begin{matrix}Hg\\ \\Hg\end{matrix}NH_2\right]I\downarrow + 7I^- + 2H_2O$$

奈氏试剂　　(棕黄色)

（2）NO_2^-和NO_3^-的鉴定

① 棕色环法：取试液加入固体 $FeSO_4$，使其大部分溶解。若加入 HAc 后，溶液呈棕色，示有NO_2^-，若加入浓 H_2SO_4，在浓 H_2SO_4与试液分界处出现棕色环，则示有NO_3^-。

$$2NO_2^- + 2H^+ = 2HNO_2 = NO + NO_2 + H_2O$$

$$NO_3^- + 3Fe^{2+} + 4H^+ = NO + 3Fe^{3+} + 2H_2O$$

$$NO + Fe^{2+} = FeNO^{2+}\ (棕色)$$

② 对氨基苯磺酸-α-萘胺法：NO_2^-在醋酸溶液中，能使对氨基苯磺酸重氮化，然后与α-萘胺生成粉红色偶氮染料。当NO_2^-浓度大时，粉红色很快褪去，并产生褐色沉淀。

$$H_2N-C_6H_4-SO_3H + NO_2^- + 2HAc \longrightarrow N\equiv N^+-C_6H_4-SO_3H + 2Ac^- + 2H_2O$$

$$N{\equiv}N^+\text{-}C_6H_4\text{-}SO_3H + C_{10}H_7NH_2(\alpha\text{-萘胺}) \longrightarrow HO_3S\text{-}C_6H_4\text{-}N{=}N\text{-}C_{10}H_6\text{-}NH_2$$

如果NO_2^-不存在时，要鉴定NO_3^-，可先将NO_3^-用Zn粉还原为NO_2^-，再用上法检验。

③ 二苯胺法：将试液以H_2SO_4酸化，加二苯胺的浓H_2SO_4溶液，有NO_3^-存在时，溶液变深蓝色。

$$C_6H_5\text{-}NH\text{-}C_6H_5 \xrightarrow{HNO_3} C_6H_5\text{-}NH\text{-}C_6H_4\text{-}C_6H_4\text{-}NH\text{-}C_6H_5 \longrightarrow C_6H_5\text{-}N{=}C_6H_4{=}C_6H_4{=}N\text{-}C_6H_5$$

此反应受NO_2^-的干扰，必须先除去。可以用尿素$CO(NH_2)_2$或铵盐NH_4Cl破坏NO_2^-。反应式为：

$$2NO_2^- + 2H^+ + CO(NH_2)_2 = 2N_2\uparrow + CO_2\uparrow + 3H_2O$$

$$HNO_2 + NH_4^+ = N_2\uparrow + H^+ + 2H_2O$$

（3）PO_4^{3-}、$(PO_3)_n^{n-}$和$P_2O_7^{4-}$的鉴定

① 钼酸铵法：PO_4^{3-}、$(PO_3)_n^{n-}$和$P_2O_7^{4-}$在酸性溶液中都能与$(NH_4)_2MoO_4$生成黄色的磷钼酸铵沉淀。

$$PO_4^{3-} + 3NH_4^+ + 12MoO_4^{2-} + 24H^+ = (NH_4)_3PO_4\cdot 12MoO_3\cdot 6H_2O\downarrow + 6H_2O$$

SiO_3^{2-}也发生类似反应，生成黄色硅钼酸铵沉淀（可溶于浓HNO_3）。强还原剂可将钼酸根还原成低价态的蓝色产物。大量的Cl^-可与Mo(Ⅵ)配位，从而降低鉴定的灵敏性。因此，上述反应必须在加热以及钼酸铵过量几倍和浓HNO_3存在下进行较为有利。

② Ag^+盐法：PO_4^{3-}与Ag^+形成黄色的Ag_3PO_4沉淀；$(PO_3)_n^{n-}$和$P_2O_7^{4-}$都与Ag^+形成白色的$AgPO_3$和$Ag_4P_2O_7$沉淀。但是，由于$(PO_3)_n^{n-}$的强吸水性，可破坏蛋白质的水化膜，能使蛋白质沉聚，而$P_2O_7^{4-}$则不能。

【仪器和药品】

温度计，酒精灯，离心机。

（除特别注明外，试剂浓度单位为$mol\cdot L^{-1}$）$NaNO_2$(0.1)，$NaNO_3$(0.1)，$AgNO_3$(0.1)，Na_3PO_4(0.1)，NaH_2PO_4(0.1)，Na_2HPO_4(0.1)，$Na_4P_2O_7$(0.1)，Na_3PO_3(0.1)，KI(0.1)，$CuSO_4$(0.1)，$FeSO_4$(0.1，s)，$CaCl_2$(0.1)，HNO_3(浓，2)，H_2SO_4(浓，1)，HAc(2)，HCl(2)，$NH_3\cdot H_2O$(2)，$KMnO_4$(0.01)，奈氏试剂，尿素，NH_4Cl(s)，$(NH_4)_2SO_4$(s)，$(NH_4)_2Cr_2O_7$(s)，$NaNO_3$(s)，$Cu(NO_3)_2$(s)，$AgNO_3$(s)，Zn片，红磷，I_2(s)，pH试纸，石蕊试纸，对氨基苯磺酸，*α*-萘胺，二苯胺，钼酸铵试剂。

【实验内容】

解释实验现象时，若涉及化学反应，均要求书写反应方程式。

1．铵盐和硝酸盐的热分解

（1）铵盐的热分解　在水平放置的干燥试管中放入约1g NH_4Cl(s)，并用干的玻璃棒将其

压紧，在试管口贴一条湿润的红色石蕊试纸，并将试管夹稳，均匀加热试管底部，观察试纸颜色的变化。在较冷的试管壁上附着的白霜物质是什么？解释现象。

分别用$(NH_4)_2SO_4$(s)和$(NH_4)_2Cr_2O_7$(s)代替NH_4Cl(s)重复上述实验，观察比较它们的热分解产物。

（2）硝酸盐的热分解　分别试验$NaNO_3$(s)、$Cu(NO_3)_2$(s)和$AgNO_3$(s)的热分解，用火柴余烬检验反应生成的气体，说明这3种硝酸盐热分解产物的异同。

2．硝酸的氧化性

取一小块Zn片于试管中，加入数滴$2mol\cdot L^{-1}$ HNO_3溶液，放置几分钟，取出少量反应的溶液，设法检验反应产生的NH_4^+。

3．亚硝酸盐的氧化还原性

（1）取数滴KI溶液于试管中，用稀H_2SO_4酸化，然后逐滴加入$NaNO_2$溶液，观察现象。微热试管，有何变化。

（2）用$KMnO_4$溶液代替KI溶液，按上法试验，观察现象。

4．NO_2^-和NO_3^-的鉴定

（1）在点滴板上加1滴试液，用$2mol\cdot L^{-1}$ HAc酸化，依次加入对氨基苯磺酸和α-萘胺各1滴，生成粉红色染料，示有NO_2^-。取10滴$0.1mol\cdot L^{-1}$ $NaNO_3$于试管中，加入一小粒$FeSO_4$(s)，振荡，溶解后，沿着试管壁加入浓H_2SO_4，不要振荡，观察浓H_2SO_4和试液交界处有无棕色环出现，有棕色环，则有NO_3^-存在。

（2）各取4滴$NaNO_2$和$NaNO_3$溶液于同一试管中，加入少量尿素（s），用稀H_2SO_4酸化，于沸水浴中加热5min，待溶液冷却后，检查NO_2^-是否除尽，若已除净，则取该溶液分别用二苯胺法和棕色环法检验NO_3^-的存在。

5．正磷酸盐的性质

（1）分别检验Na_3PO_4、Na_2HPO_4、NaH_2PO_4水溶液的pH值。

（2）分别取上述溶液各5滴，加入等量的$AgNO_3$溶液，并检验沉淀后溶液的pH值变化，解释之，试验沉淀是否溶于稀HNO_3。

（3）分别取上述溶液各5滴，加入等量的$CaCl_2$溶液若干滴，观察现象。滴加$2mol\cdot L^{-1}$ $NH_3\cdot H_2O$，各有何变化？再滴加$2mol\cdot L^{-1}$ HCl，又有何变化？比较$Ca_3(PO_4)_2$、$CaHPO_4$、$Ca(H_2PO_4)_2$的溶解性及其相互转化的条件。

（4）取5滴NaH_2PO_4溶液于试管中，加入0.5mL浓HNO_3和约1mL钼酸铵试剂，水浴加热2～3min，观察现象。

【选做部分】

1．磷酸及焦磷酸盐的配位性

（1）在两支试管中各取0.5mL $FeSO_4$溶液，其中一支加入1mL 1∶1 H_2SO_4；另一支加入1mL 1∶1 H_3PO_4溶液，然后分别滴加10滴$KMnO_4$溶液，比较两溶液的颜色。并解释之。

（2）取$CuSO_4$溶液2～3滴，逐滴加入$Na_4P_2O_7$溶液，观察浅蓝色沉淀的生成，继续滴加$Na_4P_2O_7$溶液至沉淀溶解。反应式为：

$$2Cu^{2+} + P_2O_7^{4-} \longrightarrow Cu_2P_2O_7\downarrow$$

$$Cu_2P_2O_7 + 3P_2O_7^{4-} \longrightarrow 2[Cu(P_2O_7)_2]^{6-}$$

2．PO_4^{3-}，$(PO_3)_n^{n-}$和$P_2O_7^{4-}$的鉴别

有 3 瓶试剂Ⅰ、Ⅱ、Ⅲ，可能是PO_4^{3-}，$(PO_3)_n^{n-}$和$P_2O_7^{4-}$的钠盐，试设法将它们鉴别出来。

3．碘化磷的生成和水解性质（在通风橱内进行实验）

在干燥的大试管中装有粉状的碘及干燥的红磷（I_2∶P 为 1∶6），稍微加热试管，并向试管内滴加少量水，检验是否有 HI 气体生成。

【问题与讨论】

1．指出下列分子，在气相中扩散速率快慢的次序：$NH_3(g)$、$CO_2(g)$、HCl(g)。说明为什么？

2．试比较磷酸钠盐（正盐和酸式盐）与 Ag^+和 Ca^{2+}沉淀的异同点。

3．怎样消除 NO_2^-对鉴定 NO_3^-的影响？

4．试举出氮化合物歧化反应的两三例（用方程式表示）。

5．设计 3 种区别亚硝酸钠和硝酸钠的方案。

6．为什么磷酸盐能防止锅炉形成水垢？

7．用钼酸铵检验PO_4^{3-}时，为什么强调必须在加热及有过量几倍的沉淀剂和浓 HNO_3 存在下进行反应？

【安全知识】

1．除 N_2O 外，所有氮的氧化物都有毒，其中尤以 NO_2 为甚。在大气中 NO_2 的允许含量为每升空气不得超过 0.005mg。目前 NO_2 中毒尚无特效药物治疗，一般只能输入氧气以帮助呼吸和血液循环。NO_2 主要对人体造成黏膜损害引起肿胀充血；呼吸系统损害引起各种炎症；神经系统损害引起眩晕、无力、痉挛、面部发绀等；造血系统损害破坏血红素等。吸入高浓度的氮氧化物将迅速出现窒息以致死亡。因此，凡涉及氮氧化物生成的反应均应在通风橱内进行。

2．实验室常见的磷有白磷和红磷。红磷毒性较小，白磷为蜡状结晶体，燃点为 45℃，在空气中易氧化，毒性很大，常保存于水中或煤油中。磷化氢是无色恶臭剧毒的气体。PCl_3(l)、PCl_5(s)都有腐蚀性，使用时应注意。

Experiment 21　Nitrogen and Phosphorus

【Objectives】

1. To evaluate and understand the properties of some important compounds of nitrogen and phosphorus.

2．To learn how to identify NH_4^+，NO_3^-，NO_2^-，PO_4^{3-}，$P_2O_7^{4-}$，and $(PO_3)_n^{n-}$.

【Introduction】

When studying the compounds of nitrogen and phosphorus, compare them with that of oxygen and sulfur. Pay particular attention to their positions in the periodic table and their atomic structures.

1．NH_3, N_2H_4 and H_2O, H_2O_2

NH_3 and N_2H_4 are similar to H_2O and H_2O_2. They all have intermolecular hydrogen bonds, so their boiling points are higher than other similar compounds in the same group. NH_3 and N_2H_4 are weak bases. H_2O_2 is a weak acid. H_2O is neutral and usually used as a solvent. NH_3 and H_2O have strong coordination ability and solvability. N_2H_4 and H_2O_2 have similar redox properties, they act as both oxidizing agents and reducing agents, and they easily disproportionate and decompose, especially when heated.

$$3N_2H_4 \xlongequal{250^\circ C} 4NH_3 + N_2\uparrow$$

$$2H_2O_2 = 2H_2O + O_2\uparrow$$

2. Ammonium salt

(1) Ammonium salts are similar to potassium salts and rubidium salts, because NH_4^+, K^+ and Rb^+ have the same oxidation number, and a similar ionic radius (143pm, 133pm, and 148pm, respectively). For example, they have similar crystals and solubilities.

(2) All ammonium salts will hydrolyze:

$$2NH_4^+ + SO_4^{2-} + H_2O \rightleftharpoons 2NH_3\cdot H_2O + HSO_4^- + H^+$$

$$2NH_4^+ + CO_3^{2-} + H_2O \rightleftharpoons 2NH_3\cdot H_2O + H_2CO_3$$

$$NH_4^+ + AlO_2^- + H_2O \rightleftharpoons NH_3\uparrow + Al(OH)_3\downarrow$$

(3) Ammonium salts easily decompose when heated. Different ammonium salts have different products. For example, ammonium chloride will decompose into volatile HCl and ammonia. When cooled down, the two products will react to produce ammonium chloride again. If the products are H_3PO_4 or H_2SO_4, which are not volatile acids, the produced ammonia will be volatilized. If the products are HNO_3, HNO_2 and $H_2Cr_2O_7$, the produced ammonia will usually be oxidized into nitrogen gas.

3. Oxyacids and salts of nitrogen

Nitrous acid (HNO_2) and nitric acid (HNO_3) are both strong oxidizing agents. Nitric acid can be reduced to different products. Dilute HNO_3 is usually reduced to NO, or NH_4^+ by active metals. Concentrated HNO_3 is reduced to NO_2. Cold concentrated HNO_3 can make some metals (Ti, V, Cr, Al, Fe, Co, Ni, etc) passivated. HNO_2 is very unstable, and easily decomposes. This is a disproportionation reaction:

$$2HNO_2 = NO_2\uparrow + NO\uparrow + H_2O$$

Nitrous acid or nitrite can be oxidized by strong oxidizing agents.

$$2MnO_4^- + 5HNO_2 + H^+ = 2Mn^{2+} + 5NO_3^- + 3H_2O$$

$$2MnO_4^- + NO_2^- + 2OH^- = 2MnO_4^{2-} + NO_3^- + H_2O$$

Nitrites are more stable than nitrates when heated.

4. Oxyacids and salts of phosphorous

Polyphosphoric acid can be formed by the condensation reaction of phosphoric acids. The polyphosphoric acid chain can be written as $H_{n+2}P_nO_{3n+1}$ (n=16～90).

PO_4^{3-} has strong coordination (chelating) ability. For example, some colorless coordination compounds, such as $H_3[Fe(PO_4)_2]$ and $H[Fe(HPO_4)_2]$, can mask Fe(Ⅲ). Add NaH_2PO_4 to the water in a boiler, and NaH_2PO_4 reacts with Ca^{2+} and Mg^{2+} to form soluble chelates, so boiler scale can be avoided. Pyrophosphate is an important plating solution because of its strong coordination ability to metals.

$$\left[-O-\overset{\overset{O}{\uparrow}}{\underset{\underset{O}{|}}{P}}-O-\overset{\overset{O}{\uparrow}}{\underset{\underset{O}{|}}{P}}-O- \right]_n \quad \text{(the two lower O atoms bridged by M)}$$

M=Ca、Mg、Fe、Cu

n=15~45

5．Identification of ions

(1) Identification of NH_4^+

Place the test solution on a watch glass, and add concentrated NaOH. Put a moistened red litmus paper or a filter paper dipped with Nessler's reagent ($K_2[HgI_4]$ +KOH) on another watch glass. Cover the first watch glass immediately with the second watch glass. If the litmus paper changes into blue, or the filter paper changes into brown, NH_4^+ is present in the test solution. The reactions are:

$$NH_4^+ + OH^- = NH_3\uparrow + H_2O$$

$$NH_3 + 2[HgI_4]^{2-} + 3OH^- = \left[O\begin{matrix} Hg \\ \\ Hg \end{matrix}NH_2 \right] I\downarrow + 7I^- + 2H_2O$$

Nessler's reagent (brown)

(2) Identification of NO_2^- and NO_3^-

① Brown ring method: Add solid $FeSO_4$ to the test solution and dissolve it. To identify NO_2^-, add HAc. If the test solution turns brown, NO_2^- is present in the test solution. To identify NO_3^-, add concentrated H_2SO_4. If a brown ring appears at the partitioning of H_2SO_4 and the test solution, NO_3^- is present in the test solution.

$$2NO_2^- + 2H^+ = 2HNO_2 = NO + NO_2 + H_2O$$

$$NO_3^- + 3Fe^{2+} + 4H^+ = NO + 3Fe^{3+} + 2H_2O$$

$$NO + Fe^{2+} = FeNO^{2+} \text{ (brown)}$$

② *p*-aminobenzene sulfonic acid-*α*-naphthylamine method: In acetic acid solution, *p*-aminobenzene sulfonic acid can be diazotized by NO_2^-, and produces a pink azo dye with *α*-naphthylamine. The pink fades quickly if the concentration of NO_2^- is high, and a brown precipitate will be formed.

$$H_2N-C_6H_4-SO_3H + NO_2^- + 2HAc \longrightarrow N{\equiv}N^+-C_6H_4-SO_3H + 2Ac^- + 2H_2O$$

$$N{\equiv}N^+-C_6H_4-SO_3H + C_{10}H_7NH_2 \longrightarrow HO_3S-C_6H_4-N{=}N-C_{10}H_6-NH_2$$

To identify NO_3^-, reduce it to NO_2^- by using zinc first, and then use the above method.

③ Diphenylamine method: Acidify the test solution with H_2SO_4. Then add diphenylamine, which dissolves in concentrated H_2SO_4. If the solution turns dark blue, the presence of NO_3^- is confirmed.

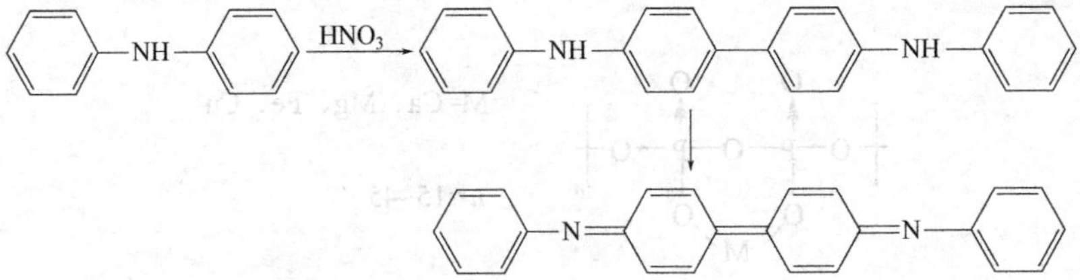

This reaction can be interfered by NO_2^-, which can be removed by $CO(NH_2)_2$ or NH_4Cl.

$$2NO_2^- + 2H^+ + CO(NH_2)_2 = 2N_2\uparrow + CO_2\uparrow + 3H_2O$$

$$HNO_2 + NH_4^+ = N_2\uparrow + H^+ + 2H_2O$$

(3) Identification of PO_4^{3-}, $(PO_3)_n^{n-}$ and $P_2O_7^{4-}$

① Ammonium molybdate method: PO_4^{3-}, $(PO_3)_n^{n-}$ and $P_2O_7^{4-}$ can all react with $(NH_4)_2MoO_4$ to produce yellow precipitates.

$$PO_4^{3-} + 3NH_4^+ + 12MoO_4^{2-} + 24H^+ = (NH_4)_3PO_4 \cdot 12MoO_3 \cdot 6H_2O\downarrow + 6H_2O$$

SiO_3^{2-} can also react with $(NH_4)_2MoO_4$ to produce yellow precipitate which can be dissolved in concentrated HNO_3. Because strong reducing agents can reduce MoO_4^{2-} to a blue product, and a large amount of Cl^- can coordinate with Mo(Ⅵ), the sensitivity of the reaction is low. So excessive $(NH_4)_2MoO_4$ and concentrated HNO_3 should be used, and the reaction should be heated.

② Ag^+ method: Yellow precipitate (Ag_3PO_4) can be formed by the reaction of PO_4^{3-} and Ag^+. Both $(PO_3)_n^{n-}$ and $P_2O_7^{4-}$ can react with Ag^+ to form white precipitates ($AgPO_3$ and $Ag_4P_2O_7$).

【Apparatus and Chemicals】

Thermometer, alcohol burner, centrifuge.

(The following reagents utilize the concentration unit $mol \cdot L^{-1}$)

$NaNO_2$ (0.1), $NaNO_3$ (0.1), $AgNO_3$ (0.1), Na_3PO_4 (0.1), NaH_2PO_4 (0.1), Na_2HPO_4 (0.1), $Na_4P_2O_7$ (0.1), Na_3PO_3 (0.1), KI (0.1), $CuSO_4$ (0.1), $FeSO_4$ (0.1, solid), $CaCl_2$ (0.1), HNO_3 (concentrated, 2), H_2SO_4 (concentrated, 1), HAc (2), HCl (2), $NH_3 \cdot H_2O$ (2), $KMnO_4$ (0.01), nessler's reagent, urea, NH_4Cl (s), $(NH_4)_2SO_4$ (s), $(NH_4)_2Cr_2O_7$ (s), $NaNO_3$ (s), $Cu(NO_3)_2$ (s), $AgNO_3$ (s), Zn, red phosphorus, I_2 (s), pH test paper, litmus paper, *p*-aminobenzene sulfonic acid, *α*-naphthylamine, diphenylamine, ammonium molybdate.

【Experimental Procedure】

Fill in the appropriate table and write the chemical equations in your report.

1. Thermal decomposition of ammonium salts and nitrates

(1) Thermal decomposition of ammonium salts Place about 1g of solid NH_4Cl in a dry test tube, and put a moistened red litmus test paper at the opening of the test tube. Use a test tube clamp to hold the test tube, and heat the bottom of the test tube slightly. Observe the color change of the test paper. What is the white compound in the test tube after being cooled down? Record and explain in your report.

Use $(NH_4)_2SO_4$ (s) and $(NH_4)_2Cr_2O_7$ (s) to replicate the above experiments. Observe and compare the products.

(2) Thermal decomposition of nitrates Heat $NaNO_3$ (s), $Cu(NO_3)_2$ (s) and $AgNO_3$ (s) respectively, and test the produced gas with a match ember. Compare their products and explain.

2. Oxidizability of nitric acid

Put a small piece of zinc in a test tube, and add several drops of $2mol \cdot L^{-1}$ HNO_3 solution. After a few minutes, use a small amount of the solution to identify the produced NH_4^+.

3. Oxidizability and reducibility of nitrite

(1) Put several drops of KI solution in a test tube, and acidify with dilute H_2SO_4. Then add $NaNO_2$ solution drop by drop. Observe the reaction. Heat the test tube slightly. Observe and explain in your report.

(2) Use $KMnO_4$ solution to replace KI solution to do the above experiment. Observe and record in your report.

4. Identification of NO_2^- and NO_3^-

(1) Add one drop of test solution on a spot plate, and acidify with $2mol \cdot L^{-1}$ HAc. First add one drop of *p*-aminobenzene sulfonic acid. Then add one drop of *α*-naphthylamine. The pink color confirms the presence of NO_2^-. Add 10 drops of $0.1mol \cdot L^{-1}$ $NaNO_3$ in a test tube, then add a small amount of solid $FeSO_4$. After dissolution of $FeSO_4$, add concentrated H_2SO_4 along the wall of the test tube. Do not shake the test tube. Observe the brown ring at the partitioning of H_2SO_4 and the test solution. The brown ring confirms that NO_3^- is present.

(2) Add four drops of $NaNO_2$ and four drops of $NaNO_3$ in the same test tube. Then add a small amount of solid $CO(NH_2)_2$, and acidify with dilute H_2SO_4. Heat the test tube in a boiled water bath for five minutes. After cooling the test tube, confirm that NO_2^- has been removed. Next, identify NO_3^- by the diphenylamine and brown ring methods.

5. Properties of phosphates

(1) Check the pH values for Na_3PO_4, Na_2HPO_4 and NaH_2PO_4 solutions.

(2) Add five drops of Na_3PO_4, Na_2HPO_4 and NaH_2PO_4 solutions to three test tubes respectively, and add the same amount of $AgNO_3$ solution to each test tube. Observe the precipitates and test the pH values. Then test if the precipitates dissolve in dilute HNO_3. Record and explain your findings in your report.

(3) Add five drops of Na_3PO_4, Na_2HPO_4 and NaH_2PO_4 solutions to three test tubes respectively, and add the same amount of $CaCl_2$ solution to each test tube. Observe whether there will be precipitates. Then add $2mol \cdot L^{-1}$ $NH_3 \cdot H_2O$ to each test tube, what do you observe? Then, add $2mol \cdot L^{-1}$ HCl, what do you observe? Compare the solubilities of $Ca_3(PO_4)_2$, $CaHPO_4$ and $Ca(H_2PO_4)_2$, and the conditions of their transformation into each other.

(4) Add five drops of NaH_2PO_4 in a test tube, and then add 0.5mL of concentrated HNO_3 and about 1mL of $(NH_4)_2MoO_4$. Heat the test tube in a water bath for two to three minutes. Record your observations and explain.

【Questions and Discussion】

1. List the gases, NH_3(g), CO_2(g) and HCl(g), in sequence according to their diffusion rates. Why?

2. Compare the similarities and differences of the reactions of Na_3PO_4, Na_2HPO_4 and NaH_2PO_4 with Ag^+ and Ca^{2+}.

3. When identifying NO_3^-, how do we remove NO_2^-?

4. Write three disproportionation reactions of the compounds of nitrogen.

5. Design three methods to identify $NaNO_2$ and $NaNO_3$.

6. How can the phosphates prevent the boiler from forming boiler scale?

7. When identifying PO_4^{3-} with $(NH_4)_2MoO_4$, why should excessive $(NH_4)_2MoO_4$ and concentrated HNO_3 be used? Why should the reaction be heated?

实验 22 ds 区元素的化合物

【实验目的】

1．试验并了解 ds 区元素的氢氧化物（或氧化物）的酸碱性及对热稳定性。

2．了解铜、银、锌、镉、汞的金属离子形成配合物的特征。

3．了解 Cu(Ⅱ)与 Cu(Ⅰ)，Hg(Ⅱ)与 Hg(Ⅰ)的相互转化条件。

4．了解铜、银、锌、镉、汞的离子鉴定。

【实验提要】

ds 区元素包括铜、银、金、锌、镉和汞。它们的价电子层结构分别为$(n-1)d^{10}ns^1$ 和$(n-1)d^{10}ns^2$。在化合物中常见的氧化值，铜为+2 和+1，银为+1，锌和镉为+2，汞为+2 和+1。这些元素的简单阳离子具有或接近 18 电子的构型。在化合物中与某些阳离子有较强的相互极化作用，成键的共价成分较大。多数化合物较难溶于水，对热稳定性较差，易形成配位化合物，化合物常显不同的颜色。

例如，这些元素的氢氧化物均较难溶于水，且易脱水变成氧化物。银和汞的氢氧化物极不稳定。常温下即失水变成 Ag_2O（棕黑色）和 HgO（黄色）。黄色 HgO 加热则生成橘红色 HgO 变体。

$Cu(OH)_2$、$Zn(OH)_2$ 和 $Cd(OH)_2$ 在常温下较稳定，但受热亦会失水成氧化物。浅蓝色 $Cu(OH)_2$ 在 80℃失水成棕黑色 CuO，白色 $Zn(OH)_2$ 在 125℃开始失水成黄色（冷却后为白色）的 ZnO，白色 $Cd(OH)_2$ 在 250℃变成棕红色的 CdO。

$Zn(OH)_2$ 是典型的两性氢氧化物，$Cu(OH)_2$ 呈较弱的两性（偏碱），$Cd(OH)_2$ 和 $Hg(OH)_2$（或 HgO)呈碱性，而 AgOH 为强碱性。

Cu^{2+}、Ag^+、Zn^{2+}、Cd^{2+}、Hg^{2+}与 Na_2S 溶液反应都生成难溶的硫化物，即 CuS（黑色）、Ag_2S（黑色）、ZnS（白色）、CdS（黄色）和 HgS（黑色）。其中 HgS 可溶于过量的 Na_2S，与 S^{2-}生成无色的$[HgS_2]^{2-}$。若在此溶液中加入盐酸又生成黑色 HgS 沉淀。此反应可作为分离 HgS 的方法。根据 ZnS、CdS、Ag_2S、CuS 和 HgS 溶度积大小，ZnS 可溶于稀酸，CdS 溶于 $6mol \cdot L^{-1}$ HCl 溶液，Ag_2S 和 CuS 溶于氧化性的 HNO_3，而 HgS 溶于王水。

ds 区元素阳离子都有较强的接受配体的能力，易与 H_2O、NH_3、X^-、CN^-、SCN^-和 en 等形成配离子。例如$[Cu(en)_2]^{2+}$、$[Ag(SCN)_2]^-$、$[Zn(H_2O)_4]^{2+}$、$[Cd(NH_3)_4]^{2+}$和$[HgCl_4]^{2-}$等。

Hg^{2+}与 I^-反应先生成橘红色 HgI_2 沉淀，加入过量的 I^-则生成无色的$[HgI_4]^{2-}$，它和 KOH 的混合溶液称为奈斯勒试剂，该试剂能有效地检验铵盐的存在。

Cu^{2+}、Ag^+、Zn^{2+}、Cd^{2+}与氨水反应生成$[Cu(NH_3)_4]^{2+}$（深蓝色）、$[Ag(NH_3)_2]^+$(无色)、$[Zn(NH_3)_4]^{2+}$(无色)、$[Cd(NH_3)_4]^{2+}$(无色)等配离子。Hg^{2+}只有在过量的铵盐存在下才与 NH_3 生成配离子。当铵盐不存在时，则生成氨基化合物沉淀。如：

$$HgCl_2+2NH_3 \Longrightarrow HgNH_2Cl\downarrow+NH_4Cl$$

Hg_2^{2+} 在 $NH_3 \cdot H_2O$ 中不生成配离子，而发生歧化反应。

$$2Hg_2(NO_3)_2+4NH_3+H_2O \Longrightarrow HgO\cdot HgNH_2NO_3\downarrow+2Hg\downarrow+3NH_4NO_3$$

难溶物和配合物的形成，可以改变元素的电极电位，影响元素的性质。以铜为例加以说明，铜的部分电位图（单位：V）为：

$$Cu^{2+} \xrightarrow{+0.153} Cu^{+} \xrightarrow{+0.521} Cu$$

$$Cu^{2+} \xrightarrow{+0.533} CuCl \xrightarrow{+0.137} Cu$$

$$[Cu(NH_3)_4]^{2+} \xrightarrow{-0.010} [Cu(NH_3)_2]^{+} \xrightarrow{-0.12} Cu$$

$$Cu^{2+} \xrightarrow{+0.86} CuI \xrightarrow{-0.185} Cu$$

$$CuS \xrightarrow{-0.54} Cu_2S \xrightarrow{-0.93} Cu$$

在水溶液中，Cu^+难于存在，它易发生歧化

$$2Cu^+ = Cu + Cu^{2+}$$

要使上述平衡向左移动，可使Cu^+生成难溶盐或配合物

$$2Cu^{2+} + 4I^- = 2CuI\downarrow + I_2$$

$$Cu^{2+} + Cu + 4X^- = 2[CuX_2]^-$$

$$[CuX_2]^- \rightleftharpoons CuX\downarrow + X^-$$

$$[Cu(NH_3)_4]^{2+} + Cu \rightleftharpoons 2[Cu(NH_3)_2]^+$$

在$CuSO_4$溶液中加入过量的Na_2SO_3溶液能将Cu^{2+}还原为Cu^+，同时形成$[Cu(SO_3)_3]^{5-}$：

$$2Cu^{2+} + 7SO_3^{2-} + H_2O = 2[Cu(SO_3)_3]^{5-} + SO_4^{2-} + 2H^+$$

如果移去Cu^+难溶盐的阴离子或配合物的配体，Cu^+又发生歧化分解。

Hg^{2+}与Hg_2^{2+}在一定条件下能相互转化。它们有如下平衡：

$$Hg^{2+} + Hg \rightleftharpoons Hg_2^{2+} \qquad K = 166$$

反应的方向将取决于对反应条件的控制。例如，在配位剂存在下或当pH≥3时，Hg_2^{2+}发生歧化反应：

$$Hg_2^{2+} + 4I^- = [HgI_4]^{2-} + Hg\downarrow$$

$$Hg_2^{2+} + H_2O = HgO\downarrow + 2H^+ + Hg\downarrow$$

$$Hg_2^{2+} + 2OH^- = Hg\downarrow + HgO\downarrow + H_2O$$

若Hg^{2+}形成难溶沉淀物或稳定的配合物，上述平衡右移，Hg_2^{2+}发生歧化。

【离子的鉴定】

（1）Cu^{2+}　Cu^{2+}与黄血盐$K_4[Fe(CN)_6]$反应，生成红棕色$Cu_2[Fe(CN)_6]$沉淀，方法灵敏。Fe^{3+}有干扰。

（2）Zn^{2+}　Zn^{2+}与硫氰合汞酸铵$(NH_4)_2[Hg(SCN)_4]$生成白色的$Zn[Hg(SCN)_4]$沉淀。

（3）Cd^{2+}　Cd^{2+}与S^{2-}生成黄色沉淀。若要消除其它金属离子的干扰，可在KSCN存在时鉴定。

（4）Hg^{2+}和Hg_2^{2+}　Hg^{2+}可被$SnCl_2$分步还原，还原产物由白色（Hg_2Cl_2）变为灰色或黑色（Hg）沉淀。

【仪器和药品】

离心机，点滴板。

（除特别注明外，试剂浓度单位为$mol \cdot L^{-1}$）$CuSO_4$ (0.1)，$AgNO_3$ (0.1)，$ZnSO_4$ (0.1)，$CdSO_4$ (0.1)，$Hg(NO_3)_2$ (0.1)，Na_2S (0.1)，Na_2SO_3 (0.5)，NaCl (0.1)，KBr (0.1)，KI (0.1)，$Na_2S_2O_3$ (0.1)，$CuCl_2$ (饱和)，NaOH (2、6)，HCl (浓)，$NH_3 \cdot H_2O$ (2，6，浓)，H_2SO_4 (1，2)，HNO_3 (1，6，浓)，乙二胺 (0.1)，EDTA (0.1)，$K_4[Fe(CN)_6]$ (0.1)，KSCN (质量分数为25%)，葡萄糖（质量

分数为 10%)，汞，铜屑，TAA。

【实验内容】

请用表格报告实验结果。解释实验现象时，若涉及化学反应，均要求写出反应方程式。实验废液回收在指定容器中。

1．氢氧化物（或氧化物）的生成和性质

在数滴 $CuSO_4$ 溶液中，滴加适量的 2mol·L^{-1} NaOH 溶液，生成沉淀后离心分离。将沉淀分成 3 份，其中 1 份加热，试验其对热的稳定性；其它两份分别试验沉淀在 6mol·L^{-1} NaOH 和 2mol·L^{-1} H_2SO_4 溶液中溶解的情况。

分别用 $ZnSO_4$、$CdSO_4$、$AgNO_3$ 和 $Hg(NO_3)_2$ 溶液代替 $CuSO_4$ 溶液，将制得的沉淀分别试验其对热、对稀酸和强碱作用的情况。

2．配合物

（1）氨合物　在 $AgNO_3$ 溶液中滴加 2mol·L^{-1} $NH_3·H_2O$，观察沉淀的生成与溶解。再用沉淀溶解后的溶液试验其对热稳定性和与酸、碱作用的情况。

分别用 $CuSO_4$、$ZnSO_4$ 和 $CdSO_4$ 溶液代替 $AgNO_3$ 溶液重复上述实验。

（2）铜的其它配合物

① 取几滴 $CuSO_4$ 溶液于试管中，先滴加 2mol·L^{-1} $NH_3·H_2O$ 至生成的沉淀溶解，转加乙二胺溶液，观察溶液颜色的变化，再滴加 EDTA 溶液，溶液颜色又有何变化？比较以上 3 种铜的配合物的稳定性并解释之。

② 试验 $CuSO_4$ 溶液与 $K_4[Fe(CN)_6]$溶液的作用，观察沉淀的颜色。此反应常用于鉴定 Cu^{2+}。

（3）配位反应与沉淀反应　利用实验提供的下列试剂：$AgNO_3$、NaCl、KBr、KI、$Na_2S_2O_3$ 和 2mol·L^{-1} $NH_3·H_2O$ 等溶液设计试管实验，比较 AgCl、AgBr、AgI 的溶解度和$[Ag(NH_3)_2]^+$、$[Ag(S_2O_3)_2]^{3-}$稳定性的大小。

3．Cu(Ⅱ)与 Cu(Ⅰ)的相互转化

（1）氧化亚铜的生成和性质　在数滴 $CuSO_4$ 溶液中加入过量的 6mol·L^{-1} NaOH 溶液，使生成的沉淀溶解后再加入数滴葡萄糖溶液。摇匀，水浴微热，观察现象。将沉淀离心分离，弃去清液，沉淀用蒸馏水洗涤后加入 2mol·L^{-1} H_2SO_4 溶液，水浴加热，观察沉淀溶解的情况，溶液的颜色，剩余沉淀是何物？

（2）碘化亚铜的形成　取数滴 $CuSO_4$ 溶液于离心试管中，滴加 KI 溶液，观察现象。CuI 是什么颜色？如何消除 I_2 颜色的干扰？

（3）CuCl 的形成　在小烧杯中，加入约 1mL 饱和的 $CuCl_2$ 溶液和约 2mL 浓 HCl，再加入少许铜屑，小火加热片刻，此时溶液颜色加深，继续加热微沸片刻，待溶液颜色由深变浅时，将溶液倾入约 100mL 水中，观察产物的颜色和状态。

【选做部分】

1．汞的化合物

（1）HgS 的生成与性质　于两支离心试管中，各加入 2 滴 $Hg(NO_3)_2$ 溶液，分别滴加 TAA 溶液，观察沉淀的颜色。离心分离，弃去清液，将沉淀洗涤后，于一支试管加入 Na_2S 溶液，观察现象。于另一支试管中加入几滴 6mol·L^{-1} HNO_3 溶液，搅匀、观察沉淀是否溶解？若不溶解再滴加几滴王水[$V(HNO_3)$: V(浓 HCl)＝1 : 3]，搅匀，此时有何变化？

（2）Hg^{2+}配合物的生成及其应用

① 在 2 滴 $Hg(NO_3)_2$ 溶液中，逐滴加入 KI 溶液，观察沉淀的颜色。继续加入 KI，观察

沉淀的溶解，再加入几滴 2mol·L^{-1} NaOH 溶液并摇匀。然后试验上述溶液与 NH_4Cl 溶液作用的情况。

② 取 2 滴 $Hg(NO_3)_2$ 溶液于试管中，加入数滴 KSCN 溶液，观察现象。将溶液分成两份，分别试验该溶液与硫酸锌溶液和氯化钴溶液作用的情况。此反应常用来鉴定 Zn^{2+}和 Co^{2+}。

2．Hg(Ⅱ)和 Hg(Ⅰ)的相互转化

（1）取 2 滴 $Hg(NO_3)_2$ 溶液于试管中，加入数滴 NaCl 溶液，观察现象。再加入 2mol·L^{-1} $NH_3·H_2O$，有何变化？

（2）取 4 滴 $Hg(NO_3)_2$ 溶液于试管中，加入 1 滴汞（小心取用，切勿洒出瓶外！），振荡试管。将清液转移至另外两支试管（余下的汞要回收！），于其中一支加入数滴 NaCl 溶液，观察现象。于另一支加入几滴 2mol·L^{-1} $NH_3·H_2O$，观察现象。并与实验（1）做比较。

3．用下列试剂设计 Cu^+的歧化和反歧化实验。

$CuSO_4$(0.1mol·L^{-1})，Na_2SO_3(0.5mol·L^{-1})，H_2SO_4(2mol·L^{-1})和浓 $NH_3·H_2O$。

4．现有 Cu^{2+}、Ag^+、Zn^{2+}和 Hg^{2+}4 种离子的混合液，根据氯化物和硫化物的性质，将它们分离并检出。

混合液取 0.1mol·L^{-1} 的下列试剂：$CuSO_4$、$AgNO_3$、$ZnSO_4$ 和 $Hg(NO_3)_2$ 溶液各 3 滴配成。

【问题与讨论】

1．久置的$[Ag(NH_3)_2]^+$碱性溶液，有产生氮化银 Ag_3N 的危险，应采用什么办法来破坏$[Ag(NH_3)_2]^+$。

2．总结 Cu^{2+}-Cu^+，Hg^{2+}-Hg_2^{2+} 相互转化的条件。

3．试分析为什么在 $CuSO_4$ 溶液中加入 KI 即产生 CuI 沉淀，而加入 KCl 溶液时却不出现 CuCl 沉淀。

4．Hg 和 Hg^{2+}有剧毒，试验时应注意些什么？

Experiment 22 Elements in ds-block and Their Compounds

【Objectives】

1．To evaluate and understand the acidic and basic properties of hydroxides and oxides of ds-block elements, and their thermostability.

2．To understand the coordination characteristics of Cu, Ag, Zn, Cd and Hg.

3．To understand the transformation between Cu(Ⅱ) and Cu(Ⅰ), Hg(Ⅱ) and Hg(Ⅰ).

4．To identify Cu^{2+}, Ag^+, Zn^{2+}, Cd^{2+} and Hg^{2+}.

【Introduction】

Cu, Ag, Au, Zn, Cd and Hg are ds-block elements. Their outer electron configurations can be expressed as $(n-1)d^{10}ns^1$ and $(n-1)d^{10}ns^2$. The common oxidation numbers for Cu are +2 and +1, for Ag +1, for Zn and Cd +2, for Hg +2 and +1. Their simple cations are or close to 18-electron configurations. Many of their compounds are insoluble in water, and have poor thermostabilities. Elements in ds-block are easy to form coordination compounds which usually have different colors.

The hydroxides of ds-block elements are insoluble, and easy to form oxides by removing water. The hydroxides of Ag and Hg are very unstable. They will lose water to form Ag_2O (brownish black) and HgO (yellow) at room temperature. Yellow HgO can be heated to produce orange HgO.

$Cu(OH)_2$, $Zn(OH)_2$ and $Cd(OH)_2$ are stable at room temperature, but they can form oxides by heating. Blue $Cu(OH)_2$ will change to black CuO at 80℃. White $Zn(OH)_2$ will change to yellow ZnO at 125℃, and yellow ZnO will turn white after cooling down. White $Cd(OH)_2$ will change to reddish brown CdO at 250℃.

$Zn(OH)_2$ is a typical amphoteric hydroxide. $Cu(OH)_2$ is a weak base. $Cd(OH)_2$ and $Hg(OH)_2$ (or HgO) are both bases. AgOH is a strong base.

Cu^{2+}, Ag^+, Zn^{2+}, Cd^{2+} and Hg^{2+} all react with Na_2S solution to form insoluble sulfides, i.e. CuS (black), Ag_2S (black), ZnS (white), CdS (yellow) and HgS (black). HgS can be dissolved into excessive Na_2S, forming colorless $[HgS_2]^{2-}$ with S^{2-}. But adding HCl to the solution, black HgS precipitate will be formed again. This is the method to separate HgS. ZnS can be dissolved in dilute acid, and CdS can be dissolved in $6mol \cdot L^{-1}$ HCl solution. Ag_2S and CuS can be dissolved in HNO_3, but HgS can only be dissolved in aqua regia.

Cations of the ds-block all have strong abilities to accept ligands, such as H_2O, NH_3, X^-, CN^-, SCN^- and en, forming $[Cu(en)_2]^{2+}$, $[Ag(SCN)_2]^-$, $[Zn(H_2O)_4]^{2+}$, $[Cd(NH_3)_4]^{2+}$ and $[HgCl_4]^{2-}$, etc.

Hg^{2+} reacts with I^- to form an orange precipitate HgI_2. $[HgI_4]^{2-}$ complex ion will be form if excessive I^- is used. The mixture of $[HgI_4]^{2-}$ and KOH is called Nessler's reagent, which is used to identify ammonium salts.

Cu^{2+}, Ag^+, Zn^{2+} and Cd^{2+} can react with aqueous ammonia to produce $[Cu(NH_3)_4]^{2+}$ (dark blue), $[Ag(NH_3)_2]^+$ (colorless), $[Zn(NH_3)_4]^{2+}$ (colorless) and $[Cd(NH_3)_4]^{2+}$ (colorless) complex ions. Hg^{2+} can only produce complex ions with NH_3 when excessive ammonium salts are present. Without ammonium salts, a precipitate will be formed. For example:

$$HgCl_2+2NH_3 = HgNH_2Cl\downarrow +NH_4Cl$$

Hg_2^{2+} will react with $NH_3 \cdot H_2O$, but this is a disproportionated reaction.

$$2Hg_2(NO_3)_2+4NH_3+H_2O = HgO \cdot HgNH_2NO_3\downarrow +2Hg\downarrow +3NH_4NO_3$$

The produced insoluble compounds and complexes can change the electrode potentials of the elements. Take Cu as an example(V).

$$Cu^{2+} \xrightarrow{+0.153} Cu^+ \xrightarrow{+0.521} Cu$$

$$Cu^{2+} \xrightarrow{+0.533} CuCl \xrightarrow{+0.137} Cu$$

$$[Cu(NH_3)_4]^{2+} \xrightarrow{-0.010} [Cu(NH_3)_2]^+ \xrightarrow{-0.12} Cu$$

$$Cu^{2+} \xrightarrow{+0.86} CuI \xrightarrow{-0.185} Cu$$

$$CuS \xrightarrow{-0.54} Cu_2S \xrightarrow{-0.93} Cu$$

Cu^+ is unstable and disproportionates in solution:

$$2Cu^+ = Cu+Cu^{2+}$$

To shift the above equilibrium to the left, produce insoluble salts or complexes of Cu^+.

$$2Cu^{2+}+4I^- = 2CuI\downarrow +I_2$$

$$Cu^{2+}+Cu+4X^- = 2[CuX_2]^-$$

$$[CuX_2]^- \rightleftharpoons CuX\downarrow +X^-$$

$$[Cu(NH_3)_4]^{2+}+Cu \rightleftharpoons 2[Cu(NH_3)_2]^+$$

Excessive Na_2SO_3 solution can reduce Cu^{2+} to Cu^+, and form $[Cu(SO_3)_3]^{5-}$ complex ions.

$$2Cu^{2+}+7SO_3^{2-}+H_2O = 2[Cu(SO_3)_3]^{5-}+SO_4^{2-}+2H^+$$

If removing the anions of insoluble salts, or removing the ligands of the complexes, Cu^+ will disproportionate again.

Hg^{2+} and Hg_2^{2+} can transform into each other under certain conditions.

$$Hg^{2+}+Hg \rightleftharpoons Hg_2^{2+} \qquad K=166$$

The direction of the reaction depends on the reaction conditions. For example, when pH≥3 or there are ligands, Hg_2^{2+} disproportionates:

$$Hg_2^{2+}+4I^- = [HgI_4]^{2-}+Hg\downarrow$$

$$Hg_2^{2+}+H_2O = HgO\downarrow+2H^++Hg\downarrow$$

$$Hg_2^{2+}+2OH^- = Hg\downarrow+HgO\downarrow+H_2O$$

【Identification of Ions】

(1) Cu^{2+} Cu^{2+} reacts with $K_4[Fe(CN)_6]$ to form a reddish brown $Cu_2[Fe(CN)_6]$ precipitate. This is a sensitive method to identify Cu^{2+}. Fe^{3+} will interfere with the reaction.

(2) Zn^{2+} Zn^{2+} reacts with $(NH_4)_2[Hg(SCN)_4]$ to form a white $Zn[Hg(SCN)_4]$ precipitate.

(3) Cd^{2+} Cd^{2+} reacts with S^{2-} to form a yellow precipitate. To eliminate the interference of other metal ions, identify Cd^{2+} with the presence of KSCN.

(4) Hg^{2+} and Hg_2^{2+} Hg^{2+} can be reduced by $SnCl_2$ step by step. The products change from white (Hg_2Cl_2) to grey or black precipitates (Hg).

【Apparatus and Chemicals】

Centrifuge, spot plate.

(The following reagents utilize the concentration unit $mol \cdot L^{-1}$)

$CuSO_4$ (0.1), $AgNO_3$ (0.1), $ZnSO_4$ (0.1), $CdSO_4$ (0.1), $Hg(NO_3)_2$ (0.1), Na_2S (0.1), Na_2SO_3 (0.5), NaCl (0.1), KBr (0.1), KI (0.1), $Na_2S_2O_3$ (0.1), $CuCl_2$ (saturated), NaOH (2、6), HCl (concentrated), $NH_3 \cdot H_2O$ (2, 6, concentrated), H_2SO_4 (1, 2), HNO_3 (1, concentrated), ethylenediamine (0.1), EDTA (0.1), $K_4[Fe(CN)_6]$ (0.1), KSCN (25%, wt/wt), glucose (10%, wt/wt), Hg, Cu, TAA.

【Experimental Procedure】

Fill in the appropriate table and write the chemical equations in your report.

1. Properties of hydroxides and oxides

Add an appropriate amount of $2mol \cdot L^{-1}$ NaOH solution to a few drops of $CuSO_4$ solution. Centrifuge and separate the precipitates. Divide the precipitates to three equal portions. Heat the first portion and test its thermostability. Add $6mol \cdot L^{-1}$ NaOH solution and $2mol \cdot L^{-1}$ H_2SO_4 solution to the other two portions respectively, and test the solubilities of the precipitates.

Use $ZnSO_4$, $CdSO_4$, $AgNO_3$ and $Hg(NO_3)_2$ to replace $CuSO_4$ solution to replicate the above experiments.

2. Coordination compounds

(1) Ammine Add $2mol \cdot L^{-1}$ $NH_3 \cdot H_2O$ to $AgNO_3$ solution, observe the formation and dissolution of the precipitates. Use the final solution to test its thermostability, and its reactions with

acid and base.

Use $CuSO_4$, $ZnSO_4$ and $CdSO_4$ to replace $AgNO_3$ solution to replicate the above experiments.

(2) Other complexes

① Place a few drops of $CuSO_4$ solution into a test tube, and add $2mol \cdot L^{-1}$ $NH_3 \cdot H_2O$ until precipitates form. Then add ethylenediamine solution, and observe the color change. Next add EDTA solution, what happens? Compare the stabilities of these three complexes and explain your observations.

② Test the reaction of $CuSO_4$ solution with $K_4[Fe(CN)_6]$ solution, and observe the color of the precipitate. This reaction is used to identify Cu^{2+}.

(3) Coordination reaction and precipitate reaction Use $AgNO_3$, NaCl, KBr, KI, $Na_2S_2O_3$ and $2mol \cdot L^{-1}$ $NH_3 \cdot H_2O$ to design experiments to compare the solubilities of AgCl, AgBr and AgI, and the stabilities of $[Ag(NH_3)_2]^+$ and $[Ag(S_2O_3)_2]^{3-}$.

3. Transformation between Cu(Ⅱ) and Cu(Ⅰ)

(1) Cu_2O Add excessive $6mol \cdot L^{-1}$ NaOH to a few drops of $CuSO_4$ solution. After the produced precipitate is dissolved, add a few drops of glucose solution. Mix well and heat in a water bath. Centrifuge and discard the supernatant. Wash the precipitate with distilled water, then add $2mol \cdot L^{-1}$ H_2SO_4 solution, and heat in a water bath. Observe the dissolution of the precipitate. What is the color of the solution? What is the final precipitate?

(2) CuI Place a few drops of $CuSO_4$ solution in a centrifuge tube, and add KI solution. What is the color of CuI? How do we eliminate the interference of the I_2 color?

(3) CuCl In a small beaker, add about 1mL of saturated $CuCl_2$ solution and about 2mL of concentrated HCl. Then add a few pieces of copper, and heat. Continue heating until the color of the solution changes from dark to light. Decant the solution to about 100mL of water, and observe the products.

【Questions and Discussion】

1. If the basic solution of $[Ag(NH_3)_2]^+$ is stored for a long time, it may produce dangerous Ag_3N. How do we decompose $[Ag(NH_3)_2]^+$?

2. Summarize the conditions of the transformation between Cu^{2+} and Cu^+, Hg^{2+} and Hg_2^{2+}.

3. Adding KI to $CuSO_4$ solution can form CuI precipitate, but adding KCl to $CuSO_4$ solution can't form CuCl precipitate. Why?

4. Hg and Hg^{2+} are toxic. What precautions should you take when doing experiments with them?

实验23 钛、钒、铬和锰的化合物

【实验目的】

试验并了解钛、钒、铬和锰的某些重要化合物的性质。

【实验提要】

1．钛、钒、铬、锰在元素周期表中的位置

钛、钒、铬、锰在元素周期表中的位置如表5-3所示。

表5-3 钛、钒、铬、锰在元素周期表中的位置

族 / 周期	ⅣB	VB	ⅥB	ⅦB
	钛分族	钒分族	铬分族	锰分族
4	22 Ti $[Ar]3d^2 4s^2$	23 V $[Ar]3d^3 4s^2$	24 Cr $[Ar]3d^5 4s^1$	25 Mn $[Ar]3d^5 4s^2$

2．氢氧化物和盐类

这些元素的氢氧化物的酸碱性符合一般规律。$M(OH)_2$都是难溶的中强碱；$M(OH)_3$是弱碱且酸性依次增强；$Cr(OH)_3$是典型的两性氢氧化物。Cr(Ⅲ)既能形成阳离子盐也能形成阴离子盐，高氧化值时，亲氧性很强，因此即使在强酸性溶液中也不存在简单阳离子（水合）Ti^{4+}、V^{4+}、V^{5+}、Cr^{6+}、Mn^{6+}和Mn^{7+}，其中Ti(Ⅳ)、V(Ⅳ)和V(Ⅴ)主要的存在形式是其金属氧基（酰基）离子TiO^{2+}、VO^{2+}和VO_2^+；Cr(Ⅵ)、Mn(Ⅵ)和Mn(Ⅶ)则只形成阴离子盐。Cr(Ⅵ)的阴离子盐在水溶性方面类似于硫酸盐。例如Ca^{2+}、Sr^{2+}、Ba^{2+}、Pb^{2+}及Ag^+盐均难溶于水，因此可借助于$BaCrO_4$（柠檬黄色）、$PbCrO_4$（黄色）或Ag_2CrO_4（砖红色）沉淀的形成来鉴定CrO_4^{2-}（或$Cr_2O_7^{2-}$）。

此外，周期表中ⅤB和ⅥB元素的含氧酸常缩合成偏酸和多酸，例如铬酸根CrO_4^{2-}（黄色）在pH<5时缩合的主要产物为二铬酸根，即重铬酸根$Cr_2O_7^{2-}$（橙色），在更强的酸性溶液中还可以形成三铬酸根$Cr_3O_{10}^{2-}$、四铬酸根$Cr_4O_{13}^{2-}$等。

对于难溶盐来说，多酸盐的溶解度比单酸盐大。因此，加Ba^{2+}、Pb^{2+}或Ag^+于CrO_4^{2-}和$Cr_2O_7^{2-}$等的混合溶液中，总是析出难溶的铬酸盐沉淀。

3．配位化合物

在这些元素的阳离子所形成的配合物中，以Cr^{3+}的配合物为多，也较为重要。例如，$CrCl_3·6H_2O$有3种颜色的晶体，对应3种不同的配合物（Cr^{3+}的配位数为6）：

$[Cr(H_2O)_6]Cl_3$（蓝紫色）　　$[Cr(H_2O)_5Cl]Cl_2·H_2O$（蓝绿色）　　$[Cr(H_2O)_4Cl_2]Cl·2H_2O$（灰绿色）

因此，Cr^{3+}的水溶液往往因其水合程度不同而显不同的颜色。

钛和钒的亲氧能力较强，因此在水溶液中多以较稳定的氧基（酰基）离子或含氧酸根存在，不易形成其它较稳定的配合物，只有X^-在较浓时才形成若干卤合离子，例如$[TiCl_2]^{2+}$、$[TiCl_3]^+$或$[Ti(OH)Cl_5]^{2-}$、$[TiCl_6]^{2-}$及$[TiF_6]^{2-}$等。此外在酸性溶液中，Ti(Ⅳ)、V(Ⅳ)、V(Ⅴ)和Cr(Ⅵ)可以与H_2O_2形成有特征颜色的配合物：

$$TiO^{2+} + H_2O_2 \rightleftharpoons [TiO(H_2O_2)]^{2+}$$
（橘黄色）

$$VO^{2+} + H_2O_2 = [VO(H_2O_2)]^{2+}$$
（红棕色）

$$VO_2^+ + 2H_2O_2 = [VO_2(H_2O_2)_2]^+$$
（黄色）

$$Cr_2O_7^{2-} + 4H_2O_2 + 6H^+ = 2[CrO_2(H_2O_2)_2]^{2+} + 3H_2O$$

$$Cr_2O_7^{2-} + 4H_2O_2 + 2H^+ = 2CrO(O_2)_2 + 5H_2O$$
（蓝色）

向$[TiO(H_2O_2)]^{2+}$溶液中加入氨水，形成过氧钛酸H_4TiO_5 [即$TiO(H_2O_2)(OH)_2$]黄色沉淀，此法鉴定钛很灵敏：

$$[TiO(H_2O_2)]^{2+} + 2NH_3 \cdot H_2O = H_4TiO_5 + 2NH_4^+$$

Cr(Ⅵ)的H_2O_2配合物既可鉴定H_2O_2，也用于鉴定Cr(Ⅵ)，此配合物易溶于乙醚，在乙醚中较稳定。故鉴定时加乙醚萃取可提高灵敏度。

4．氧化还原性

Ti、V、Cr、Mn的标准电极电位$E_A^\ominus$图（方括号内为碱性介质的电位，单位：V）如下：

$$TiO^{3+} \xrightarrow{+0.1} TiO^{2+} \xrightarrow{-0.37} Ti^{2+} \xrightarrow{-1.63} Ti$$
（无色）（紫色）（褐色）（银灰色）

$$VO_2^+ \xrightarrow{+1.0} VO^{2+} \xrightarrow{+0.361} V^{3+} \xrightarrow{-0.255} V^{2+} \xrightarrow{-1.18} V$$
（浅黄色）（蓝色）（绿色）（紫色）（浅灰色）

$$Cr_2O_7^{2-}\text{（橙红）} \xrightarrow[{[-0.13]}]{+1.33} Cr^{3+}\text{（蓝紫色）} \xrightarrow[{[0.8]}]{-0.41} Cr^{2+}\text{（天蓝）} \xrightarrow[{[-1.4]}]{-0.91} Cr$$

$$[Cr_2O_7^{2-}]\text{（黄色）} \quad [CrO_2^-]\text{（绿色）} \quad [Cr(OH)_2]$$

$$MnO_4^-\text{（紫红）} \xrightarrow[{[+0.564]}]{+0.564} MnO_4^{2-}\text{（绿色）} \xrightarrow[{[+0.60]}]{+2.26} MnO_2\text{（棕黑）} \xrightarrow[{[-0.2]}]{+0.95}$$

$$[MnO_4^{2-}]\text{（绿色）} \quad [MnO_2]$$

$$Mn^{3+}\text{（红色）} \xrightarrow[{[+0.1]}]{+1.51} Mn^{2+}\text{（浅粉色）} \xrightarrow[{[-1.55]}]{-1.18} Mn$$

$$[Mn(OH)_3] \quad [Mn(OH)_2]\text{（白色）}$$

在酸性溶液中除Mn^{2+}外，M^{2+}都是相当强的还原剂，都能从水溶液中还原出H_2：

$$2M^{2+} + 2H^+ = 2M^{3+} + H_2\uparrow$$

Cr^{3+}是比较稳定的，其余M^{3+}均不稳定，Ti^{3+}和V^{3+}易被空气中的O_2氧化，而Mn^{3+}却发生歧化反应：

$$4M^{3+} + O_2 + 2H_2O = 4MO^{2+} + 4H^+ \quad (M = Ti，V)$$

$$2Mn^{3+} + 2H_2O = MnO_2\downarrow + Mn^{2+} + 4H^+$$

锰除了Mn(Ⅲ)容易发生歧化反应外，Mn(Ⅵ)也发生歧化反应。另外，除钛外，它们的高氧化态都是较强的氧化剂，而且氧化性依次增强。

【仪器和药品】

坩埚，pH试纸，$Pb(Ac)_2$试纸。

（除特别注明外，试剂浓度单位为 $mol \cdot L^{-1}$）新鲜配制溶液：Cl_2 水，H_2O_2 (质量分数为 3%)，Na_2S (0.5)。

H_2SO_4 (3，浓)，HCl (6，浓)，HNO_3 (6，浓)，$MnSO_4$ (0.1)，$Fe(NO_3)_3$ (0.1)，MnO_2(s)，$Cr_2(SO_4)_3$ (0.1)，$TiOSO_4$ (0.1)，$KMnO_4$ (0.01)，Na_2SO_3 (0.1)，NaOH (0.1，6)，$NH_3 \cdot H_2O$ (6)，KSCN(饱和)，K_2CrO_4 (0.1)，$K_2Cr_2O_7$ (0.1)，$AgNO_3$ (0.1)，$Pb(NO_3)_2$ (0.1)，$FeSO_4$ (0.1)，Fe(粉)，NH_4VO_3(s)，Na_2CO_3 (0.1)，Zn(粒)，$KMnO_4$(s)，$BaCl_2$ (0.1)。

【实验内容】

1．Cr 的化合物

（1）$Cr(OH_3)$的生成与性质　向 $Cr_2(SO_4)_3$ 溶液中滴加 $0.1mol \cdot L^{-1}$ NaOH 溶液，观察沉淀的生成。离心分离，沉淀分别做下面实验：

① 与稀 H_2SO_4 作用；

② 与 $6mol \cdot L^{-1}$ 的 NaOH 溶液作用；

③ 与 H_2O_2 溶液作用。

（2）Cr^{3+}的水解　检验 $Cr_2(SO_4)_3$ 溶液的 pH 值，然后分别向盛有少量 $Cr_2(SO_4)_3$ 溶液的试管中滴加 Na_2S 溶液和 Na_2CO_3 溶液，观察现象，通过试验证实所得沉淀均为 $Cr(OH)_3$。

（3）CrO_4^{2-} 与 $Cr_2O_7^{2-}$ 的相互转化　观察并比较 K_2CrO_4 和 $K_2Cr_2O_7$ 溶液的颜色，然后分别向 K_2CrO_4 溶液中滴加 $6mol \cdot L^{-1}$ 的 HNO_3，向 $K_2Cr_2O_7$ 溶液中滴加 $6mol \cdot L^{-1}$ NaOH 溶液，再比较其颜色变化。

（4）难溶性铬酸盐　分别向 K_2CrO_4 和 $K_2Cr_2O_7$ 溶液中滴加 $BaCl_2$ 溶液，观察并比较沉淀的颜色。

用 $AgNO_3$ 和 $Pb(NO_3)_2$ 代替 $BaCl_2$ 分别做同样的试验。

（5）Cr(Ⅵ)的氧化性　分别向 K_2CrO_4 和 $K_2Cr_2O_7$ 溶液中滴加 $FeSO_4$ 溶液，再用稀 H_2SO_4 酸化，观察现象。

2．Mn 的化合物

（1）Mn(Ⅱ)的化合物及其性质　检验 $MnSO_4$ 溶液的 pH 值，分别取少量 $MnSO_4$ 溶液做下列试验：

① 与适当的试剂反应获得 $Mn(OH)_2$ 沉淀，观察沉淀在空气中的变化；

② 与 $KMnO_4$ 溶液作用，观察现象；

③ 滴加适量的 Na_2S 溶液，观察沉淀的生成，试设法验证沉淀是 MnS。

（2）Mn(Ⅵ)的生成和性质　取少量 MnO_2 固体加入适量 $6mol \cdot L^{-1}$ NaOH 溶液，再滴加 $KMnO_4$ 溶液，微热，观察颜色变化，把溶液分成 3 份，分别试验溶液与氯水、Na_2SO_3 溶液和稀 H_2SO_4 的作用，观察并比较其颜色变化。

（3）Mn(Ⅶ)的氧化性　用 3 支试管，各取少量的 $KMnO_4$ 溶液，其中 1 支用 $3mol \cdot L^{-1}$ H_2SO_4 酸化，1 支用的 $6mol \cdot L^{-1}$ 的 NaOH 碱化，然后分别滴加 Na_2SO_3 溶液，观察比较 $KMnO_4$ 在不同介质中与 Na_2SO_3 的作用情况。

3．Ti 的化合物

（1）$Ti(OH)_4$ 的生成和性质　取适量的 $TiOSO_4$ 溶液，加入 $6mol \cdot L^{-1}$ 氨水，观察沉淀的颜色，将沉淀分成两份，分别试验沉淀在 $3mol \cdot L^{-1}$ H_2SO_4 和 $6mol \cdot L^{-1}$ NaOH 溶液中的溶解情况。

（2）TiO^{2+}的氧化还原性质　取适量 $TiOSO_4$ 溶液，加入少量铁粉，观察紫色的出现。另取一试管，分别滴加 KSCN 溶液和 $Fe(NO_3)_3$ 溶液各 1 滴，把上述紫色溶液滴加到 Fe^{3+}-SCN^- 溶液中，观察颜色变化。

4．V 的化合物

（1）钒(Ⅴ)的氧化性　取饱和 NH_4VO_3 溶液 1~2mL（自配，用 NH_4VO_3 固体和 6mol·L^{-1} HCl 溶液配成）。加入一小颗 Zn 粒，放置，并仔细观察溶液颜色的变化。

（2）V_2O_5 的生成和性质　于坩埚中盛少量 NH_4VO_3 固体，小火加热并不断搅拌，根据固体颜色的变化来判断反应的发生和完成。冷却，将固体分成 3 份，一份与浓 H_2SO_4 作用，另一份与 6mol·L^{-1} NaOH 溶液作用（加热），第三份加入浓 HCl，观察固体的溶解及颜色变化，检查 Cl_2 的生成。

【选做部分】

分别试验 H_2O_2 与 $TiOSO_4$、NH_4VO_3、$KMnO_4$ 和 K_2CrO_4（或 $K_2Cr_2O_7$）溶液的作用情况，注意不同介质对反应产物的影响。

【问题与讨论】

1．怎样证明 $Cr_2(SO_4)_3$ 与 Na_2S 溶液作用所生成的沉淀不是硫化物？

2．怎样配制 NH_4VO_3 饱和溶液？

3．酸化的 $TiOSO_4$ 溶液与锌粒作用得到紫色溶液，此紫色溶液在空气中放置一段时间后紫色又褪去，为什么？

4．根据 E^s 数据说明在酸性溶液中，处于中间氧化值的钛、钒能否发生歧化反应？

5．总结实验中有关反应，指出哪些反应可用于鉴定 Ti(Ⅳ)、V(Ⅴ)和 Cr(Ⅵ)，写出反应方程式，注明反应条件，并指出钛、铬、钒在鉴定反应产物中的氧化值。

6．哪些常用氧化剂能把 Mn^{2+} 氧化成 MnO_4^-？要确保 Mn^{2+} 被氧化成 MnO_4^- 而不出现任何副反应，反应条件应怎样控制？

Experiment 23　Compounds of Ti, V, Cr and Mn

【Objectives】

To evaluate and understand the properties of some important compounds of Ti, V, Cr and Mn.

【Introduction】

1．Electron configurations of Ti, V, Cr and Mn in the periodic table

Table5-3 lists the electron configurations of Ti, V, Cr and Mn in the periodic table.

Table 5-3　Electron configurations of Ti, V, Cr and Mn in the periodic table

Period	ⅣB	ⅤB	ⅥB	ⅦB
4	22 Ti [Ar]$3d^2 4s^2$	23 V [Ar]$3d^3 4s^2$	24 Cr [Ar]$3d^5 4s^1$	25 Mn [Ar]$3d^5 4s^2$

2．Hydroxides and salts

For Ti, V, Cr and Mn, their hydroxides $M(OH)_2$ are all insoluble and medium strong bases, and their hydroxides $M(OH)_3$ are weak bases. $Cr(OH)_3$ is an amphoteric hydroxide. Cr(Ⅲ) can form both cation salts and anion salts. When their oxidation numbers are high, they bond strongly to oxygen. So Ti^{4+}，V^{4+}，V^{5+}，Cr^{6+}，Mn^{6+} and Mn^{7+} can't be present even in strong acid solutions. Therefore, TiO^{2+}, VO^{2+} and VO_2^+ are the forms present for Ti(Ⅳ), V(Ⅳ) and V(Ⅴ). Cr(Ⅵ), Mn(Ⅵ) and Mn(Ⅶ) can only form anion salts. The solubilities of anion salts of Cr(Ⅵ) are similar to that of sulfates in water. For example, the sulfates of Ca^{2+}, Sr^{2+}, Ba^{2+}, Pb^{2+} and Ag^+ are insoluble. Therefore, the formation of precipitates of $BaCrO_4$ (lemon yellow), $PbCrO_4$ (yellow) and Ag_2CrO_4 (brick red) can be used to identify CrO_4^{2-} or $Cr_2O_7^{2-}$.

The oxyacids of Group ⅤB and ⅥB usually condense to meta-acids and polyacids. For example, yellow CrO_4^{2-} can turn into orange $Cr_2O_7^{2-}$ when the pH is less than 5. It can turn into $Cr_3O_{10}^{2-}$ and $Cr_4O_{13}^{2-}$ when the pH is lower.

The solubilities of $Cr_2O_7^{2-}$ salts are greater than that of CrO_4^{2-} salts. So precipitates of CrO_4^{2-} salts will form if Ba^{2+}, Pb^{2+} or Ag^+ is added to the mixture of CrO_4^{2-} and $Cr_2O_7^{2-}$.

3. Coordination compounds

The coordination compounds of Cr^{3+} are important. For example, $CrCl_3 \cdot 6H_2O$ represents three complexes, which are three crystals with different colors. The coordination number of Cr^{3+} is 6.

$[Cr(H_2O)_6]Cl_3$	$[Cr(H_2O)_5Cl]Cl_2 \cdot H_2O$	$[Cr(H_2O)_4Cl_2]Cl \cdot 2H_2O$
(purple blue)	(blue green)	(greyish green)

So, the aqueous solutions of Cr^{3+} show different colors.

Ti and V bond strongly to oxygen, so TiO^{2+}, VO^{2+} and VO_2^+ are stable in aqueous solutions. It is not easy for Ti and V to form stable coordination compounds. But they can form some complex ions with halogen ions which have high concentrations, such as $[TiCl_2]^{2+}$, $[TiCl_3]^+$, $[Ti(OH)Cl_5]^{2-}$, $[TiCl_6]^{2-}$, $[TiF_6]^{2-}$, etc. In acidic solution, Ti(Ⅳ), V(Ⅳ), V(Ⅴ) and Cr(Ⅵ) can react with H_2O_2 to form complex ions with typical colors.

$$TiO^{2+} + H_2O_2 \rightleftharpoons [TiO(H_2O_2)]^{2+}$$
（orange）

$$VO^{2+} + H_2O_2 = [VO(H_2O_2)]^{2+}$$
（red brown）

$$VO_2^+ + 2H_2O_2 = [VO_2(H_2O_2)_2]^+$$
（yellow）

$$Cr_2O_7^{2-} + 4H_2O_2 + 6H^+ = 2[CrO_2(H_2O_2)_2]^{2+} + 3H_2O$$

$$Cr_2O_7^{2-} + 4H_2O_2 + 2H^+ = 2\,CrO(O_2)_2 + 5H_2O$$
（blue）

Add ammonia water to $[TiO(H_2O_2)]^{2+}$ solution, and yellow H_4TiO_5 [or $TiO(H_2O_2)(OH)_2$] will be formed. This reaction is used to identify Ti.

$$[TiO(H_2O_2)]^{2+} + 2NH_3 \cdot H_2O = H_4TiO_5 + 2\,NH_4^+$$

Cr(Ⅵ) reacts with H_2O_2 to form a complex which can be used to identify both H_2O_2 and Cr(Ⅵ). This complex dissolves and is stable in diethyl ether. So diethyl ether is used to improve the sensitivity.

4. Oxidation-reduction properties

The standard electrode potentials ($E_A^{\ominus}$/V) of Ti、V、Cr、Mn are as follows (data or compounds in syuare brackets are the standard potentials $E_B^{\ominus}$/V or compounds in basic solution):

$$\underset{\text{(colorless)}}{TiO^{3+}} \xrightarrow{+0.1} \underset{\text{(purple)}}{TiO^{2+}} \xrightarrow{-0.37} \underset{\text{(brown)}}{Ti^{2+}} \xrightarrow{-1.63} \underset{\text{(silver grey)}}{Ti}$$

$$\underset{\text{(light yellow)}}{VO_2^{+}} \xrightarrow{+1.0} \underset{\text{(blue)}}{VO^{2+}} \xrightarrow{+0.361} \underset{\text{(green)}}{V^{3+}} \xrightarrow{-0.255} \underset{\text{(purple)}}{V^{2+}} \xrightarrow{-1.18} \underset{\text{(light grey)}}{V}$$

$$\underset{[Cr_2O_7^{2-}]\text{(yellow)}}{Cr_2O_7^{2-}\text{(orange)}} \xrightarrow[{[-0.13]}]{+1.33} \underset{[CrO_2^{-}]\text{(green)}}{Cr^{3+}\text{(purple blue)}} \xrightarrow[{[0.8]}]{-0.41} \underset{[Cr(OH)_2]}{Cr^{2+}\text{(sky blue)}} \xrightarrow[{[-1.4]}]{-0.91} Cr$$

$$MnO_4^{-}\text{(purple)} \xrightarrow[{[+0.564]}]{+0.564} \underset{[MnO_4^{2-}]\text{(green)}}{MnO_4^{2-}\text{(green)}} \xrightarrow[{[+0.60]}]{+2.26} \underset{[MnO_2]}{MnO_2\text{(brown black)}} \xrightarrow[{[-0.2]}]{+0.95}$$

$$\underset{[Mn(OH)_3]}{Mn^{3+}\text{(red)}} \xrightarrow[{[+0.1]}]{+1.51} \underset{[Mn(OH)_2]\text{(white)}}{Mn^{2+}\text{(pink)}} \xrightarrow[{[-1.55]}]{-1.18} Mn$$

In acidic solution, all M^{2+} ions are strong reducing agents except Mn^{2+}.

$$2M^{2+} + 2H^{+} \Longrightarrow 2M^{3+} + H_2\uparrow$$

Cr^{3+} is stable, but other M^{3+} ions are not. Ti^{3+} and V^{3+} are easily oxidized by O_2 in the air, but Mn^{3+} disproportionates:

$$4M^{3+} + O_2 + 2H_2O \Longrightarrow 4MO^{2+} + 4H^{+} \quad (M = Ti, V)$$

$$2Mn^{3+} + 2H_2O \Longrightarrow MnO_2\downarrow + Mn^{2+} + 4H^{+}$$

【Apparatus and Chemicals】

Crucible, pH test paper, $Pb(Ac)_2$ test paper.

(The following reagents utilize the concentration unit $mol \cdot L^{-1}$)

Freshly prepared solutions: chlorine water, H_2O_2 (3%, wt/wt), Na_2S (0.5).

H_2SO_4 (3, concentrated), HCl (concentrated), HNO_3 (6, concentrated), $MnSO_4$ (0.1), $Fe(NO_3)_3$ (0.1), MnO_2(s), $Cr_2(SO_4)_3$ (0.1), $TiOSO_4$ (0.1), $KMnO_4$ (0.01), Na_2SO_3 (0.1), NaOH (0.1, 6), $NH_3 \cdot H_2O$(6), KSCN(saturated), K_2CrO_4 (0.1), $K_2Cr_2O_7$ (0.1), $AgNO_3$ (0.1), $Pb(NO_3)_2$ (0.1), $FeSO_4$ (0.1), Fe powder, NH_4VO_3(s), Na_2CO_3 (0.1), Zn, $KMnO_4$(s), $BaCl_2$ (0.1).

【Experimental Procedure】

1. Compounds of Cr

(1) $Cr(OH_3)$　Add 0.1mol·L^{-1} NaOH solution to $Cr_2(SO_4)_3$ solution. Observe the formation of the precipitate. Centrifuge and separate the precipitate. Use the precipitate to do the following experiments:

① React with dilute H_2SO_4.

② React with 6mol·L^{-1} NaOH.

③ React with H_2O_2 solution.

(2) Hydrolysis of Cr^{3+}　Test the pH of $Cr_2(SO_4)_3$ solution. Put $Cr_2(SO_4)_3$ solution in two test

tubes. Then add Na_2S solution to one test tube, and Na_2CO_3 solution to the other. Prove the precipitates are $Cr(OH)_3$.

(3) Transformation between CrO_4^{2-} and $Cr_2O_7^{2-}$ Observe and compare the colors of K_2CrO_4 solution and $K_2Cr_2O_7$ solution. Then add $6mol \cdot L^{-1}$ HNO_3 to K_2CrO_4 solution, adding $6mol \cdot L^{-1}$ NaOH to $K_2Cr_2O_7$ solution. Observe and compare the color changes.

(4) Insoluble chromate Add $BaCl_2$ solution to K_2CrO_4 and $K_2Cr_2O_7$ respectively, and compare the colors of the precipitates. Use $AgNO_3$ and $Pb(NO_3)_2$ to replace $BaCl_2$ solution to replicate the above experiments.

(5) Oxidizability of Cr(Ⅵ) Add $FeSO_4$ solution to K_2CrO_4 and $K_2Cr_2O_7$ respectively, and add dilute H_2SO_4. Observe and record your observations.

2. Compounds of Mn

(1) Compounds of Mn(Ⅱ) Test the pH of $MnSO_4$ solution, and then do the following experiments.

① Use a small amount of $MnSO_4$ solution to react with the appropriate agent to produce $Mn(OH)_2$ precipitate. Observe the changes of the precipitate in the air.

② Use a small amount of $MnSO_4$ solution to react with $KMnO_4$ solution. Observe and record your observations.

③ Add an appropriate amount of Na_2S solution to $MnSO_4$ solution. Observe the precipitate, and test if the precipitate is MnS.

(2) Properties of Mn(Ⅵ) Use a small amount of solid MnO_2 to react with an appropriate amount of $6mol \cdot L^{-1}$ NaOH, then add $KMnO_4$ to the solution. Slightly heat and observe the color changes. Test the reaction of the solution with ammonia water, Na_2SO_3 solution and dilute H_2SO_4 respectively. Compare the color changes.

(3) Oxidizability of Mn(Ⅶ): Place a small amount of $KMnO_4$ solution into three test tubes. Acidify the first solution with $3mol \cdot L^{-1}$ H_2SO_4, and add $6mol \cdot L^{-1}$ NaOH to the second one. Then add Na_2SO_3 solution to each test tube. Observe and compare the reaction of $KMnO_4$ in the different media.

3. Compounds of Ti

(1) $Ti(OH)_4$ Add $6mol \cdot L^{-1}$ ammonia water to an appropriate amount of $TiOSO_4$ solution. Observe the color of the precipitate. Test the dissolution of the precipitate in $3mol \cdot L^{-1}$ H_2SO_4 and $6mol \cdot L^{-1}$ NaOH respectively.

(2) Oxidation and reduction of TiO^{2+} Add Fe powder to an appropriate amount of $TiOSO_4$ solution. Observe the purple color. Put one drop of KSCN solution and one drop of $Fe(NO_3)_3$ solution to a test tube, and add the above purple solution to Fe^{3+}-SCN^- solution. Observe the color change.

4. Compounds of V

(1) Oxidizability of V In a test tube, transfer 1 to 2mL of saturated NH_4VO_3 solution which is prepared by using solid NH_4VO_3 and $6mol \cdot L^{-1}$ HCl solution. Add a small grain of Zn to the prepared solution and observe the color change.

(2) V_2O_5 Put a small amount of solid NH_4VO_3 in a clean crucible. Heat and stir constantly.

Determine the completion of the reaction according to the color change of the solid. Allow the solid to cool. Divide the solid into three portions. React the first portion with concentrated H_2SO_4, the second with $6mol \cdot L^{-1}$ NaOH solution while heating, and the third with concentrated HCl. Observe the dissolution of the solid and the color changes of the solutions. Test if Cl_2 is produced.

【Questions and Discussion】

1. How do we prove the precipitate produced by the reaction of $Cr_2(SO_4)_3$ and Na_2S is not sulfide?

2. How do we prepare saturated NH_4VO_3 solution?

3. Acidified $TiOSO_4$ solution reacts with Zn to produce a purple solution. The purple color fades after it is in the air for a while. Why?

4. According to the data of E^s, can the disproportionation reactions of Ti and V occur in an acidic solution?

5. What reactions can be used to identify Ti(Ⅳ), V(Ⅴ) and Cr(Ⅵ)? Record the chemical equations. Point out the oxidation number of Ti, V and Cr in the products.

6. Record the commonly used oxidants which can oxidize Mn^{2+} to MnO_4^-. How do we control the reaction to make sure that Mn^{2+} is oxidized to MnO_4^-?

实验24 铁、钴、镍的化合物

【实验目的】

试验并了解铁、钴、镍的某些重要化合物性质。

【实验提要】

铁组元素属于Ⅷ族，包括铁、钴和镍3种元素。它们是同一周期的元素，原子结构相似（$[Ar]3d^{6\sim8}4s^2$），原子半径相近（115~117 pm）。

铁组元素常见的盐类是Fe(Ⅲ)、Fe(Ⅱ)、Co(Ⅱ)和Ni(Ⅱ)盐，其中硝酸盐、硫酸盐和卤化物均易溶于水。这些阳离子水合时，通常发生颜色变化。例如 Fe^{2+}由白色变成浅绿色，Co^{2+}由蓝色变成粉红色，Ni^{2+}由黄色变成亮绿色。它们的盐从水中结晶析出时，常形成带结晶水的晶体。随着结晶水量的变化，其颜色也发生变化。例如：

$$\underset{\text{(粉红色)}}{CoCl_2\cdot 6H_2O} \underset{}{\overset{52℃}{\rightleftharpoons}} \underset{\text{(紫红色)}}{CoCl_2\cdot 2H_2O} \overset{90℃}{\rightleftharpoons} \underset{\text{(蓝紫色)}}{CoCl_2\cdot H_2O} \overset{120℃}{\rightleftharpoons} \underset{\text{(蓝色)}}{CoCl_2}$$

$CoCl_2$的这一性质常用于实验室显示干燥剂（硅胶）吸水的情况。

Fe^{3+}盐与Cr^{3+}盐、Al^{3+}盐很类似。例如，MCl_3均可双聚形成共边双四面体分子M_2Cl_6，且易升华；其硫酸盐与碱金属元素的硫酸盐类均能形成$M_2SO_4\cdot M_2(SO_4)_3\cdot 24H_2O$的“矾”，如与通常净水用的明矾$K_2SO_4\cdot Al_2(SO_4)_3\cdot 24H_2O$类似的有钾铁矾$K_2SO_4\cdot Fe_2(SO_4)_3\cdot 24H_2O$等。

铁组元素常见的重要难溶盐见表5-4。

表5-4 铁组元素常见的难溶盐

阳离子	阴离子			
	S^{2-}	CO_3^{2-}	CrO_4^{2-}	$[Fe(CN)_6]^{4-}$
Fe^{3+}	黑色	—	—	蓝色
Fe^{2+}	黑色	白色	白色	白色
Co^{2+}	黑色	粉红色	浅粉红色	绿色
Ni^{2+}	黑色	浅绿色	浅绿色	浅绿色

铁组元素的阳离子是形成配合物的较好形成体，能形成很多的配合物。唯 Fe^{3+}、Fe^{2+}与OH^-结合的能力较强，它们难以在氨水中形成稳定的氨合离子。常见的稳定配合物见表5-5。由于配位后发生溶解度和颜色等的改变，常用于离子的分离和鉴定。

表5-5 铁组元素常见的配合物

中心离子	配体				
	H_2O	CN^-	NH_3	SCN^-	F^-
Fe^{3+}	浅紫色	浅黄色	—	血红色	无色
Fe^{2+}	浅绿色	黄色	—	无色	—
Co^{3+}	—	—	红色	—	—
Co^{2+}	粉红色	紫色	橙色	蓝色	—
Ni^{2+}	绿色	黄色	深蓝色	—	—

铁组元素氢氧化物的酸碱性完全符合一般规律。

M(Ⅵ)的氢氧化物为强酸，故无论在酸性还是碱性溶液中均以酸根MO_4^{2-}的形式存在。其中较为重要的是高铁酸盐。FeO_4^{2-}与MnO_4^-一样呈紫红色，与MnO_4^{2-}、CrO_4^{2-}和SO_4^{2-}等类似，

能使 Ba^{2+} 沉淀。

铁组元素的 $M(OH)_3$ 和 $M(OH)_2$ 均为碱，但碱性不强。新沉淀出来的 $Fe(OH)_3$ 还有微弱的酸性，可溶于热的浓 KOH 溶液中，形成铁酸钾 $KFeO_2$。因此铁组元素的可溶性盐均发生水解，且 M^{3+} 盐比 M^{2+} 盐的水解度要大些，尤以 Fe^{3+} 盐突出。Fe^{3+} 盐的水解产物常使其溶液变成棕黄色或棕红色：

$$\mathrm{Fe^{3+}} + n\mathrm{H_2O} \rightleftharpoons [\mathrm{Fe(OH)}_n]^{3-n} + n\mathrm{H^+} \qquad (n = 1，2，3)$$

Fe^{3+} 仅存在于强酸性溶液（pH＜2）中。稀释或提高溶液的 pH 值，会析出胶状的红棕色 $Fe(OH)_3$ 沉淀。

铁组元素的电位图 $E_A^{\ominus}$ /V（方括号内注明的是碱性溶液中的状态及 $E_B^{\ominus}$/V）为：

$$\mathrm{FeO_4^{2-}} \xrightarrow[{[+0.9]}]{+1.9} \underset{[\mathrm{Fe(OH)_3}]}{\mathrm{Fe^{3+}}} \xrightarrow[{[-0.56]}]{+0.771} \underset{[\mathrm{Fe(OH)_2}]}{\mathrm{Fe^{2+}}} \xrightarrow[{[-0.887]}]{-0.440} \mathrm{Fe}$$

$$\mathrm{CoO_2} \xrightarrow[{[+0.7]}]{\geqslant +1.8} \underset{[\mathrm{Co(OH)_3}]}{\mathrm{Co^{3+}}} \xrightarrow[{[+0.17]}]{+1.82} \underset{[\mathrm{Co(OH)_2}]}{\mathrm{Co^{2+}}} \xrightarrow[{[-0.73]}]{-0.277} \mathrm{Co}$$

$$\mathrm{NiO_4^{2-}} \xrightarrow[{[>+0.4]}]{+1.8} \mathrm{NiO_2} \xrightarrow[{[+0.49]}]{+1.68} \underset{[\mathrm{Ni(OH)_2}]}{\mathrm{Ni^{2+}}} \xrightarrow[{[-0.72]}]{-0.250} \mathrm{Ni}$$

氧化值大于或等于+4 的铁组元素化合物是很强的氧化剂。在酸性条件下它们的氧化性比 MnO_4^- 还强；在碱性介质中，它们的氧化性大为降低；在强碱中用强氧化剂（NaClO 等）就可以将低氧化值的铁组元素氧化。例如：

$$\mathrm{2Fe(OH)_3 + 3ClO^- + 4OH^- = 2\,FeO_4^{2-} + 3Cl^- + 5H_2O}$$

$$\mathrm{Fe_2O_3 + 3KNO_3 + 4KOH \xlongequal{熔融} 2K_2FeO_4 + 3KNO_2 + 2H_2O}$$

M(Ⅲ)在酸性介质中也是氧化剂，其中 Fe^{3+} 为中强氧化剂；Co^{3+} 为强氧化剂，能氧化水和盐酸；Ni^{3+} 的氧化性更强。在酸性介质中，Fe^{3+} 能被 Fe、Cu、Sn^{2+}、S^{2-}、I^- 或 SO_2 等还原。

若有配体或沉淀剂存在，M(Ⅲ)被配位或形成难溶盐而稳定存在。例如(单位：V)：

$$\mathrm{Fe^{3+}} \xrightarrow{+0.771} \mathrm{Fe^{2+}} \qquad \mathrm{Co^{3+}} \xrightarrow{+1.842} \mathrm{Co^{2+}}$$

$$[\mathrm{FeF_6}]^{3-} \xrightarrow{-0.135} \mathrm{Fe^{2+}} \qquad [\mathrm{Co(NH_3)_6}]^{3+} \xrightarrow{+0.1} [\mathrm{Co(NH_3)_6}]^{2+}$$

$$[\mathrm{Fe(CN)_6}]^{3-} \xrightarrow{+0.36} [\mathrm{Fe(CN)_6}]^{4-} \qquad [\mathrm{Co(CN)_6}]^{3-} \xrightarrow{+0.81} [\mathrm{Co(CN)_6}]^{4-}$$

综上所述，铁组元素 M(Ⅲ)、M(Ⅱ)和 M(0)氧化还原性及相应氢氧化物的酸碱性变化规律为：

在酸性溶液中还原性增强（←）

Fe（银白色）	Co（银白色）	Ni（银白色）
Fe^{2+}（浅绿色）	Co^{2+}（粉红色）	Ni^{2+}（亮绿色）
Fe^{3+}（浅紫色）	Co^{3+}（绿色）	Ni^{3+}（粉红色）

还原性增强（↑，左侧）；氧化性增强（↓，右侧）

氧化性增强（→）

在碱性溶液中还原性增强

还原性增强，碱性增强 | | | 氧化性增强，碱性减弱

Fe	Co	Ni
(银白色)	(银白色)	(银白色)
$Fe(OH)_2$	$Co(OH)_2$	$Ni(OH)_2$
(白色)	(粉红色)	(浅绿色)
$Fe(OH)_3$	$Co(OH)_3$	$Ni(OH)_3$
(红棕色)	(暗棕色)	(灰黑色)

氧化性增强

【铁组离子鉴定】

（1）Fe^{3+}　在弱酸性介质中 Fe^{3+} 与 KSCN 作用形成血红色配离子 $[Fe(SCN)_n]^{3-n}$。

$$Fe^{3+} + nKSCN = [Fe(SCN)_n]^{3-n} + nK^+$$

此法无离子干扰。若其它重金属离子浓度不太高，也可借助 Fe^{3+} 与黄血盐 $K_4[Fe(CN)_6]$ 作用生成深蓝色的普鲁士蓝 $[KFe(CN)_6Fe]_x$ 沉淀来鉴定：

$$xFe^{3+} + xK^+ + x[Fe(CN)_6]^{4-} = [KFe(CN)_6Fe]_x\downarrow$$

（2）Fe^{2+}　Fe^{2+} 与赤血盐 $K_3[Fe(CN)_6]$ 作用生成深蓝色的滕氏蓝 $[KFe(CN)_6Fe]_x$ 沉淀。

$$xFe^{2+} + xK^+ + x[Fe(CN)_6]^{3-} = [KFe(CN)_6Fe]_x\downarrow$$

当重金属离子含量不高时，此法的灵敏度和选择性均较高。

（3）Co^{2+}　Co^{2+} 与 KSCN 作用生成蓝色配离子 $[Co(SCN)_4]^{2-}$。

$$Co^{2+} + 4SCN^- = [Co(SCN)_4]^{2-}$$

该配离子在丙酮中较稳定，故鉴定时需加入丙酮或戊醇。若有 Fe^{3+}，可加入 NaF 以形成无色的 $[FeF_6]^{3-}$ 掩蔽之。

（4）Ni^{2+}　Ni^{2+} 在弱碱性（$NH_3\cdot H_2O$）介质中与丁二酮肟生成鲜红色螯合物沉淀，是鉴定 Ni^{2+} 的灵敏反应。

$$Ni^{2+} + 2\begin{matrix} CH_3-C=NOH \\ | \\ CH_3-C=NOH \end{matrix} + 2NH_3 = \begin{matrix} & O\cdots H-O & \\ & | \quad\quad | & \\ CH_3-C=N & & N=C-CH_3 \\ | & Ni & | \\ CH_3-C=N & & N=C-CH_3 \\ & | \quad\quad | & \\ & O-H\cdots O & \end{matrix} + 2NH_4^+$$

【仪器和药品】

（除特别注明外，试剂浓度单位为 $mol\cdot L^{-1}$）HCl (浓)，H_2S (饱和，新配)，NaOH (s，6，新配)，H_2SO_4(稀)，氨水 (6)，NaClO (新配)，$FeSO_4$ (0.1)，NH_4SCN (饱和)，NH_4F (2)，$Fe(NO_3)_3$ (0.1)，$CoCl_2$ (0.1)，Na_2S (0.1)，KI (0.1)，$NiSO_4$ (0.1)，$BaCl_2$ (0.5)，$K_3[Fe(CN)_6]$ (0.1)，$K_4[Fe(CN)_6]$ (0.1)，氯水(新配)，溴水，Fe^{3+}、Co^{2+}、Ni^{2+} 混合溶液 4 组[1]，H_2O_2 (质量分数为 3%)，CCl_4，戊醇，铁粉，KI-淀粉试纸，丁二酮肟试剂。

[1] 混合液由实验室提供，共 4 组。每组溶液含有待测阳离子 2～3 种，可由 $0.1mol\cdot L^{-1}$ 的单一盐溶液混合而成。学生应记录混合液编号和检验结果。

【实验内容】

用表格报告实验结果，解释现象时，若涉及化学反应，均要求写出反应方程式。

1．氢氧化物的生成和性质

（1）$M(OH)_2$的生成和性质　观察$FeSO_4$、$CoCl_2$和$NiSO_4$溶液的颜色，检验其 pH 值，由这 3 种盐溶液和其它合适试剂作用获得$Fe(OH)_2$、$Co(OH)_2$和$Ni(OH)_2$，观察沉淀的颜色及它们的酸碱性。将沉淀放置一段时间再观察比较其颜色变化。

提示：$Fe(OH)_2$极易被溶液中的氧气氧化，必须很小心地操作才能观察到白色$Fe(OH)_2$的生成。制备时，可以在一支试管中加入 5mL $6mol \cdot L^{-1}$新配 NaOH 溶液，小心煮沸，冷却备用；用另一支试管取适量$FeSO_4$溶液，用稀H_2SO_4酸化，加入少量铁粉，煮沸，用一细长滴管吸取冷却的 NaOH 溶液，小心插入$FeSO_4$溶液中，缓慢放出 NaOH。

（2）$M(OH)_3$的生成和性质　分别由$FeSO_4$、$CoCl_2$和$NiSO_4$溶液和其它合适的试剂作用，获得$Fe(OH)_3$、$Co(OH)_3$和$Ni(OH)_3$。观察比较它们的状态和颜色，将沉淀离心分离。分别在沉淀中加入浓 HCl，摇荡，微热，并用湿润的 KI-淀粉试纸检验是否有氧化性气体逸出。

提示：制备$M(OH)_3$时要注意依据 M(Ⅱ)还原性的强弱选择不同氧化能力的氧化剂，包括空气中的氧、溴水、H_2O_2、氯水、NaClO 溶液等。

2．硫化物的生成和性质

（1）Fe(Ⅱ)的硫化物　由$FeSO_4$溶液、饱和H_2S溶液及其它合适试剂制备 FeS，观察 FeS 的状态和颜色，试验 FeS 与稀 HCl 的作用情况。

（2）Co(Ⅱ)、Ni(Ⅱ)的硫化物　分别由$CoCl_2$、$NiSO_4$溶液代替$FeSO_4$做同样的试验。

（3）Fe(Ⅲ)在不同介质中与S^{2-}的作用

① 取适量饱和H_2S溶液，逐滴加入$Fe(NO_3)_3$溶液，摇荡，观察沉淀的状态和颜色。

② 用Na_2S溶液代替饱和H_2S溶液做上述试验。

3．配合物的生成与性质

（1）氨的配合物　分别向$Fe(NO_3)_3$、$FeSO_4$、$CoCl_2$和$NiSO_4$溶液中滴加$6mol \cdot L^{-1}$的氨水，观察沉淀的生成，继续滴加氨水观察沉淀是否溶解。比较颜色变化。

（2）与SCN^-形成的配合物　用NH_4SCN溶液代替氨水做上述试验，观察比较颜色变化。

（3）配合物的生成对氧化还原性质的影响　取少量$Fe(NO_3)_3$溶液，滴加 2~3 滴CCl_4，逐滴加入 KI 溶液，摇荡，观察CCl_4层的颜色变化，向此混合溶液中滴加NH_4F溶液，摇荡，观察CCl_4层的颜色变化。

（4）配合物的生成在离子鉴定中的应用

① 与$K_3[Fe(CN)_6]$的作用：在点滴板上，各取少量的$Fe(NO_3)_3$、$FeSO_4$、$CoCl_2$和$NiSO_4$溶液，向各溶液中逐滴加入$K_3[Fe(CN)_6]$试剂。观察并比较状态和颜色变化（本实验可直接在滤纸上做）。

② 与$K_4[Fe(CN)_6]$的作用：用$K_4[Fe(CN)_6]$代替$K_3[Fe(CN)_6]$做上述同样的试验，观察现象，总结Fe^{3+}、Fe^{2+}、Co^{2+}、Ni^{2+} 4 种离子分别与$[Fe(CN)_6]^{3-}$和$[Fe(CN)_6]^{4-}$作用的情况。

③ 从Fe^{3+}、Co^{2+}、Ni^{2+}混合液中检出单一离子：取 1mL 混合液放入试管中，加入NH_4SCN溶液 0.5mL，观察溶液颜色变化，缓慢加入NH_4F溶液，然后加入 1mL 戊醇，摇荡后静置，观察戊醇层颜色变化；用吸管吸取试管下层（水相）溶液 2～3 滴，放在点滴板上，再滴加 1～

2 滴氨水，最后加入 1~2 滴丁二酮肟溶液，观察沉淀的颜色变化。根据试验结果判断 Fe^{3+}、Co^{2+}、Ni^{2+}的存在。

【选做部分】

（1）设计实验 试验$[Co(NH_3)_6]^{2+}$、$[Ni(NH_3)_6]^{2+}$的还原性，比较它们还原性的相对强弱，并解释实验结果。

（2）制备 $BaFeO_4$ 于 100mL 的烧杯中注入 5mL NaClO 溶液，加少量 NaOH 固体（约 0.2g），加热至沸，然后在搅拌下滴加 0.5mL $Fe(NO_3)_3$ 溶液，加完后煮沸片刻，观察溶液颜色。取少量上述溶液离心分离，取出上清液，加入少量 $BaCl_2$ 溶液中，观察沉淀的颜色。

【问题与讨论】

1．制备 $Fe(OH)_2$ 时，有关溶液均需煮沸并避免振荡，为什么？

2．本实验中检验 Co^{2+}时为什么需加入 NH_4F 溶液和戊醇？在什么情况下可不加 NH_4F？指出戊醇层中蓝色物质的化学组成。该物质是在加入戊醇之前形成的还是在加入戊醇之后形成的？

3．怎样从 $Fe(OH)_3$ 制备 $FeCl_2$？写出反应方程式。

4．比较 $Fe(OH)_3$、$Al(OH)_3$、$Cr(OH)_3$ 的性质。怎样利用这些性质把 Fe^{3+}、Al^{3+}、Cr^{3+}从溶液中分离出来？

Experiment 24 Compounds of Fe, Co and Ni

【Objectives】

To evaluate and understand the properties of some important compounds of Fe, Co and Ni

【Introduction】

Fe, Co and Ni have similar configurations ($[Ar]3d^{6\sim8}4s^2$) and similar atomic radii (115~117 pm).

The salts of Fe(Ⅲ), Fe(Ⅱ), Co(Ⅱ) and Ni(Ⅱ) are common. Their nitrates, sulfates and halides are all soluble in water. The hydration of these cations results in color changes. For example, Fe^{2+} changes from white to light green; Co^{2+} changes from blue to pink; Ni^{2+} changes from yellow to bright green. These salts usually exist as a series of compounds that differ in their degree of hydration, and have different colors. For example:

$$\underset{\text{(pink)}}{CoCl_2\cdot 6H_2O} \xrightleftharpoons{52℃} \underset{\text{(red purple)}}{CoCl_2\cdot 2H_2O} \xrightleftharpoons{90℃} \underset{\text{(purple blue)}}{CoCl_2\cdot H_2O} \xrightleftharpoons{120℃} \underset{\text{(blue)}}{CoCl_2}$$

This property of $CoCl_2$ is usually used in the laboratory to show the effects of desiccants (silica gel).

The salts of Fe^{3+} are similar to that of Cr^{3+} and Al^{3+}. For example, MCl_3 can form dimer M_2Cl_6, with tetracoordinate M. Their sulfates and the sulfates of alkali metals can form alums with the general formula $M_2SO_4\cdot M_2(SO_4)_3\cdot 24H_2O$. For example, $K_2SO_4\cdot Al_2(SO_4)_3\cdot 24H_2O$ and $K_2SO_4\cdot Fe_2(SO_4)_3\cdot 24H_2O$.

Table 5-4 shows some important insoluble salts of Fe, Co and Ni.

Table 5-4　Common insoluble salts of Fe, Co and Ni

Cations	Anions			
	S^{2-}	CO_3^{2-}	CrO_4^{2-}	$[Fe(CN)_6]^{4-}$
Fe^{3+}	black	—	—	blue
Fe^{2+}	black	white	white	white
Co^{2+}	black	pink	light pink	green
Ni^{2+}	black	light green	light green	light green

The cations of Fe, Co and Ni can form a lot of coordination compounds. Fe^{3+} and Fe^{2+} react strongly with OH^-. So they can't form stable complex ions in ammonia water. Table 5-5 shows some common and stable coordination compounds of Fe, Co and Ni. These coordination compounds have different solubilities and colors which can be used to identify Fe, Co and Ni.

Table 5-5　Common coordination compounds of Fe, Co and Ni

Metal ions	Ligands				
	H_2O	CN^-	NH_3	SCN^-	F^-
Fe^{3+}	light purple	light yellow	—	blood red	colorless
Fe^{2+}	light green	yellow	—	colorless	—
Co^{3+}	—	—	red	—	—
Co^{2+}	pink	purple	orange	blue	—
Ni^{2+}	green	yellow	dark blue	—	—

The acidic and basic strength of the hydroxides of Fe, Co and Ni abides by the general rules as follows:

Because the hydroxides of M(Ⅵ) are strong acids, their general formula is MO_4^{2-} no matter in acidic or basic solutions. The color of FeO_4^{2-} is purple, the same as MnO_4^-. Similar to MnO_4^{2-}, CrO_4^{2-} and SO_4^{2-}, FeO_4^{2-} can precipitate Ba^{2+}.

The hydroxide formula for Fe, Co and Ni is $M(OH)_3$ or $M(OH)_2$, and they are not strong bases. Freshly prepared $Fe(OH)_3$ is weakly acidic, and can be dissolved in hot concentrated KOH solution to form $KFeO_2$. The soluble salts of Fe, Co and Ni all hydrolyze. The hydrolysis of M^{3+} is stronger than that of M^{2+}, especially for Fe^{3+}. The hydrolysis products of Fe^{3+} will make the solution brown or red brown.

$$Fe^{3+} + nH_2O \rightleftharpoons [Fe(OH)_n]^{3-n} + nH^+ \qquad (n = 1, 2, 3)$$

Fe^{3+} can only exist in strong acid solutions (pH<2). Diluting the solution or increasing the pH value of the solution can produce red brown $Fe(OH)_3$ precipitate.

The standard potentials $E_B^\ominus$/V for compounds of Fe, Co and Ni are as follows (data or compounds in square brackets are the standard potentials $E_B^\ominus$/V or compounds in the basic solution):

$$FeO_4^{2-} \xrightarrow[{[+0.9]}]{+1.9} \underset{[Fe(OH)_3]}{Fe^{3+}} \xrightarrow[{[-0.56]}]{+0.771} \underset{[Fe(OH)_2]}{Fe^{2+}} \xrightarrow[{[-0.887]}]{-0.440} Fe$$

$$CoO_2 \xrightarrow[{[+0.7]}]{>+1.8} \underset{[Co(OH)_3]}{Co^{3+}} \xrightarrow[{[+0.17]}]{+1.82} \underset{[Co(OH)_2]}{Co^{2+}} \xrightarrow[{[-0.73]}]{-0.277} Co$$

$$NiO_4^{2-} \xrightarrow[{[>+0.4]}]{+1.8} NiO_2 \xrightarrow[{[+0.49]}]{+1.68} \underset{[Ni(OH)_2]}{Ni^{2+}} \xrightarrow[{[-0.72]}]{-0.250} Ni$$

For the compounds of Fe, Co and Ni, if the oxidation numbers of Fe, Co and Ni are equal to or greater than +4, they are very strong oxidants. In acidic solutions, their oxidizing strength is greater than MnO_4^-. But in basic solutions, their oxidizing strength will be greatly reduced. The compounds of Fe, Co and Ni, which have small oxidation numbers, can be oxidized by strong oxidants, such as NaClO, in strong bases. For example:

$$2Fe(OH)_3+3ClO^-+4OH^- = 2FeO_4^{2-}+3Cl^-+5H_2O$$

$$Fe_2O_3+3KNO_3+4KOH \xrightarrow{\text{melting}} 2K_2FeO_4+3KNO_2+2H_2O$$

M(Ⅲ) is an oxidant in acidic solutions. Fe^{3+} is a medium strong oxidant. Co^{3+} is a strong oxidant, and can oxidize H_2O and HCl. Ni^{3+} is the strongest oxidant of the three. In acidic solutions, Fe^{3+} can be reduced by Fe, Cu, Sn^{2+}, S^{2-}, I^- and SO_2.

M(Ⅲ) can form stable coordination compounds and insoluble salts. For example:

$$Fe^{3+} \xrightarrow{+0.771} Fe^{2+} \qquad Co^{3+} \xrightarrow{+1.842} Co^{2+}$$

$$[FeF_6]^{3-} \xrightarrow{-0.135} Fe^{2+} \qquad [Co(NH_3)_6]^{3+} \xrightarrow{+0.1} [Co(NH_3)_6]^{2+}$$

$$[Fe(CN)_6]^{3-} \xrightarrow{+0.36} [Fe(CN)_6]^{4-} \qquad [Co(CN)_6]^{3-} \xrightarrow{+0.81} [Co(CN)_6]^{4-}$$

For Fe, Co and Ni, the redox properties of M(Ⅲ), M(Ⅱ) and M(0), and the acidic or basic strength of their hydroxides are generalized in the following diagrams.

Increased reducibility in acidic solution

Fe	Co	Ni
(silver)	(silver)	(silver)
Fe^{2+} (light green)	Co^{2+} (pink)	Ni^{2+} (bright green)
Fe^{3+} (light purple)	Co^{3+} (green)	Ni^{3+} (pink)

Increased reducibility (left, upward); Increased oxidizability (right, downward)

Increased oxidizability

Increased reducibility in basic solution

Fe	Co	Ni
(silver)	(silver)	(silver)
$Fe(OH)_2$ (white)	$Co(OH)_2$ (pink)	$Ni(OH)_2$ (light green)
$Fe(OH)_3$ (red brown)	$Co(OH)_3$ (dark brown)	$Ni(OH)_3$ (black grey)

Increased reducibility, Increased alkalinity (left, upward); Increased oxidizability, Reduced alkalinity (right, downward)

Increased oxidizability

【Identification of Fe, Co and Ni】

(1) Fe^{3+} In weak acid solution, Fe^{3+} reacts with KSCN to form blood red complex ion $[Fe(SCN)_n]^{3-n}$.

$$Fe^{3+} + nKSCN \xlongequal{} [Fe(SCN)_n]^{3-n} + nK^+$$

Other ions won't interfere with this reaction. If the concentrations of heavy metal ions are not high, Fe^{3+} reacts with $K_4[Fe(CN)_6]$ to from a dark blue complex called prussian blue $[KFe(CN)_6Fe]_x$:

$$xFe^{3+} + xK^+ + x[Fe(CN)_6]^{4-} \xlongequal{} [KFe(CN)_6Fe]_x\downarrow$$

(2) Fe^{2+} Fe^{2+} reacts with $K_3[Fe(CN)_6]$ to form dark blue $[KFe(CN)_6Fe]_x$.

$$xFe^{2+} + xK^+ + x[Fe(CN)_6]^{3-} \xlongequal{} [KFe(CN)_6Fe]_x\downarrow$$

If the concentrations of heavy metal ions are not high, this reaction is both high sensitivity and selectivity.

(3) Co^{2+} Co^{2+} reacts with KSCN to form blue $[Co(SCN)_4]^{2-}$.

$$Co^{2+} + 4SCN^- \xlongequal{} [Co(SCN)_4]^{2-}$$

This complex ion is stable in acetone, so acetone or pentanol should be added when identifying Co^{2+}. If Fe^{3+} is present, it can be masked by reacting with NaF to form colorless $[FeF_6]^{3-}$.

(4) Ni^{2+} With the presence of $NH_3\cdot H_2O$, Ni^{2+} reacts with dimethylglyoxime to form a bright red chelate. This is a sensitive reaction to identify Ni^{2+}.

$$Ni^{2+} + 2\,\begin{matrix} CH_3-C=NOH \\ | \\ CH_3-C=NOH \end{matrix} + 2NH_3 \xlongequal{} Ni(C_4H_7N_2O_2)_2 + 2NH_4^+$$

【Apparatus and Chemicals】

(The following reagents utilize the concentration unit $mol\cdot L^{-1}$) HCl (concentrated), H_2S (saturated, freshly prepared), NaOH (s, 6, freshly prepared), H_2SO_4(dilute)ammonia water (6), NaClO (freshly prepared), $FeSO_4$ (0.1), NH_4SCN (saturated), NH_4F (2), $Fe(NO_3)_3$ (0.1), $CoCl_2$ (0.1), Na_2S (0.1), KI (0.1), $NiSO_4$ (0.1), $BaCl_2$ (0.5), $K_3[Fe(CN)_6]$ (0.1), $K_4[Fe(CN)_6]$ (0.1), chlorine water(freshly prepared), bromine water, four sets of Fe^{3+}、Co^{2+}、Ni^{2+} mixed solutions, H_2O_2 (3%, wt/wt), CCl_4, pentanol, iron powder, KI-starch test paper, dimethylglyoxime reagent.

【Experimental Procedure】

Fill in the appropriate table and write the chemical equations in your report.

1. Formation and properties of hydroxide

(1) Formation and properties of $M(OH)_2$ Observe the colors of $FeSO_4$ solution, $CoCl_2$ solution and $NiSO_4$ solution, and test their pH values. Use the above solutions and other appropriate reagents to prepare $Fe(OH)_2$, $Co(OH)_2$ and $Ni(OH)_2$. Observe the colors of the precipitates and test their pH values. After a while, observe the precipitates again and compare.

Tips: $Fe(OH)_2$ is easily oxidized by oxygen. So the formation of white $Fe(OH)_2$ must be performed very carefully. Place 5mL of freshly prepared 6mol·L^{-1}NaOH into a test tube, then heat and boil carefully. Let the solution cool and it will be ready to use. Put an appropriate amount of $FeSO_4$ solution in another test tube, and add dilute H_2SO_4 and a small amount of iron powder. Then, boil the solution. Use a long thin pipet to draw NaOH, and put the tip of the pipet into $FeSO_4$ solution carefully. Then release NaOH slowly.

(2) Formation and properties of $M(OH)_3$ Use $FeSO_4$ solution, $CoCl_2$ solution and $NiSO_4$ solution, and other appropriate reagents to prepare $Fe(OH)_3$, $Co(OH)_3$ and $Ni(OH)_3$. Observe and compare their states and colors. Centrifuge to get the precipitates. Add concentrated HCl to each precipitate, and heat gently. Test if gases are produced with moistened KI-starch test paper.

Tips: According to the reductivity of M(Ⅱ), choose appropriate oxidants to prepare $M(OH)_3$, such as oxygen in the air, bromine water, H_2O_2, chlorine water, NaClO solution, etc.

2．Formation and properties of sulfides

(1) Sulfides of Fe(Ⅱ) Use $FeSO_4$ solution, saturated H_2S solution and other appropriate reagents to prepare FeS. Observe the state and the color of FeS. Evaluate the reaction of FeS with dilute HCl.

(2) Sulfides of Co(Ⅱ) and Ni(Ⅱ) Replace $FeSO_4$ solution with $CoCl_2$ solution and $NiSO_4$ solution to replicate the above experiments.

(3) Reactions of Fe(Ⅲ) and S^{2-} in different media

① Add $Fe(NO_3)_3$ solution drop by drop to an appropriate amount of saturated H_2S solution. Shake well and observe the precipitate.

② Use Na_2S solution to replace saturated H_2S solution to replicate the above experiment.

3．Formation and properties of coordination compounds

(1) Coordination compounds containing ammonia Add 6mol·L^{-1} ammonia water to $Fe(NO_3)_3$, $FeSO_4$, $CoCl_2$ and $NiSO_4$ respectively. Observe the formation of precipitates. Continue adding ammonia water, and observe whether the precipitates dissolve. Compare the color changes.

(2) Coordination compounds formed with SCN^- Replace ammonia water with NH_4SCN solution to replicate the above experiments, and compare the color changes.

(3) Effects of the formation of coordination compounds on redox properties Add 2 to 3 drops of CCl_4 to a small amount of $Fe(NO_3)_3$ solution, then add KI solution drop by drop. Shake well and observe the color of the CCl_4 layer. Continue adding NH_4F solution. Shake well and observe the color of the CCl_4 layer again.

(4) Identification of ions by coordination reactions

① Reaction with $K_3[Fe(CN)_6]$: Add $Fe(NO_3)_3$, $FeSO_4$, $CoCl_2$ and $NiSO_4$ solutions on a spot plate respectively. Then add $K_3[Fe(CN)_6]$ drop by drop to the four solutions. Observe and compare the states and colors. (The experiments can be performed on filter paper.)

② Reaction with $K_4[Fe(CN)_6]$: Use $K_4[Fe(CN)_6]$ to replace $K_3[Fe(CN)_6]$ to replicate the above experiments. Summarize the reactions of four ions, Fe^{3+}, Fe^{2+}, Co^{2+} and Ni^{2+}, with $[Fe(CN)_6]^{3-}$ and $[Fe(CN)_6]^{4-}$ respectively.

③ Identification of Fe^{3+}, Co^{2+} and Ni^{2+} in a mixed solution: Transfer 1mL of mixed solution in a test tube, and add 0.5mL of NH_4SCN solution. Observe the color of the solution. Then slowly add NH_4F solution and 1mL of pentanol. Shake well and observe the color of the pentanol layer. Use a pipet to draw 2 to 3 drops of the solution from the lower layer (aqueous phase) onto a spot plate. Next add 1 to 2 drops of ammonia water, and 1 to 2 drops of dimethylglyoxime solution. Observe the color change of the precipitate. Identify the presence of Fe^{3+}, Co^{2+} and Ni^{2+} according to the results of the experiments.

【Questions and Discussion】

1．When preparing $Fe(OH)_2$, the solutions should be boiled and shaking should be avoided. Why?

2．To identify Co^{2+} in this experiment, why should NH_4F solution and pentanol be added? When is NH_4F not needed? What is the blue compound in the pentanol layer? Does the formation of the blue compound occur before or after the pentanol is added?

3．How do we prepare $Fe(OH)_3$ from $FeCl_2$? Record the chemical equations.

4．Compare the properties of $Fe(OH)_3$, $Al(OH)_3$ and $Cr(OH)_3$. How do we separate Fe^{3+}, Al^{3+} and Cr^{3+} by their properties?

实验25 配 合 物

【实验目的】

1．比较并解释配离子的稳定性。

2．理解配位平衡。

3．了解配合物的一些应用。

【实验提要】

配合物的组成一般可分为外界和内界两个部分，中心离子和配体组成配合物的内界，其余离子处于外界。如$[Cu(NH_3)_4]SO_4$中的Cu^{2+}和4个NH_3组成内界，SO_4^{2-}处于外界。在水溶液中主要以$[Cu(NH_3)_4]^{2+}$和SO_4^{2-}两种离子存在。

配离子在水溶液中会发生解离，也就是说配离子在溶液中同时存在着配位过程和解离过程，即存在配位平衡。如：

$$Cu^{2+} + 4\,NH_3 \rightleftharpoons [Cu(NH_3)_4]^{2+}$$

配位反应广泛应用于离子鉴定、分离，在冶金工业中广泛应用于提炼、分离金属等。

【仪器和药品】

（除特别注明外，试剂浓度单位为$mol \cdot L^{-1}$）HCl（1），$NH_3 \cdot H_2O$（6），KI（0.1），KBr（0.1），$K_4[Fe(CN)_6]$（0.1），NaCl（0.1），Na_2S（0.1），$Na_2S_2O_3$（0.1），EDTA二钠盐（0.1），KSCN（0.1），NH_4SCN（饱和），$(NH_4)_2C_2O_4$（饱和），NH_4F（2），$AgNO_3$（0.1），$CuSO_4$（0.1），$FeCl_3$（0.1），$NiSO_4$（0.1），Fe^{3+}和Co^{2+}混合试液，碘水，锌粉，丁二酮肟（1%），戊醇。

【实验内容】

1．配离子的稳定性

（1）往盛有2滴$0.1mol \cdot L^{-1}$ $FeCl_3$溶液的试管中，加1滴$0.1mol \cdot L^{-1}$ KSCN溶液，有何现象？然后再逐滴加入饱和$(NH_4)_2C_2O_4$溶液，观察溶液颜色有何变化？写出有关反应方程式，并比较Fe^{3+}的两种配离子的稳定性大小。

（2）在盛有10滴$0.1\ mol \cdot L^{-1}$ $AgNO_3$溶液的试管中，加入10滴$0.1mol \cdot L^{-1}$ NaCl溶液。倾去上层清液，然后在试管中依次进行下列试验：

① 滴加$6mol \cdot L^{-1}$氨水（不断摇动试管）至沉淀刚好溶解；

② 加10滴$0.1mol \cdot L^{-1}$ KBr溶液，有何沉淀生成？

③ 倾去上层清液，滴加$0.1mol \cdot L^{-1}$ $Na_2S_2O_3$溶液至沉淀溶解；

④ 滴加$0.1mol \cdot L^{-1}$ KI溶液，又有何沉淀生成？

写出以上各反应的方程式，并根据实验现象比较：

① $[Ag(NH_3)_2]^{+}$、$[Ag(S_2O_3)_2]^{3-}$的稳定性大小；

② AgCl、AgBr、AgI的K_{sp}的大小。

（3）在0.5mL碘水中，逐滴加入$0.1mol \cdot L^{-1}$ $K_4[Fe(CN)_6]$溶液振荡，有何现象？写出反应式。

2．配位平衡

在盛有5mL $0.1\ mol \cdot L^{-1}$ $CuSO_4$溶液的小烧杯中加入$6mol \cdot L^{-1}$氨水，直至最初生成的碱式

盐 $Cu_2(OH)_2SO_4$ 沉淀又溶解为止。

现欲破坏该配离子，请按下述要求，自己设计实验步骤进行实验，并写出有关反应式。

（1）利用酸碱反应来破坏配离子；

（2）利用沉淀反应来破坏配离子；

（3）利用氧化还原反应来破坏配离子；

提示：$[Cu(NH_3)_4]^{2+} + 2e^- \longrightarrow Cu + 4NH_3 \quad E^\ominus = -0.02V$

$[Zn(NH_3)_4]^{2+} + 2e^- \longrightarrow Zn + 4NH_3 \quad E^\ominus = -1.02V$

（4）利用生成更稳定配合物（如螯合物）的方法破坏配离子。

3．配合物的某些应用

（1）利用生成有色配合物来定性鉴定某些离子　Ni^{2+}与丁二酮肟作用生成鲜红色螯合物沉淀：在白色点滴板上加入 Ni^{2+}试液 1 滴、6mol·L^{-1}氨水 1 滴和 1%丁二酮肟溶液 1 滴，有鲜红色沉淀生成表示有 Ni^{2+}存在。

（2）利用生成配合物掩蔽干扰离子　取 Fe^{3+}和 Co^{2+}混合试液 2 滴于一试管中，加 8~10 滴饱和 NH_4SCN 溶液，有何现象产生？逐滴加入 2mol·L^{-1}NH_4F 溶液，并摇动试管，有何现象？继续滴加至溶液变为淡红色，然后加戊醇 6 滴，振荡试管，静置，观察戊醇层的颜色。

【问题与讨论】

1．可用哪些不同类型的反应，使$[FeSCN]^{2+}$配离子的红色褪去？

2．请用适当的方法将下列各组化合物逐一溶解：

①AgCl，AgBr，AgI；②$Mg(OH)_2$，$Zn(OH)_2$，$Al(OH)_3$；③CuC_2O_4，CuS。

Experiment 25　Coordination Compounds

【Objectives】

1. To compare and explain the stabilities of complex ions.

2. To understand the complex-ion equilibrium.

3. To understand the applications of coordination compounds.

【Introduction】

A coordination compound usually consists of a complex ion and counter ion. A number of ligands bonded to the transition metal ion form a coordination sphere, which can be a cation, an anion or a molecule. For example, $[Cu(NH_3)_4]SO_4$ is a coordination compound: the four NH_3 molecules are the ligands, and $[Cu(NH_3)_4]^{2+}$ is the coordination sphere or the complex ion. $[Cu(NH_3)_4]^{2+}$ and SO_4^{2-} are the main forms present in aqueous solution.

The complex ions dissociate in aqueous solution. Complex-ion equilibrium is reached when the forward reaction rate is equal to the reverse reaction rate. For example:

$$Cu^{2+} + 4NH_3 \rightleftharpoons [Cu(NH_3)_4]^{2+}$$

Coordination compounds are widely used for the identification and separation of ions both in the laboratory and in industry.

【Apparatus and Chemicals】

（The fallowing reagents utilize the concentration unit mol·L^{-1}）HCl（1）, $NH_3 \cdot H_2O$（6）, KI

（0.1）, KBr（0.1）, $K_4[Fe(CN)_6]$（0.1）, NaCl（0.1）, Na_2S（0.1）, $Na_2S_2O_3$（0.1）, disodium EDTA（0.1）, KSCN（0.1）, NH_4SCN（saturated）, $(NH_4)_2C_2O_4$（saturated）, NH_4F（2）, $AgNO_3$（0.1）, $CuSO_4$（0.1）, $FeCl_3$（0.1）, $NiSO_4$（0.1）, mixed solution of Fe^{3+} and Co^{2+}, iodine water, zinc powder, dimethylglyoxime, pentanol.

【Experimental Procedure】

1. Stabilities of complex ions

(1) Place two drops of $0.1mol \cdot L^{-1}$ $FeCl_3$ solution in a test tube, and add one drop of $0.1mol \cdot L^{-1}$ KSCN solution. What will happen? Then add saturated $(NH_4)_2C_2O_4$ solution drop by drop to the test tube. Observe the color change and record the chemical equations. Compare the stabilities of the two complex ions of Fe^{3+}.

(2) Put ten drops of $0.1mol \cdot L^{-1}$ $AgNO_3$ solution into a test tube, and add ten drops of $0.1mol \cdot L^{-1}$ NaCl solution. Decant the supernatant, and use the precipitate to do the following experiments successively:

① Add $6mol \cdot L^{-1}$ ammonia water till the precipitate dissolves (shaking the test tube continuously);

② Add ten drops of $0.1mol \cdot L^{-1}$ KBr solution. What precipitate should be formed?

③ Decant the supernatant, and add $0.1mol \cdot L^{-1}$ $Na_2S_2O_3$ solution till the precipitate dissolves;

④ Add $0.1mol \cdot L^{-1}$ KI solution. What precipitate should be formed?

Write all the chemical equations, and compare the following:

① The stabilities of $[Ag(NH_3)_2]^+$ and $[Ag(S_2O_3)_2]^{3-}$.

② The K_{sp} of AgCl, AgBr and AgI.

(3) In a test tube, add 0.5mL of iodine water, and then add $0.1mol \cdot L^{-1}$ $K_4[Fe(CN)_6]$ drop by drop. Observe and record the chemical equation.

2. Complex-ion equilibria

Put 5mL of $0.1mol \cdot L^{-1}$ $CuSO_4$ in a beaker, and add $6mol \cdot L^{-1}$ ammonia water until the produced $Cu_2(OH)_2SO_4$ precipitate dissolves.

To make the obtained complex ion dissociate, design and do your experiments according to the following instructions:

(1) Use an acid or a base;

(2) Produce a precipitate;

(3) Use redox reaction;

Tips: $[Cu(NH_3)_4]^{2+} + 2e^- \rightleftharpoons Cu + 4NH_3 \quad E^\ominus = -0.02V$

$[Zn(NH_3)_4]^{2+} + 2e^- \rightleftharpoons Zn + 4NH_3 \quad E^\ominus = -1.02V$

(4) Produce a more stable complex ion (such as a chelate).

3. Applications of coordination compounds

(1) Identification of an ion by the color of the produced complex ion Ni^{2+} reacts with dimethylglyoxime to produce a red precipitate: add one drop of test solution on a white spot plate, and add one drop of $6mol \cdot L^{-1}$ ammonia water and one drop of 1% dimethylglyoxime. The red precipitate confirms the presence of Ni^{2+}.

(2) Masking the interfering ions by formation of complex ions Put two drops of Fe^{3+} and Co^{2+} mixed solution in a test tube. Add 8 to 10 drops of saturated NH_4SCN solution. What do you observe? Add $2mol \cdot L^{-1}NH_4F$ solution drop by drop, and shake the test tube. What do you observe? Continue adding NH_4F solution until the solution is light pink, and then add six drops of pentanol. Shake the test tube and observe the color of the pentanol layer.

【Questions and Discussion】

1. To make the red color of $[FeSCN]^{2+}$ fade, what kinds of reactions can be used?

2. Use proper methods to dissolve the compounds one by one in each group:

①AgCl，AgBr，AgI；②$Mg(OH)_2$，$Zn(OH)_2$，$Al(OH)_3$；③CuC_2O_4 and CuS.

第6章　离子的分离与鉴定
Chapter 6　Separation and Identification of Ions

实验26　混合溶液中阳离子的分析（Ⅰ）

【实验目的】

1．将Al^{3+}、Zn^{2+}、Cr^{3+}、Mn^{2+}、Fe^{3+}、Co^{2+}、Ni^{2+}从混合溶液中分离检出，并掌握检出的条件。

2．熟悉以上离子的有关性质。

【实验提要】

参见“附录10常见阳离子的分析”。

【仪器和药品】

离心机。

（除特别注明外，试剂浓度单位为 $mol \cdot L^{-1}$）Al^{3+}、Zn^{2+}、Cr^{3+}、Mn^{2+}、Fe^{3+}、Co^{2+}、Ni^{2+}混合溶液，$(NH_4)_2[Hg(SCN)_4]$，H_2O_2(质量分数为3%)，NaOH (6, 0.1)，HNO_3 (6)，$NaBiO_3$(s)，NH_4SCN (饱和)，$NH_3 \cdot H_2O$ (2)，HAc (6)，H_2SO_4 (1)，NH_4F(2)，Na_2S (0.5)，戊醇，丁二酮肟，铝试剂。

【实验内容】

1．沉淀分离

取试液1mL于离心试管中，滴加6$mol \cdot L^{-1}$ NaOH 2mL，并过量3滴，加质量分数为3%的H_2O_2 10~20滴，于沸水浴中加热并不断搅拌3~4min，离心分离，吸出清液于另一试管中（留做第4步用）。沉淀用0.1$mol \cdot L^{-1}$ NaOH 1mL搅拌洗涤一次，再以热蒸馏水洗涤两次（每次搅拌洗涤），经离心分离，将洗出液弃去。

2．沉淀溶解

将上述沉淀加2$mol \cdot L^{-1}$ HNO_3 2mL、H_2O_2 8~10滴，水浴加热，搅拌使沉淀溶解，继续水浴加热2~3min，使过量的H_2O_2分解。

3．沉淀溶解后的离子检验

沉淀溶解后的溶液中，可能含有Fe^{3+}、Co^{2+}、Ni^{2+}、Mn^{2+}。

（1）取上述清液5滴，加6$mol \cdot L^{-1}$ HNO_3 1mL，加入少量的固体$NaBiO_3$，搅拌，放置1~2min，若显紫红色，示有Mn^{2+}。

（2）取上述清液5滴，加1mL水，加饱和NH_4SCN 2滴，加2 $mol \cdot L^{-1}$ NH_4F至溶液由血红色转为无色，示有Fe^{3+}。

（3）取1mL上述清液，加饱和NH_4SCN 0.5mL，缓慢加2$mol \cdot L^{-1}$ NH_4F，加1mL戊醇，摇荡后静置，若戊醇呈蓝色，示有Co^{2+}。

（4）用吸管取（3）管中下层（水相）溶液2~3滴于点滴板上，加5滴2$mol \cdot L^{-1}$氨水，加1~2滴丁二酮肟，若显红色，示有Ni^{2+}。

4．清液的离子检验

用第一步得到的清液分别鉴定Al^{3+}、Zn^{2+}、Cr^{3+}

（1）取试液10滴，加$6mol \cdot L^{-1}$ HAc酸化，加1~2滴铝试剂，加$2mol \cdot L^{-1}$氨水碱化至微碱性，微热有鲜红色絮状沉淀，示有Al^{3+}。

（2）取试液10滴，加戊醇10滴，滴加$2mol \cdot L^{-1}$ H_2SO_4酸化，滴加质量分数为3%的H_2O_2，摇匀，戊醇层呈蓝色，示有Cr^{3+}。

（3）①取上述清液10滴，加$2mol \cdot L^{-1}$ H_2SO_4酸化，滴加$(NH_4)_2[Hg(SCN)_4]$2滴，摩擦试管壁，有白色沉淀，示有Zn^{2+}（若无沉淀生成，可加1滴质量分数为0.02%的$CuSO_4$溶液）。

②取上述清液10滴，滴加$0.5mol \cdot L^{-1}$ Na_2S溶液，产生白色沉淀，示有Zn^{2+}存在。

综上所述，这些阳离子的分离与检出如图6-1所示。

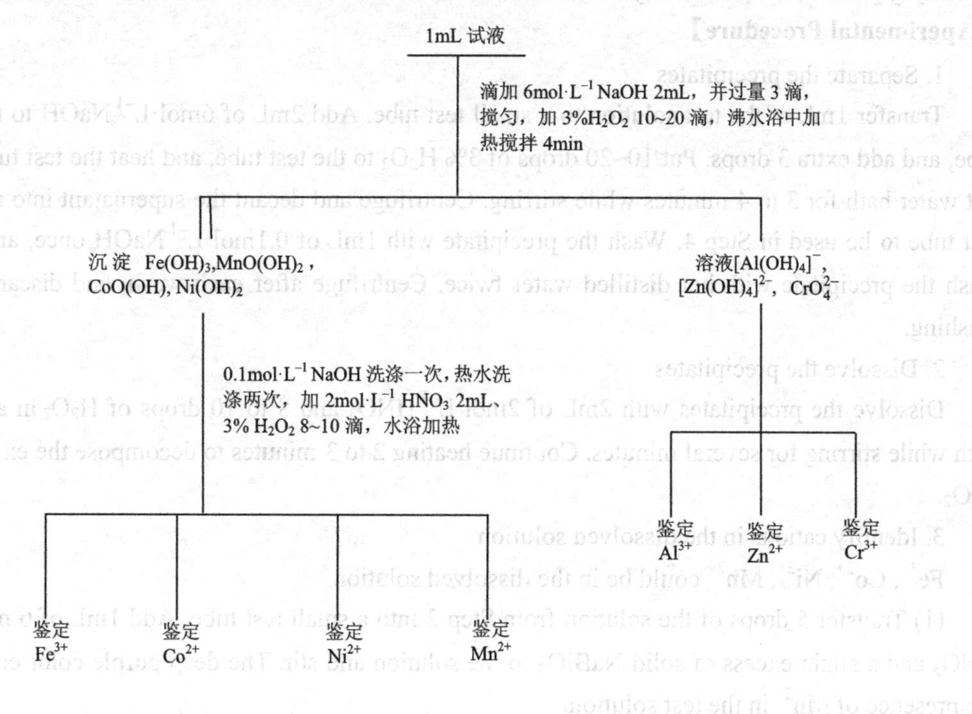

图6-1 Al^{3+}, Zn^{2+}, Cr^{3+}, Mn^{2+}, Fe^{3+}, Co^{2+}, Ni^{2+}混合液分离与检出示意图

【问题与讨论】

1．本实验分离过程第一步为何要加过量NaOH，还要加3%H_2O_2？而反应后，为什么要使H_2O_2完全分解？

2．在第二步，为什么又要加3%H_2O_2？H_2O_2在这里起什么作用？

3．写出有关反应的化学反应式。

Experiment 26 Qual Ⅰ. Al^{3+}, Zn^{2+}, Cr^{3+}, Mn^{2+}, Fe^{3+}, Co^{2+}, Ni^{2+}

【Objectives】

1. To separate and identify the presence of one or more of the cations, Al^{3+}, Zn^{2+}, Cr^{3+}, Mn^{2+},

Fe^{3+}, Co^{2+}, Ni^{2+}, in an aqueous solution.

2. To observe and utilize the chemical and physical properties of those cations.

【Introduction】

See Appendix 10 *Analysis of common cations*.

【Apparatus and Chemicals】

Centrifuge.

(The following reagents utilize the concentration unit $mol \cdot L^{-1}$)Al^{3+}, Zn^{2+}, Cr^{3+}, Mn^{2+}, Fe^{3+}, Co^{2+}, Ni^{2+} test solution, $(NH_4)_2[Hg(SCN)_4]$, 3% H_2O_2, NaOH (6, 0.1), HNO_3 (6), $NaBiO_3$ (s), NH_4SCN (saturated), $NH_3 \cdot H_2O$ (2), HAc (6), H_2SO_4 (1), NH_4F(2), Na_2S (0.5), pentanol, dimethylglyoxime, aluminon reagent.

【Experimental Procedure】

1. Separate the precipitates

Transfer 1mL of the test solution to a small test tube. Add 2mL of $6mol \cdot L^{-1}$ NaOH to the test tube, and add extra 3 drops. Put 10~20 drops of 3% H_2O_2 to the test tube, and heat the test tube in a hot water bath for 3 to 4 minutes while stirring. Centrifuge and decant the supernatant into a small test tube to be used in Step 4. Wash the precipitate with 1mL of $0.1mol \cdot L^{-1}$ NaOH once, and then wash the precipitate with hot distilled water twice. Centrifuge after each wash, and discard each washing.

2. Dissolve the precipitates

Dissolve the precipitates with 2mL of $2mol \cdot L^{-1}$ HNO_3 and 8 to 10 drops of H_2O_2 in a water bath while stirring for several minutes. Continue heating 2 to 3 minutes to decompose the excessive H_2O_2.

3. Identify cations in the dissolved solution

Fe^{3+}, Co^{2+}, Ni^{2+}, Mn^{2+} could be in the dissolved solution.

(1) Transfer 5 drops of the solution from Step 2 into a small test tube. Add 1mL of 6 $mol \cdot L^{-1}$ HNO_3 and a slight excess of solid $NaBiO_3$ to the solution and stir. The deep purple color confirms the presence of Mn^{2+} in the test solution.

(2) Transfer 5 drops of the solution from Step 2 into a small test tube, and add 1mL of water to it. Then add 2 drops of saturated NH_4SCN solution. Adding $2mol \cdot L^{-1}$ NH_4F until the color of the solution changes from blood-red to colorless confirms the presence of Fe^{3+}.

(3) Transfer 1mL of the solution from Step 2 into a small test tube, and add 0.5mL of saturated NH_4SCN to it. Slowly add $2mol \cdot L^{-1}$ NH_4F, and then add 1mL of pentanol and shake well. The blue color of pentanol layer confirms the presence of Co^{2+}.

(4) Use a dropping pipet to transfer 2 to 3 drops of solution (lower layer) from the test tube of Step (3). Add 5 drops of $2mol \cdot L^{-1}$ammonia water, and 1 to 2 drops of dimethylglyoxime solution. Appearance of a red precipitate confirms the presence of Ni^{2+}.

4. Identify cations in the supernatant

Identify Al^{3+}、Zn^{2+}、Cr^{3+} from the supernatant of Step 1.

(1) Acidify 10 drops of the supernatant with $6mol \cdot L^{-1}$ HAc. Add 2 drops of the aluminon reagent, and add several drops of $2mol \cdot L^{-1}$ ammonia water until the solution is basic. Heat the test

tube slightly, and a red precipitate confirms the presence of Al^{3+}.

(2) Transfer 10 drops of the supernatant into a test tube and add 10 drops of pentanol. Add $2mol\cdot L^{-1}$ H_2SO_4 to acidify the solution. Then add 3% H_2O_2, and shake well. The blue color of the pentanol layer confirms the presence of Cr^{3+}.

(3) ① Acidify 10 drops of the supernatant with $2mol\cdot L^{-1}$ H_2SO_4. Add 2 drops of $(NH_4)_2[Hg(SCN)_4]$, and rub the test tube wall. A white precipitate confirms the presence of Zn^{2+}. (If there is no precipitate, add 1 drop of 0.02% $CuSO_4$ solution.)

②Transfer 10 drops of the supernatant into a test tube and add $0.5mol\cdot L^{-1}$ Na_2S solution. A white precipitate confirms the presence of Zn^{2+}.

Figure 6-1 shows the procedure of the experiment.

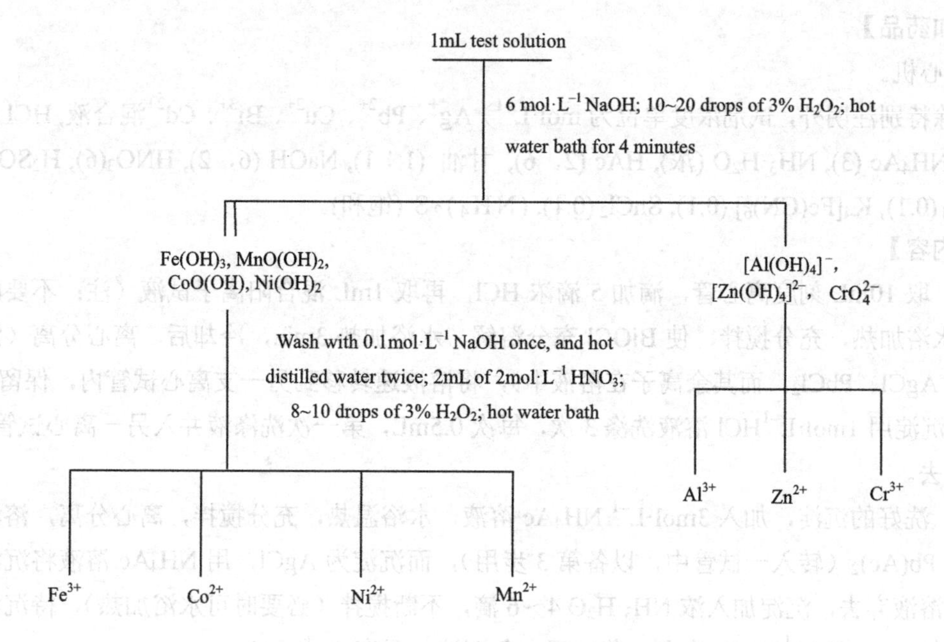

Figure 6-1 Flow diagram of Qual Ⅰ. Al^{3+}, Zn^{2+}, Cr^{3+}, Mn^{2+}, Fe^{3+}, Co^{2+}, Ni^{2+}

【Questions and Discussion】

1. Why should excessive NaOH and 3% H_2O_2 be added in the first step? Why should the excessive H_2O_2 decompose?

2. In the second step, why should 3% H_2O_2 be added again? What is the purpose of adding H_2O_2 here?

3. Record all the chemical equations in your report.

实验 27　混合溶液中阳离子的分析（Ⅱ）

【实验目的】

1．学会将 Ag^{+}、Pb^{2+}、Cu^{2+}、Bi^{3+}、Cd^{2+}从混合液中进行分离，并掌握它的分离条件和方法。

2．熟悉和巩固以上有关离子的性质。

【实验提要】

参见“附录 10 常见阳离子的分析”。

【仪器和药品】

离心机。

（除特别注明外，试剂浓度单位为 $mol \cdot L^{-1}$）Ag^{+}、Pb^{2+}、Cu^{2+}、Bi^{3+}、Cd^{2+}混合液, HCl (浓，1，2), NH_4Ac (3), $NH_3 \cdot H_2O$ (浓), HAc (2，6), 甘油 (1∶1), NaOH (6，2), HNO_3(6), H_2SO_4(1), K_2CrO_4 (0.1), $K_4[Fe(CN)_6]$ (0.1), $SnCl_2$ (0.1), $(NH_4)_2S$ (饱和)。

【实验内容】

1．取 10mL 刻度离心管，滴加 5 滴浓 HCl，再取 1mL 混合阳离子试液（注：不要颠倒加）。水浴加热，充分搅拌，使 BiOCl 充分溶解，水浴加热 3min，冷却后，离心分离（沉淀可能有 AgCl、$PbCl_2$，而其余离子在溶液中），将溶液速转移到另一支离心试管内，保留第 4 步用。沉淀用 $1mol \cdot L^{-1}$HCl 溶液洗涤 3 次，每次 0.5mL，第一次洗涤液并入另一离心试管内，其余弃去。

2．洗好的沉淀，加入 $3mol \cdot L^{-1}$ NH_4Ac 溶液，水浴温热，充分搅拌，离心分离，溶液内可能为 $Pb(Ac)_2$（转入一试管中，以备第 3 步用），而沉淀为 AgCl，用 NH_4Ac 溶液将沉淀洗一次，溶液弃去，沉淀加入浓 $NH_3 \cdot H_2O$ 4～6 滴，不断搅拌（必要时可水浴加热），待沉淀溶解后，加入 $6mol \cdot L^{-1}$ HNO_3 中和，若出现白色沉淀，示有 Ag^{+}存在。

3．往盛有溶液的试管滴 $6mol \cdot L^{-1}$ HAc，酸化，再加 $0.1mol \cdot L^{-1}$$K_2CrO_4$ 溶液，有黄色沉淀，示有 Pb^{2+}存在。

4．取第 1 步分离出的溶液加入甘油（1∶1）8 滴，充分搅拌，滴加 $6mol \cdot L^{-1}$NaOH 至溶液呈碱性，再多加 10 滴，搅拌均匀后，离心分离。溶液可能有 Cu^{2+}、Bi^{3+}的甘油化合物，转入一试管中，以备第 6 步和第 7 步用。沉淀应是 $Cd(OH)_2$，沉淀加入甘油-碱液[20 滴 $2mol \cdot L^{-1}$ NaOH 加入 5 滴甘油（1∶1）配成]，充分搅拌后，离心分离，同样洗涤 3 次，直至不含 Cu^{2+}、Bi^{3+}为止（Cu^{2+}、Bi^{3+}的存在影响 Cd^{2+}的鉴定）。

5．洗涤好后沉淀应为 $Cd(OH)_2$，加入 $2mol \cdot L^{-1}$ HCl 6~8 滴、水 20 滴，搅拌至沉淀完全溶解（若溶不完，可适当水浴加热），加饱和$(NH_4)_2S$ 1~2 滴，有黄色沉淀，示有 Cd^{2+}存在。

若第 4 步洗涤不干净，有少量剩余 Pb^{2+}、Cu^{2+}、Bi^{3+}存在，则生成对应的硫化物，呈棕褐色，此时应离心分离，溶液弃去，沉淀加 $1mol \cdot L^{-1}$ H_2SO_4 10~12 滴，在不断搅拌下，水浴加热 3min，此时 CdS 溶解，而其它硫化物仍是沉淀，离心分离，将溶液转入一小试管中，滴加饱和$(NH_4)_2S$ 溶液 2~3 滴，有黄色沉淀，示有 Cd^{2+}存在。

6．从第 4 步分离的溶液取 2 滴于点滴板上，加 1 滴 $6mol\cdot L^{-1}$ HAc 酸化，再加 $0.1mol\cdot L^{-1}$ $K_4[Fe(CN)_6]$，有红棕色沉淀，示有 Cu^{2+} 存在。

7．从第 4 步分离的溶液中取 2 滴于点滴板上，加入新配的 Na_2SnO_2 溶液（即 $0.1mol\cdot L^{-1}$ $SnCl_2$ 加过量 $2mol\cdot L^{-1}$ NaOH 至有沉淀后，沉淀溶解则为 Na_2SnO_2），有黑色沉淀，示有 Bi^{3+} 存在。

综上所述，这些阳离子的分离与检出如图 6-2 所示。

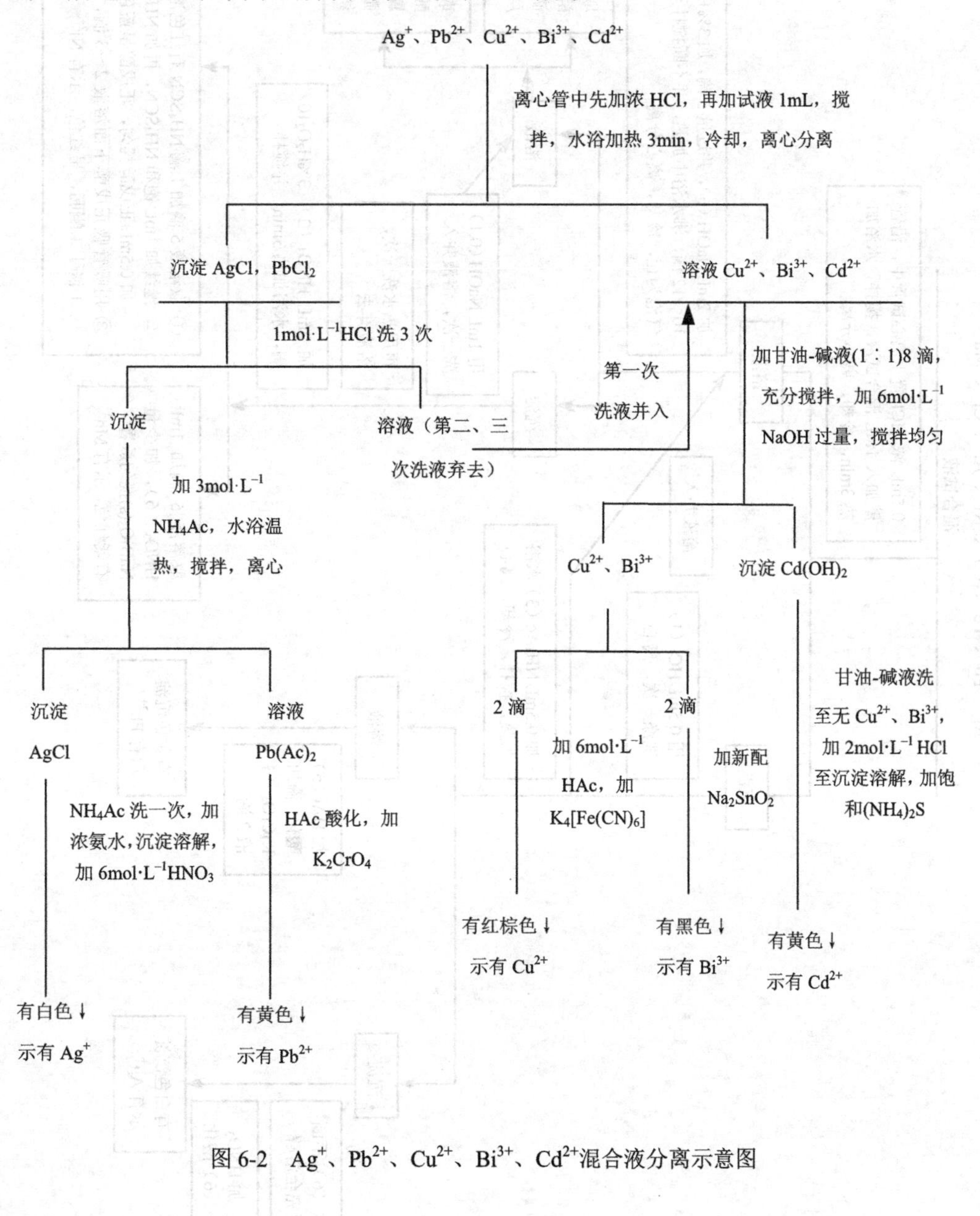

图 6-2　Ag^+、Pb^{2+}、Cu^{2+}、Bi^{3+}、Cd^{2+} 混合液分离示意图

【选做部分】

Ag^+、Pb^{2+}、Fe^{3+}、Co^{2+}、Ni^{2+}、Al^{3+}、Cr^{3+}、Mn^{2+} 混合溶液分析。

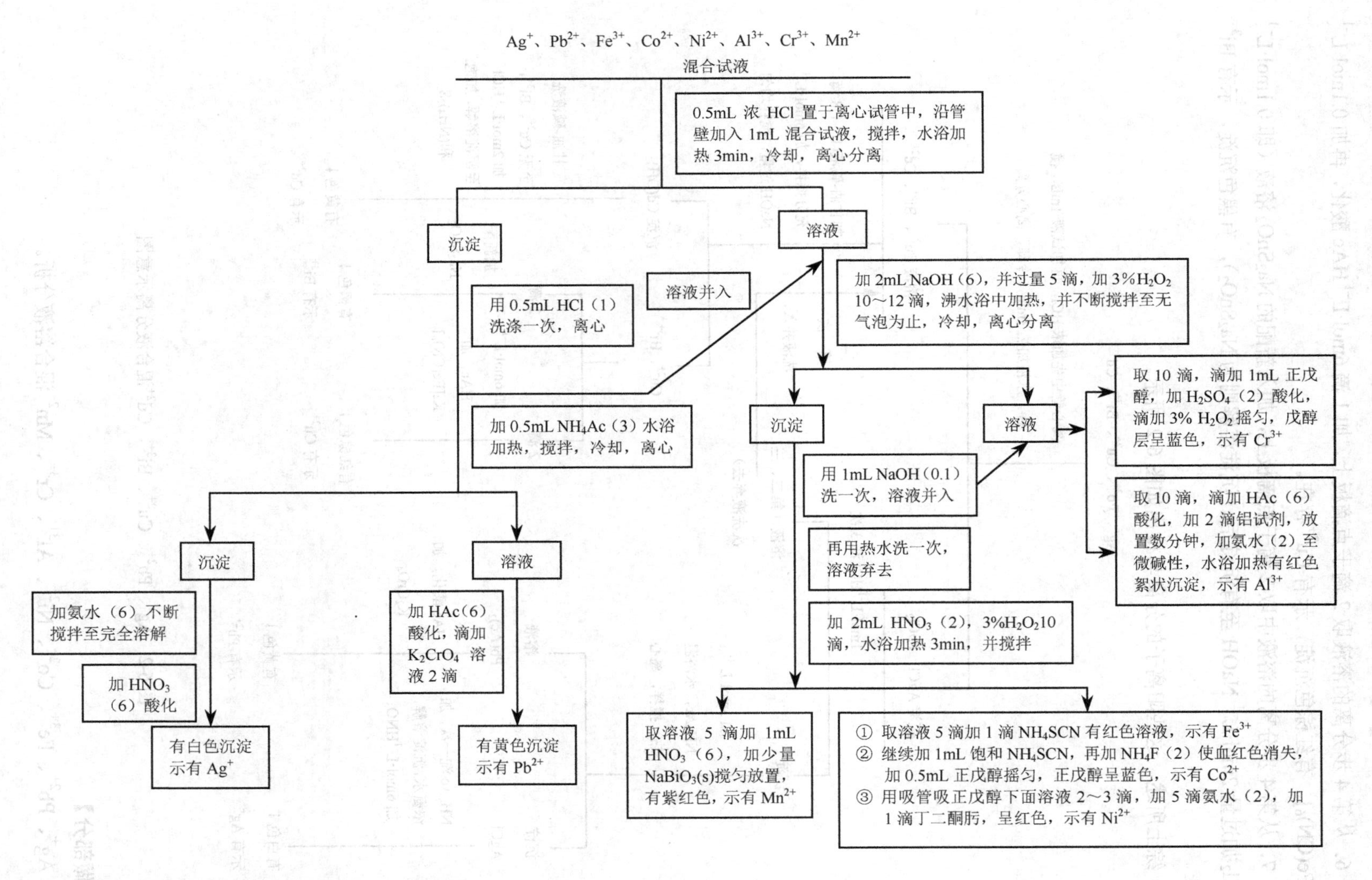

注：药品除特别注明外，试剂浓度单位均为 $mol \cdot L^{-1}$，括号中即为浓度。

【问题与讨论】

1．为什么要把试液加到浓 HCl 中？

2．用 $K_4[Fe(CN)_6]$检出 Cu^{2+} 时，为何要用 HAc 酸化？

Experiment 27 Qual Ⅱ. Ag^+, Pb^{2+}, Cu^{2+}, Bi^{3+}, Cd^{2+}

【Objectives】

1. To separate and identify the presence of one or more of the cations, Ag^+, Pb^{2+}, Cu^{2+}, Bi^{3+}, Cd^{2+}, in an aqueous solution.

2. To observe the chemical and physical properties of those cations.

【Introduction】

See Appendix 10 *Analysis of common cations*.

【Apparatus and Chemicals】

(The following reagents utilize the concentration unit $mol \cdot L^{-1}$)Ag^+, Pb^{2+}, Cu^{2+}, Bi^{3+}, Cd^{2+} test solution, HCl (concentrated, 1, 2) , NH_4Ac (3), $NH_3 \cdot H_2O$ (concentrated), HAc (6, 2), glycerin (1∶1), NaOH (6, 2), HNO_3(6), H_2SO_4(1), K_2CrO_4 (0.1), $K_4[Fe(CN)_6]$ (0.1), $SnCl_2$ (0.1), $(NH_4)_2S$ (saturated).

【Experimental Procedure】

1. In a 10mL graduated centrifuge tube Ⅰ, add 5 drops of concentrated HCl first, and then add 1mL of test solution. (*Caution*: do not add reversely.) Heat the centrifuge tube in a water bath for 3 minutes and stir thoroughly to dissolve BiOCl. Cool the centrifuge tube and centrifuge. (The precipitates may contain AgCl, $PbCl_2$. The rest of the cations are in the supernatant.) Decant the supernatant into centrifuge tube Ⅱ for Step 4. Wash the precipitates 3 times with $1mol \cdot L^{-1}$ HCl. Use 0.5mL of HCl with each wash. After the first wash, transfer the supernatant to centrifuge tube Ⅱ for Step 4. Discard the rest of the supernatant after wash.

2. Add $3mol \cdot L^{-1}$ NH_4Ac solution to the precipitates which are already washed. Heat the centrifuge tube in a warm water bath and stir thoroughly. After centrifuge, decant the supernatant, which may be $Pb(Ac)_2$, to a test tube Ⅲ for Step 3. The precipitate is AgCl. Wash the precipitate with NH_4Ac once, and discard the supernatant. Add 4 to 6 drops of concentrated $NH_3 \cdot H_2O$ and stir (If necessary, use a water bath). After the precipitate dissolves, add $6mol \cdot L^{-1}$ HNO_3. Appearance of a white precipitate confirms the presence of Ag^+.

3. Add $6mol \cdot L^{-1}$ HAc to the supernatant in test tube Ⅲ from Step 2, and add $0.1mol \cdot L^{-1}$ K_2CrO_4 solution. A yellow precipitate confirms the presence of Pb^{2+}.

4. Add 8 drops of glycerin (1∶1) to the supernatant in centrifuge tube Ⅱ from Step 1 and stir. Then add $6mol \cdot L^{-1}$ NaOH till the solution is alkaline. Add an extra 10 drops of NaOH, stir and centrifuge. Decant the supernatant, which may contain Cu^{2+} and Bi^{3+}, into test tube Ⅳ for Step 6 and 7. Wash the $Cd(OH)_2$ precipitate with glycerin-NaOH solution [mix 20 drops of $2mol \cdot L^{-1}$ NaOH and 5 drops of glycerin (1∶1)] 3 times until there are no Cu^{2+} and Bi^{3+} ions present (Cu^{2+} and Bi^{3+} will affect the identification of Cd^{2+}).

5. The precipitate should be $Cd(OH)_2$. Add 6 to 8 drops of $2mol \cdot L^{-1}$ HCl and 20 drops of H_2O, stirring until the precipitate dissolves. (If necessary, heat with a water bath.) Add 1 to 2 drops of

saturated $(NH_4)_2S$ solution, and a yellow precipitate confirms the presence of Cd^{2+}.

In Step 4, if the wash is not complete, a small amount of Pb^{2+}, Cu^{2+} and Bi^{3+} will be present in the solution and their sulfides will be formed as demonstrated by a brown color. If so, centrifuge and discard the supernatant. Then add 10 to 12 drops of $1mol \cdot L^{-1}$ H_2SO_4 to the precipitates and heat in a water bath for 3 minutes while stirring. CdS should be dissolved, and the rest of sulfide precipitates can be separated by centrifuge. Decant the supernatant into a test tube and add 2 to 3 drops of saturated $(NH_4)_2S$ solution. A yellow precipitate confirms the presence of Cd^{2+}.

6. Transfer 2 drops of supernatant in test tube Ⅳ from Step 4 to a spot plate. Acidify it with 1 drop of $6mol \cdot L^{-1}$ HAc, and then add $0.1mol \cdot L^{-1}$ $K_4[Fe(CN)_6]$. A red-brown precipitate confirms the presence of Cu^{2+}.

7. Transfer 2 drops of supernatant in test tube Ⅳ from Step 4 to a spot plate. Add freshly prepared Na_2SnO_2 solution. (Add excessive $2mol \cdot L^{-1}$ NaOH to $0.1mol \cdot L^{-1}$ $SnCl_2$ until a precipitate forms. Dissolve the precipitate and Na_2SnO_2 will be produced.) A black precipitate confirms the presence of Bi^{3+}.

Figure 6-2 shows the procedure of the experiment.

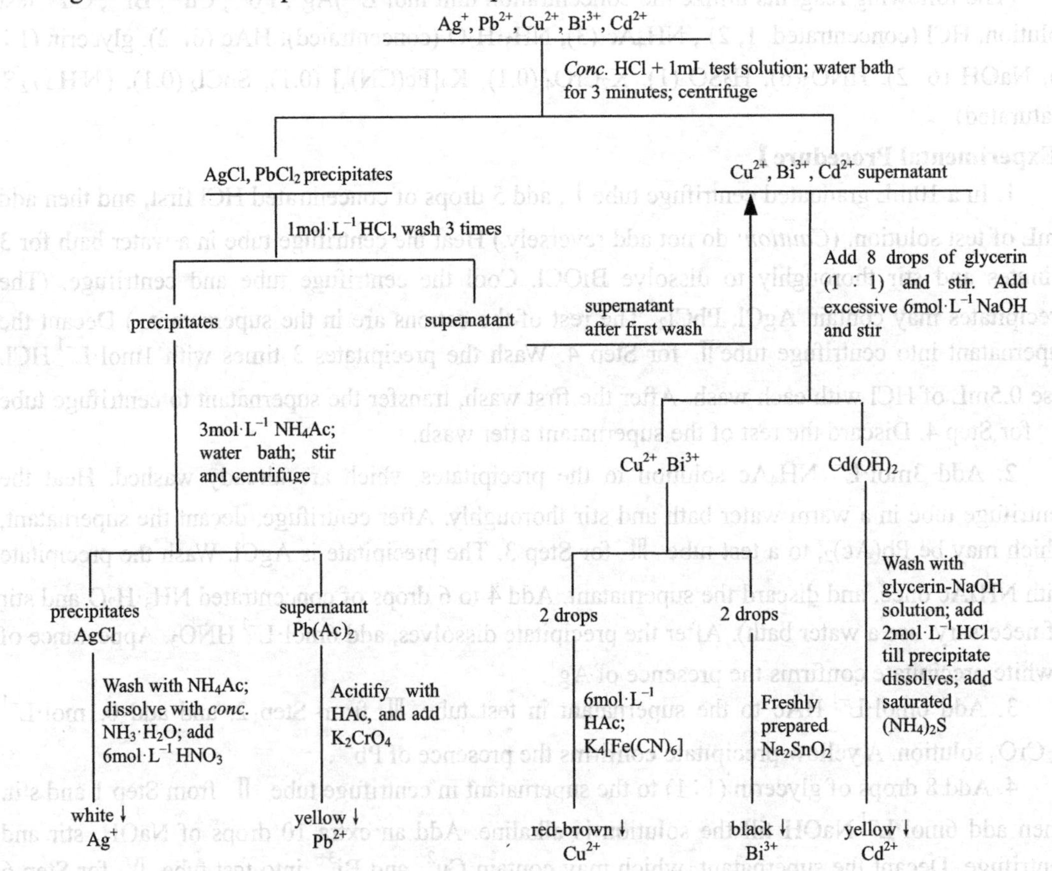

Figure 6-2 Flow diagram of Qual Ⅱ. Ag^+, Pb^{2+}, Cu^{2+}, Bi^{3+}, Cd^{2+}

【Questions and Discussion】

1. Why should the test solution be added to the concentrated HCl?

2. Why do we use HAc when identifying Cu^{2+} with $K_4[Fe(CN)_6]$?

实验 28　混合溶液中阴离子的分析

【实验目的】

1．了解分离检出 11 种常见阴离子的方法步骤和条件。

2．熟悉常见阴离子的有关性质。

3．检出未知液中的阴离子。

【实验提要】

参见“附录 11 常见阴离子的分析”。

【仪器和药品】

离心机，U 形气体瓶（自制），点滴板，离心试管。

（除特别注明外，试剂浓度单位为 $mol \cdot L^{-1}$）待检试液：CO_3^{2-}、S^{2-}、NO_3^-、PO_4^{3-}、SO_3^{2-}、Cl^-、Br^-、I^-、$S_2O_3^{2-}$、SO_4^{2-}、NO_2^- 的钠盐或钾盐溶液（均为 0.1）。

$BaCl_2$ (0.5)，$Ba(OH)_2$ (0.5)，HNO_3 (6)，HCl (6，2)，HAc (2)，$AgNO_3$ (0.1)，$KMnO_4$(0.01)，$Sr(NO_3)_2$ (0.1)，H_2O_2 (质量分数为 3%)，氯水，碘水，$(NH_4)_2CO_3$（饱和），尿素（饱和溶液或固体），H_2SO_4（浓），$Na_3[Co(NO_2)_6]$，$CdCO_3$(s)，$(NH_4)_2MoO_4$ 试剂，Zn 粉，CCl_4，对氨基苯磺酸，α-萘胺，二苯胺，淀粉溶液，I_2-淀粉溶液，KI-淀粉试纸，$Pb(Ac)_2$ 试纸，pH 试纸。

【实验内容】

分离检出常见的阴离子。

在进行初步试验的各项操作及离子的个别鉴定时，每次取试液 1~2 滴，需要分离时，取试液 8~10 滴，点滴试验时取试液 1 滴。

领取未知液一份，可能含有 CO_3^{2-}，NO_2^-，NO_3^-，Cl^-，Br^-，I^-，PO_4^{3-}，S^{2-}，SO_3^{2-}，$S_2O_3^{2-}$ 和 SO_4^{2-}。按以下步骤检出未知液中的阴离子。

（1）观察溶液的颜色并试验其酸碱性。

（2）按表 6-1 内容进行初步试验，以推测阴离子存在的可能范围。

表 6-1　阴离子初步试验[①]

试剂			SO_4^{2-}	SO_3^{2-}	$S_2O_3^{2-}$	S^{2-}	CO_3^{2-}	PO_4^{3-}	Cl^-	Br^-	I^-	NO_2^-	NO_3^-
稀 H_2SO_4		理论		+	+	+	+					+	
		试验											
$AgNO_3$ (稀 HNO_3)		理论			+	+			+	+	+		
		试验											
氯化钡	中性或碱性	理论	+	+	(+)		+	+					
		试验											
	加稀 HCl	理论	+										
		试验											
KI-淀粉 (稀 H_2SO_4)		理论										+	(+)
		试验											
I_2-淀粉 (稀 H_2SO_4)		理论		+	+	+							
		试验											

续表

试剂		SO_4^{2-}	SO_3^{2-}	$S_2O_3^{2-}$	S^{2-}	CO_3^{2-}	PO_4^{3-}	Cl^-	Br^-	I^-	NO_2^-	NO_3^-
$KMnO_4$	理论		+	+	+			(+)	+	+	+	
(稀 H_2SO_4)	试验											
可能存在的离子												
未能确定的离子												

①“+”表示有反应（正结果），“(+)”表示阴离子浓度大时才产生反应。“试验”表示未知溶液的试验结果。

（3）根据离子存在的可能范围（可能存在的和未能确定的离子），设计具体的分析方案，进行分离与检出。

有些离子之间有干扰，干扰离子必须在被干扰离子之前先行检出，当干扰离子不存在时，没有必要分离即可进行个别离子的鉴定。其它互不干扰的离子，先后鉴定均一样。分离方法与个别离子的鉴定，可参考有关实验的内容。

（4）对实验结果进行综合分析，若最后结果与初步试验有矛盾时，必须再作必要的重复试验或用多种方法加以验证。

提示：

CO_3^{2-} 的鉴定：CO_3^{2-} 与酸作用产生 CO_2 气体，可使 $Ba(OH)_2$ 溶液变浑浊。

$$CO_3^{2-} + 2H^+ \longrightarrow CO_2\uparrow + H_2O$$

$$CO_2 + Ba(OH)_2 \longrightarrow BaCO_3\downarrow + 2H_2O$$

操作：取 1~2 滴试液于自制的 U 形气体瓶中，先滴加 2 滴质量分数为 3%的 H_2O_2，然后从一头插入吸有少量 $Ba(OH)_2$ 的小滴管；另一头插入吸有稀 HCl 的小滴管，当滴加 HCl 时，若 $Ba(OH)_2$ 溶液迅速变浑浊，则表示有 CO_3^{2-} 存在。

【问题与讨论】

1. 根据阴离子的性质，你认为哪些阴离子必须先行分离才能检出？

2. 用 $Ba(OH)_2$ 溶液检验 CO_3^{2-} 时，要先加入 H_2O_2 后再加稀 HCl，能否颠倒操作？加 H_2O_2 的目的是什么？

3. 在混合液中加入 $BaCl_2$ 溶液后，再用稀 HCl 酸化，只缓慢析出白色乳状物，这是什么物质？这时是否能判断一定有 SO_4^{2-} 存在？$BaSO_4$ 沉淀的状态是怎样的？

4. 用氯水检验 Br^- 和 I^- 时，先检出 I^-，后检出 Br^-，若 I^- 含量大则会影响 Br^- 的检出，应采取什么方法先将 I^- 除去后再检出 Br^-？

5. 某一试样不溶于水，易溶于稀 HNO_3，并已证实试样中含有 Ba^{2+} 和 Ag^+，试分析哪些阴离子需要检验？哪些阴离子不需要检验？

Experiment 28　Analysis of Anions

【Objectives】

1. To understand the methods and procedures of separating and identifying eleven anions.

2. To become familiar with the properties of common anions.

3. To identify the anions in a test solution.

【Introduction】

See Appendix 11 *Analysis of common anions*.

【Apparatus and Chemicals】

Centrifuge, U-type gas bottle (self-prepared), spot plate, centrifuge tube.

Test solution: sodium salts or potassium salts of the following anions: CO_3^{2-}, S^{2-}, NO_3^-, PO_4^{3-}, SO_3^{2-}, Cl^-, Br^-, I^-, $S_2O_3^{2-}$, SO_4^{2-}, NO_2^-, all with a concentration of $0.1 mol \cdot L^{-1}$.

(The following reagents utilize the concentration unit $mol \cdot L^{-1}$) $BaCl_2$ (0.5), $Ba(OH)_2$ (0.5), HNO_3 (6), HCl (6, 2), HAc (2), $AgNO_3$ (0.1), $KMnO_4$(0.01), $Sr(NO_3)_2$ (0.1), H_2O_2 (3%, wt/wt), chlorine water, iodine water, $(NH_4)_2CO_3$ (saturated), carbamide (saturated or solid), H_2SO_4 (concentrated), $Na_3[Co(NO_2)_6]$, $CdCO_3$(s), $(NH_4)_2MoO_4$ reagent, Zn powder, CCl_4, *p*-aminobenzene sulfonic acid, *α*-naphthylamine, diphenylamine, I_2-starch solution, KI-starch test paper, $Pb(Ac)_2$ test paper, pH test paper.

【Experimental Procedure】

Separation and identification of common anions.

When you do the preliminary tests and identify each anion, use 1 to 2 drops of solution. When you do a separation, use 8 to 10 drops of solution. When you do the experiment on a spot plate, use one drop of solution.

Obtain a test solution from your instructor, which may contain CO_3^{2-}, NO_2^-, NO_3^-, Cl^-, Br^-, I^-, PO_4^{3-}, S^{2-}, SO_3^{2-}, $S_2O_3^{2-}$ and SO_4^{2-}. Identify the anions in the test solution according to the following procedures:

(1) Observe the color of the solution and test if it is acidic or basic.

(2) According to Table 6-1, do the preliminary tests and determine the anions which may be present in the test solution.

Table 6-1 Preliminary tests of anions[1]

Reagents		SO_4^{2-}	SO_3^{2-}	$S_2O_3^{2-}$	S^{2-}	CO_3^{2-}	PO_4^{3-}	Cl^-	Br^-	I^-	NO_2^-	NO_3^-
dilute H_2SO_4	Theory		+	+	+	+					+	
	Test											
$AgNO_3$ (dilute HNO_3)	Theory			+	+			+	+	+		
	Test											
$BaCl_2$ — Neutral or basic media	Theory	+	+	(+)		+	+					
	Test											
$BaCl_2$ — dilute HCl	Theory	+										
	Test											
KI-starch (dilute H_2SO_4)	Theory										+	(+)
	Test											
I_2-starch (dilute H_2SO_4)	Theory		+	+	+							
	Test											
$KMnO_4$ (dilute H_2SO_4)	Theory		+	+	+			(+)	+	+	+	
	Test											

continuous

Reagents	SO_4^{2-}	SO_3^{2-}	$S_2O_3^{2-}$	S^{2-}	CO_3^{2-}	PO_4^{3-}	Cl^-	Br^-	I^-	NO_2^-	NO_3^-
Anions may be present											
Uncertain anions											

① "+" means reactions occur (positive result). "(+)" means reactions occur when the concentration of anion is big. "Test" means the results of experiment for test solution.

(3) According to your results from the above steps, design a detailed scheme to separate and identify each anion.

Specific anion tests are subject to interferences from other anions and cations. So elimination of the interferences is necessary. Use proper methods to identify each anion.

(4) Analyze your results carefully. If your final results contradict your preliminary tests, it is necessary to repeat the experiments or confirm your results with other methods.

Tips:

Identification of CO_3^{2-} : CO_3^{2-} reacts with acid to produce CO_2, which can react with $Ba(OH)_2$ solution to form a precipitate.

$$CO_3^{2-}+2H^+ = CO_2\uparrow +H_2O$$

$$CO_2+Ba(OH)_2 = BaCO_3\downarrow+2H_2O$$

Operation: Place 1 to 2 drops of test solution to a self-prepared U-type gas bottle, and add two drops of 3% (wt/wt) H_2O_2. Then insert a pipet with $Ba(OH)_2$ to one end, and insert another pipet with HCl to the other end. Release HCl. If the $Ba(OH)_2$ solution immediately turns opaque, the presence of CO_3^{2-} is confirmed.

【Questions and Discussion】

1. Which anions should be separated before identifying them according to their properties?

2. When identifying CO_3^{2-} with $Ba(OH)_2$ solution, we have to add H_2O_2 first, then add dilute HCl. Can we do it in reverse? What is the purpose of adding H_2O_2?

3. After adding $BaCl_2$ solution to the test solution and then using dilute HCl to acidify the solution, white milky precipitate slowly forms. What is it? Can we be sure that SO_4^{2-} is present? What does the $BaSO_4$ precipitate look like?

4. When using chlorine water to identify Br^- and I^-, we identify I^- first, then Br^-. If there is a large amount of I^- in the solution, it will affect the identification of Br^-. What method should we use to remove I^- first, then identify Br^-?

5. A sample does not dissolve in water, but dissolves in dilute HNO_3. It is known that the sample contains Ba^{2+} and Ag^+. Which anions need to be further identified in the sample? Which anions do not need to be identified?

第7章 创新实验

Chapter 7 Innovative Experiments

实验29 葡萄糖酸锌的制备

【实验目的】

1．学习和掌握药用微量元素药物合成的基本方法。

2．学习并掌握葡萄糖酸锌的合成和鉴别。

3．了解锌的生物意义。

【实验提要】

锌在人体内是必需的微量元素之一，参与多种重要酶的合成，并对维持正常生理功能具有重要意义。人体需要的锌主要从日常饮食中获得。如发现缺锌现象，则必须给予适当的补充。以前常用的补锌剂是硫酸锌，而现在一般用葡萄糖酸锌（Zinc gluconate）。硫酸锌对肠胃有刺激，而葡萄糖酸锌副作用小，可制成含片应用，且易于吸收。

葡萄糖酸锌为白色粒状结晶或粉末，无臭，略有不适味，溶于水，易溶于沸水（约1∶1），15℃时饱和溶液质量分数为25%，不溶于无水乙醇、氯仿和乙醚。它可由葡萄糖酸钙和硫酸锌反应制得：

$$Ca(Glu)_2 + ZnSO_4 \longrightarrow Zn(Glu)_2 + CaSO_4 \downarrow$$

过滤除去硫酸钙沉淀，溶液经浓缩可得葡萄糖酸锌结晶。

葡萄糖酸钙[Calcium gluconate，$Ca(C_6H_{11}O_7)_2 \cdot H_2O$]本身是补钙剂，也是医药原料。

【仪器和药品】

布氏漏斗，普通漏斗。

葡萄糖酸钙，硫酸锌（$ZnSO_4 \cdot 7H_2O$），活性炭。

【实验内容】

称取葡萄糖酸钙[$Ca(C_6H_{11}O_7)_2 \cdot H_2O$] 11.2g (0.025mol)，放入100mL烧杯中，加入30mL蒸馏水。另称取硫酸锌($ZnSO_4 \cdot 7H_2O$) 7.6 g (0.026mol)，用30mL蒸馏水使之溶解，在不断搅拌下，把硫酸锌溶液逐滴加入葡萄糖酸钙溶液中，加完后在40℃水浴中保温约20min，抽滤除去$CaSO_4$沉淀，溶液转入烧杯，加热近沸，加入少量活性炭脱色，趁热过滤。滤液浓缩至原体积的1/4~1/3。静置冷却，抽滤得到粗产品结晶。粗产品用适量热水重结晶，可得供压制片剂的葡萄糖酸锌。

【选做部分】

锌含量分析：准确称取产品0.8g，溶于20mL水中（可微热），加入10mL NH_3-NH_4Cl缓冲溶液，加4滴铬黑T指示剂，用0.1mol·L^{-1} EDTA标准溶液滴定至溶液由紫红色变为蓝色，即为终点。重复上述操作2次，记录EDTA标准溶液的用量，按下式计算产品中锌的含量：

$$w_{Zn} = \frac{V(EDTA)c(EDTA) \times 65.38}{m_{产品} \times 1000} \times 100\%$$

【问题与讨论】

1．可否用如下的化合物与葡萄糖酸钙反应来制备葡萄糖酸锌？为什么？

① ZnO；② $ZnCl_2$；③ $ZnCO_3$；④ $Zn(CH_3COO)_2$。

2．试设计一方案制备葡萄糖酸亚铁。

Experiment 29　Preparation of Zinc Gluconate

【Objectives】

1. To learn the method of drug synthesis with microelements.
2. To learn how to prepare zinc gluconate.
3. To understand the biological importance of zinc.

【Introduction】

Zinc is an essential microelement of exceptional biologic and public health importance. It is the reactive center of many important enzymes in the human body. We usually obtain zinc from our daily food. But zinc supplement is necessary if zinc deficiency occurs. Zinc gluconate is often used to treat zinc deficiency.

Zinc gluconate is a white crystalline or powder substance. It is odourless, and can be dissolved in water. It dissolves more readily in hot water (about 1∶1). Zinc gluconate can not be dissolved in anhydrous ethanol, chloroform, or diethyl ether. Calcium gluconate reacts with zinc sulfate to produce zinc gluconate:

$$Ca(Glu)_2+ZnSO_4 \longrightarrow Zn(Glu)_2+CaSO_4\downarrow$$

The calcium sulfate precipitate can be removed by filtration. Then the zinc gluconate crystals will be obtained after the concentration of the solution.

Calcium gluconate, $Ca(C_6H_{11}O_7)_2 \cdot H_2O$, is a calcium supplement itself.

【Apparatus and Chemicals】

Büchner funnel, funnel.

Calcium gluconate, $ZnSO_4 \cdot 7H_2O$, activated carbon.

【Experimental Procedure】

Weigh 11.2g (0.025mol) of calcium gluconate in a 100mL beaker. Add 30mL of distilled water to dissolve it. Weigh 7.6g (0.026mol) of $ZnSO_4 \cdot 7H_2O$, and dissolve it with 30mL of distilled water. Add the zinc sulfate solution drop by drop to the calcium gluconate solution while stirring. Then maintain the mixture at 40℃ for twenty minutes. Next, remove the calcium sulfate precipitate by vacuum filtration. Transfer the filtrate to a beaker, and heat it till boiling. Add a small amount of activated carbon for decoloration purposes. Filter while the solution is hot. Finally, concentrate the filtrate to about one third of its original volume. Cool the solution till the crystals form completely. The zinc gluconate crystals can be separated by vacuum filtration. Refined crystals can be obtained by recrystallization with hot water.

【Questions and Discussion】

1. Can the following compounds be used to react with calcium gluconate to prepare zinc gluconate? Why?

①ZnO；②$ZnCl_2$；③$ZnCO_3$；④$Zn(CH_3COO)_2$.

2. Design an experiment to prepare ferrous gluconate.

实验30 从茶叶和紫菜中分离与鉴定某些元素

【实验目的】

1．了解并掌握从茶叶和紫菜中分离与鉴定某些化学元素的方法。

2．增加对探索大自然奥秘的兴趣，提高学习化学的积极性。

【实验提要】

植物是有机体，主要由 C、H、O 和 N 等元素组成，还含有 P、I 和某些金属元素。把植物烧成灰烬，经过一系列的化学处理，即可从中分离和鉴定某些元素。本实验要求从茶叶中检出钙、镁、铝、铁和磷 5 种元素，从紫菜中检出碘元素。

钙、镁、铝和铁 4 种金属离子的氢氧化物完全沉淀的 pH 范围如下：$Ca(OH)_2$ 为 pH＞13；$Mg(OH)_2$ 为 pH＞11；$Al(OH)_3$ 为 pH≥5.2；$Fe(OH)_3$ 为 pH≥4.1。而 pH＞9 时，$Al(OH)_3$ 又开始溶解。故本实验先用 $2mol \cdot L^{-1}$HCl 溶解茶叶灰，然后用浓 $NH_3 \cdot H_2O$ 将其滤液调至 pH 值为 7 左右，此时，只有铝和铁的氢氧化物完全沉淀。过滤后，Mg^{2+}和 Ca^{2+}留在滤液中，从滤液中可分别鉴定 Mg^{2+}和 Ca^{2+}。沉淀与过量的 $2mol \cdot L^{-1}$NaOH 溶液反应，由于 $Al(OH)_3$ 具有两性，因而又可将 Al^{3+}分离并鉴定之。

茶叶灰用浓 HNO_3 溶解后，从滤液中可检出磷元素。紫菜灰用醋酸溶解后，从滤液中可检出碘元素。

【仪器和药品】

酒精喷灯，研钵，蒸发皿，电子天平，烧杯（100mL）。

（除特别注明外，试剂浓度单位为 $mol \cdot L^{-1}$）茶叶，紫菜，HCl (2)，$(NH_4)_2C_2O_4$ (0.5)，$NH_3 \cdot H_2O$(浓)，镁试剂，铝试剂，KSCN (饱和)，$K_4[Fe(CN)_6]$ (0.1)，氯水，CCl_4，HNO_3(浓)，钼酸铵试剂，HAc (2)。

【实验内容】

1．茶叶中钙、镁、铝和铁元素的分离和鉴定

（1）称取 15g 干燥的茶叶，放入蒸发皿中，用酒精喷灯加热充分灰化（在通风橱中进行），并移入研钵磨细（取出少量茶叶灰以作鉴定 P 元素用）。然后将 15mL HCl 加入灰中搅拌，过滤此盐酸溶液。

（2）设计一实验方案，分离并鉴定上面滤液中 Ca、Mg、Al 和 Fe 4 种元素，并实验之。

2．茶叶中磷元素的鉴定

取一药匙茶叶灰于 100mL 烧杯中，用 2mL 浓 HNO_3 溶解，再加入 30mL 蒸馏水，过滤后得透明溶液，鉴定 PO_4^{3-} 的存在。

3．紫菜中碘元素的鉴定

（1）取 10g 左右的紫菜进行灰化（在通风橱中进行）。

（2）取一药匙紫菜灰于 100mL 烧杯中，放入 10mL HAc 溶液，稍加热溶解，过滤。

（3）在滤液中鉴定 I^-的存在。

【问题与讨论】

1．茶叶和紫菜中还含有哪些元素？可用何种方法鉴定？

2．从茶叶中分离和鉴定钙、镁、铝和铁 4 种元素时，各步反应的条件应如何控制？

Experiment 30 Separation and Identification of Elements from Tea and Laver

【Objectives】

1. To understand the methods of separation and identification of some elements from tea and laver.

2. To become more interested in nature and chemistry.

【Introduction】

Organisms, such as plants, mainly consist of C, H, O, N, P, I and other metal elements. Before separating and determining the elements, the plant sample must be treated by dry ashing and a series of other treatments. In this experiment, five elements, Ca, Mg, Al, Fe and P, will be identified from tea; the element I will be identified from laver.

To completely precipitate the hydroxides of Ca, Mg, Al and Fe, an appropriate pH range is necessary. For $Ca(OH)_2$, the pH should be more than 13; for $Mg(OH)_2$, the pH should be more than 11; for $Fe(OH)_3$, the pH should be equal to or more than 4.1; for $Al(OH)_3$, the pH should be equal to or more than 5.2. But when the pH is over 9, $Al(OH)_3$ starts to dissolve again. In this experiment, the ashes of the tea should be dissolved in $2mol \cdot L^{-1}$ HCl. Then adjust the pH of the filtrate to about 7 with concentrated $NH_3 \cdot H_2O$. At this pH, only the hydroxides of Al and Fe precipitate completely. After filtration, Mg^{2+} and Ca^{2+} are in the filtrate, so they can be identified respectively. Use the precipitates $Al(OH)_3$ and $Fe(OH)_3$ to react with $2mol \cdot L^{-1}$NaOH. Because $Al(OH)_3$ is amphoteric, it will dissolve. Therefore, we can separate and identify Al and Fe.

After dissolving the ashes of tea in concentrated HNO_3 and filtration, the element P can be identified from the filtrate. For laver, after dissolving the ashes with acetic acid, the element I can be identified from the filtrate, too.

【Apparatus and Chemicals】

Bunsen burner, mortar, evaporating dish, electronic balance, beaker (100mL).

（The following reagents utilize the concentration unit $mol \cdot L^{-1}$）Tea, laver, HCl (2), $(NH_4)_2C_2O_4$ (0.5), $NH_3 \cdot H_2O$ (concentrated), magneson, aluminon, KSCN (saturated), $K_4[Fe(CN)_6]$ (0.1), Chlorine water, CCl_4, HNO_3(concentrated), $(NH_4)_2MoO_4$, HAc (2).

【Experimental Procedure】

1. Separation and identification of Ca, Mg, Al and Fe in tea

(1) Weigh 15g of dry tea in an evaporating dish, and heat to ashes with a Bunsen burner in a fume hood. Place the ashes in a mortar and grind them into powder. Save a small amount of ashes for the identification of the element P. Then add 15mL of HCl to the ashes, stir and filter. Save the filtrate.

(2) Design an experimental scheme to separate and identify Ca, Mg, Al and Fe from the above filtrate.

2. Identification of the element P in tea

Place a small amount of ashes in a 100mL beaker, and dissolve with 2mL of concentrated HNO_3. Add 30mL of distilled water and filter. The filtrate should be transparent. Identify the presence of PO_4^{3-}.

3. Identification of the element I in tea

(1) Weigh 10g of laver, and heat to ashes in a fume hood.

(2) Place a small amount of ashes in a 100mL beaker, and dissolve with 10mL of HAc solution. Heat slightly to dissolve the ashes, and then filter.

(3) Identify I^- from the filtrate.

【Questions and Discussion】

1. What are the other elements in tea and laver? What method can be used to identify those elements?

2. When separating and identifying Ca, Mg, Al and Fe from tea, how do we control the reaction conditions in each step?

实验31 固相合成硒芳香杂环化合物

【实验目的】

1．了解固相合成方法。

2．了解硒芳香杂环化合物的合成。

【实验提要】

固相化学反应是指有固体物质直接参与的反应，它既包括经典的固-气反应和固-液反应，也包括新型的固-固反应。所有固相化学反应都是非均相反应。固相反应是有机合成中的研究热点，主要特色反应表现在钯催化的偶联反应、杂环合成、环加成、烯醇烷基化反应等方面。固相化学反应，特别是新型的固-固反应，所得到的粗产物没有受到任何溶剂的影响，大大简化了反应过程，省去了传统溶液化学反应中耗时的纯化和分离步骤，只需过滤和洗涤就可除去多余的试剂和副产物，因此固相化学反应是环境友好的绿色反应，其产率往往比相应的液相反应高。

芳香二胺类化合物与二氧化硒生成硒二唑化合物的反应称为Hinsberg反应，是合成该系列化合物的最常用、最经典的方法，也是合成有机硒的重要方法之一。可以用以下的化学反应方程式表示：

$$\text{R–}C_6H_3(NH_2)_2 + SeO_2 \longrightarrow \text{R–}C_6H_3N_2Se + 2H_2O$$

传统上，该类反应是在乙醇或稀酸的液相环境中进行，反应涉及亲核进攻、闭环、消去等多步反应，过程复杂。固相化学反应速率主要取决于固相中物质向反应位点的传递，依赖于新相的成核作用，通常物质在固态中的扩散速率远远低于液态或气态，因而无溶剂的固相反应通常显得困难。然而，Hinsberg反应却能在固相中顺利进行且纯度较高，这是十分有趣的。该类反应的产物——硒二唑化合物分子中的两个氮原子容易质子化，在酸性环境中容易产生质子化的副产物，给分离纯化带来一定的困难，而固相反应却避免了质子化副产物的形成。尽管Na_2SeO_3也可以与芳香二胺发生Hinsberg反应，但在固相反应中使用SeO_2为原料更为有利，因为反应的唯一副产物是H_2O，容易处理，这也是固相中进行的Hinsberg反应产物纯度较高的主要原因。

【仪器和药品】

ELEMENTAR Vario EL型元素分析仪（EA），Bruker Equinox 55型红外光谱仪（IR），X射线粉末衍射仪（XRD），电感耦合等离子光谱发生仪（ICP）。

芳香二胺类化合物, SeO_2, 环已烷及其余试剂均为AR级。实验用水均为石英亚沸蒸馏器所得的二次蒸馏水。

【实验内容】

在室温下，分别将研磨成细粉的邻苯二胺化合物与二氧化硒按摩尔比1∶1均匀混合于研钵中，充分研磨，在反应过程中，可以观察到明显剧烈的变化，以XRD跟踪反应进程，监测结果表明，研磨约30min，反应完成，得相应的化合物。将合成产物溶解于环已烷中，过滤（目的：除去可能存在的不溶性杂质），所得滤液用二次蒸馏水洗涤3次（用何仪器？），蒸发除去环己烷，产物无需进一步纯化，即得纯品，干燥、称量、计算产率，避光保存备用。

将反应物和固相反应的产物分别做XRD、IR实验；对固相反应所得产物用ICP和EA分别测定Se和C、H、N的含量。

【问题与讨论】

1. Na_2SeO_3和SeO_2均可以和芳香二胺发生Hinsberg反应。在水溶液中的合成反应通常选用Na_2SeO_3，而在固相反应中通常选用SeO_2，为什么？

2. 讨论硒二唑化合物中，Se和N的键合状态。

Experiment 31 Solid-Phase Synthesis of Heterocyclic Aromatic Selenium Compounds

【Objectives】

1. To understand the method of solid-phase synthesis.
2. To understand the synthesis of heterocyclic aromatic selenium compounds.

【Introduction】

If we use a solid as a reactant in a reaction, it is called a solid-phase reaction, such as solid-solid reaction, solid-gas reaction, and solid-liquid reaction. Solid-phase reactions are all heterogeneous. In chemistry, solid-phase synthesis is a method in which molecules are bound on a bead and synthesized, step by step, in a reactant solution. Compared with normal synthesis in a liquid state, it is easier to remove excess reactant or byproduct. This method is used for the synthesis of peptides, DNA, and other molecules that need to be synthesized in a certain alignment. Moreover, this method is also used in combinatorial chemistry.

The reaction between aromatic diamines and selenium dioxide is called Hinsberg reaction. It is both a classic and important method to synthesize selenadiazole compounds. Here is the chemical equation:

R–(benzene ring with two NH_2) + SeO_2 ⟶ R–(benzo ring fused with N–Se–N ring) + $2H_2O$

Hinsberg reactions usually occur in the presence of ethanol or dilute acid, and involve several complicated steps such as nucleophilic (electrophilic) attack, closed loop (ring-closure), and elimination. The rate of solid-phase reaction depends on the reaction sites, nucleation of new phase, etc. Generally speaking, the diffusion rate in a solid state is much lower than in a liquid or gas state, so a solid-phase reaction is usually more difficult. However, it is interesting that Hinsberg reactions can occur in solid phase and produce pure product. Because the two nitrogen atoms in the product selenadiazole are easily protonated to produce byproduct, it is difficult to separate and purify the product in a liquid state. Aromatic diamines can also react with Na_2SeO_3 to produce selenadiazole compounds in the Hinsberg reaction, but SeO_2 is a better reactant because the only byproduct is H_2O, which is easy to remove and the product, selenadiazole, is highly pure.

【Apparatus and Chemicals】

ELEMENTAR Vario EL elementary analyzer (EA), Bruker Equinox55 infrared spectrometer (IR), X-ray diffraction instrument (XRD), Inductive Coupled Plasma Emission Spectrometer(ICP).

Aromatic diamines, SeO_2, cyclohexane and other reagents are all analytically pure. Double distilled water.

【Experimental Procedure】

Grind *o*-phenylenediamine and SeO_2 to powder respectively under room temperature, and mix them together in a mortar at the ratio of 1 : 1 (mol/mol). Completely grind the mixture, and observe the reaction. XRD shows that the reaction should be accomplished in 30 minutes. Dissolve the product in cyclohexane, and filter to remove insoluble impurities. Then use double distilled water to rinse the filtrate 3 times. (What glassware should be used?) Finally evaporate to remove cyclohexane, and the pure product can be obtained. Dry the product and calculate the yield. Keep the product away from light.

Use XRD and IR to characterize the reactants and the product. Use ICP and EA to determine Se, C, H, and N in the product.

【Questions and Discussion】

1. Both Na_2SeO_3 and SeO_2 can be used in the Hinsberg reaction. But Na_2SeO_3 is usually used in liquid-phase synthesis, SeO_2 is usually used in solid-phase synthesis. Why?

2. Discuss the bonds between Se and N in selenadiazole compounds.

实验32 纳米硫的制备

【实验目的】

1．了解纳米硫的制备方法。

2．了解纳米粒子的表征方法。

【实验提要】

硫是一种重要的化学和生物学活性物质，熔点为112~120℃。与经典的晶态单质硫相比，纳米态单质硫（简称纳米硫）由于粒子的比表面积增大，表面原子具不饱和性，易于与其它原子相结合而趋于稳定，因而具有更高的活性。在去除重金属毒性、清除自由基、抗氧化、消毒杀菌等方面具有功效。

氨基酸有活性氨基和羧基，而硫外层有空的3d 轨道，与氨基酸中的氧、氮等原子成键时可形成较弱的d-p π配键，从而对纳米硫粒子起表面修饰和稳定作用。

本实验在超声波条件下，将升华硫溶解于无水乙醇中，过滤获得饱和的均匀溶液，并加入含硫氨基酸作为表面修饰剂和稳定剂，获得由含硫氨基酸（胱氨酸和蛋氨酸）修饰的纳米硫粒子，粒径集中在250nm以内。

【仪器和药品】

超声波清洗器，离心机，970CRT荧光分光光度计，马尔文粒度分析仪，TEANAI-10型透射电子显微镜。

碘量瓶（250mL），长颈漏斗，比色管，烧杯，滴定管，定量滤纸，聚偏氟乙烯微孔滤膜（F型）。

升华硫(纯度>99.5%)，无水乙醇(AR)，蛋氨酸，胱氨酸，实验用水均为二次蒸馏水。

【实验内容】

1．升华硫的溶解

将100mL无水乙醇与1g升华硫加入250mL碘量瓶中，密封，在超声波清洗器中超声30min，最后得一黄色混合溶液。将其迅速过滤，得一澄清溶液，即纳米硫澄清溶液。

2．氨基酸修饰的纳米硫制备

（1）配制蛋氨酸溶液：将0.2997g 蛋氨酸溶解在80mL 二次蒸馏水中(该溶液的浓度为$0.025mol\cdot L^{-1}$)，利用$0.01mol\cdot L^{-1}$盐酸溶液与$0.01mol\cdot L^{-1}$氢氧化钠溶液，调节蛋氨酸溶液的pH值为5.74，即蛋氨酸等电点。

（2）取9支离心管并编号，往9支离心管中各加入5mL纳米硫澄清溶液，并超声10 min。超声情况下依次向离心管中加入0.1mL、0.2mL、0.3mL、0.5mL、1mL、2mL、3mL、4mL、5 mL 蛋氨酸溶液，并继续超声10min。

（3）将以上9支离心管进行离心，离心条件为：25℃，$5000r\cdot min^{-1}$，10min。过滤，得最终产品。

（4）设计实验，获得胱氨酸修饰的纳米硫。

3．纳米硫的形貌表征

对氨基酸修饰的纳米硫利用透射电镜（Transmission Electron Microscope，TEM）进行形貌观察（条件允许，最好即做即测定）；利用粒度分析仪测定纳米硫的粒径及Zeta电位；利用荧光分光光度计（同步扫描，$\Delta\lambda=\lambda_{ex}-\lambda_{em}=0$）测量纳米硫的共振瑞利散射光谱（条件允许，在制备后的第 1h、5h、10h、24h、48h 进行测定）。

【问题与讨论】

1．对不同氨基酸修饰的纳米硫，其粒径、形貌有何不同？

2．除氨基酸外，还有哪些物质可作为纳米硫的修饰剂？

Experiment 32 Preparation of Sulfur Nanoparticles

【Objectives】

1. To understand the method of preparation of sulfur nanoparticles.

2. To understand the method of characterizations of nanoparticles.

【Introduction】

Sulfur is an essential element with the melting point between 112℃ and 120℃. Valuable sulfur properties have been widely applied for many years in medicine, agriculture and the chemical industry. Nanotechnology improved the applications of sulfur because sulfur nanoparticles have high activities and can be used in detoxicating heavy metals, scavenging free radicals, as antioxidants and antimicrobials, etc.

Amino acids have active amino groups and carboxyl groups. Sulfur has empty 3d orbits, so oxygen or nitrogen, which is from amino acids, can form d-p π coordination bonds with sulfur. Therefore, amino acids can both decorate and stabilize sulfur nanoparticles.

In this experiment, sublimed sulfur is dissolved in absolute ethanol with ultrasound, so that a saturated solution can be obtained. Next add a sulfur-containing amino acid, methionine or cystine, to obtain amino acid decorated sulfur nanoparticles. The diameters of the nanoparticles are around 250nm.

【Apparatus and Chemicals】

Ultrasonic cleaner, 970CRT fluorospectro photometer, Malvern Zetasizer Nano ZS instrument, TEANAI-10 transmission electron microscope (TEM).

Iodine flask (250mL), long neck funnel, color comparison cylinder, beaker, buret, quantitative filter paper, PVDF millipore filter (F type).

Sublimed sulfur possessing a purity equal to or more than 99.5%, absolute ethanol (AR), methionine, cystine, double distilled water.

【Experimental Procedure】

1. Dissolution of sublimed sulfur

Add 100mL of absolute ethanol and 1g of sublimed sulfur to an iodine flask. Seal the flask and put it in the ultrasonic cleaner for 30min. A yellow mixture should be obtained. Filter the solution with the PVDF millipore filter. A clear solution of sulfur nanoparticles is obtained.

2. Preparation of sulfur nanoparticles decorated with amino acid

(1) Preparation of methionine: Dissolve 0.2997g of methionine in 80mL of double distilled water. The concentration of the obtained solution is $0.025mol \cdot L^{-1}$. Use $0.01mol \cdot L^{-1}$ HCl and

0.01mol·L^{-1} NaOH to adjust the pH of the solution to 5.74, which is the isoelectric point of methionine.

(2) In each of 9 centrifuge tubes, add 5mL of sulfur nanoparticles solution from step 1. Then put the centrifuge tubes in the ultrasonic cleaner for 10min. With ultrasound operating, add 0.1mL, 0.2mL, 0.3mL, 0.5mL, 1mL, 2mL, 3mL, 4mL and 5mL methionine solution to the 9 tubes, respectively. Continue ultrasound operation for 10min.

(3) Centrifuge the 9 test tubes at 25℃, 5000r·min^{-1}, for 10min. Filter to obtain the final products.

(4) Design an experiment to obtain cystine decorated sulfur nanoparticles.

3. Characterizations of sulfur nanoparticles

First, characterize the sulfur nanoparticles decorated with amino acid with the transmission electron microscope (TEM). If possible, TEM characterization should be done as soon as the product is prepared. Second, use the Malvern Zetasizer Nano ZS instrument to determine the size distribution, and Zeta potential of the nanoparticles. Finally, use the fluorospectro photometer to obtain the resonance Rayleigh scattering spectra (synchronous scanning, $\Delta\lambda = \lambda_{ex} - \lambda_{em} = 0$). If possible, obtain the spectra at 1h, 5h, 10h, 24h and 48h after preparation of products.

【Questions and Discussion】

1. Use different amino acids to decorate sulfur nanoparticles. What are the differences in the diameters and morphology of nanoparticles?

2. Except for amino acids, which substances can decorate sulfur nanoparticles?

实验 33　纳米硒的制备

【实验目的】

1．了解利用抗坏血酸（维生素 C）制备纳米硒溶胶的方法。

2．了解不同修饰剂对纳米硒的修饰。

【实验提要】

硒是人体必需微量元素。硒具有抗氧化、调节免疫、抗有害重金属等重要生物功能。早在 20 世纪 30 年代，人们就发现，在富硒环境中的某些微生物如光合细菌、微藻、真菌等可以把 Se(Ⅳ) 或 Se(Ⅵ)还原为单质硒(Se^0)，也可以把 Se^0 转化为气态的甲基硒。研究表明，纳米态单质硒（简称纳米硒）是新型的硒形态，其生物利用性好，生物活性高。

在酸性水溶液中过量的维生素 C(又称抗坏血酸)可将亚硒酸（H_2SeO_3）还原为纳米硒。H_2SeO_3 由二氧化硒（SeO_2）与水分子结合而得到。其反应方程式为：

$$H_2SeO_3 \ + \ 2 \ [\text{ascorbic acid: OH, CHCH}_2\text{OH, O, O, HO, OH}] \longrightarrow Se^0 \ + \ 2 \ [\text{OH, CHCH}_2\text{OH, O, O, O, O}] \ + \ 3H_2O$$

H_2SeO_3被 Vc 还原成为纳米硒，在自然条件下，生成的纳米硒会不断发生聚集，纳米硒的颗粒不断增大。

在修饰剂存在下，纳米硒的生成和聚集都受到修饰剂的调控。通过控制反应条件和选择合适的表面修饰剂，可以对纳米硒的粒径进行调控，得到粒径小、粒径分布范围窄的液相纳米硒体系。

例如，在水溶液中，利用半胱氨酸为修饰剂获得的纳米硒，在室温下放置 7 天后粒径稍有增大但并不发生聚沉，粒径维持在 100~150 nm。利用谷胱甘肽（GSH）还原并修饰的纳米硒，粒径为 110nm±20 nm。上述两种物质均可以作用在纳米硒粒子的表面，使纳米粒子带上一定的电荷，从而使其在水相中稳定存在。

【仪器和药品】

超声波清洗器，970CRT 荧光分光光度计，BI9000AT 型激光散射仪，TEANAI-10 型透射电子显微镜。

维生素 C，SeO_2，实验用水均为二次蒸馏水。

【实验内容】

1．储备液的配制

维生素 C 配制成浓度为 $0.1mol \cdot L^{-1}$ 的储备液；Se(Ⅳ)储备液由 SeO_2 配制成浓度为 $4 \times 10^{-4} mol \cdot L^{-1}$ 的储备液。

2．纳米硒溶胶的制备

在 5 个 25mL 容量瓶中，分别依次加入 12.5mL 二次蒸馏水和一定量的 Vc 储备液，轻轻摇匀；然后分别依次加入 0.25mL、0.5mL、1.0mL、2.0mL、4.0mL 浓度为 $4 \times 10^{-4} mol \cdot L^{-1}$ 的 Se(Ⅳ)储备液，轻轻摇匀，加二次蒸馏水定容。得到一系列 Se 浓度分别为 $4\mu mol \cdot L^{-1}$、$8\mu mol \cdot L^{-1}$、$16\mu mol \cdot L^{-1}$、$32\mu mol \cdot L^{-1}$、$64\mu mol \cdot L^{-1}$ 的 Vc-纳米硒溶胶。

3．纳米硒的形貌表征

对纳米硒溶胶利用透射电镜进行形貌观察；利用激光散射仪测定纳米硒的粒径及其分布情况；利用荧光分光光度计（同步扫描，$\Delta\lambda=\lambda_{ex}-\lambda_{em}=0$）测量纳米硒的共振瑞利散射光谱。

【问题与讨论】

1．讨论抗坏血酸对纳米硒的粒径调控作用。

2．哪些物质可作为纳米硒的修饰剂？

Experiment 33 Preparation of Selenium Nanoparticles

【Objectives】

1. To understand the method of preparation of selenium nanoparticles from ascorbic acid.

2. To understand the surface decorations of selenium nanoparticles by different decorators.

【Introduction】

Selenium (Se) is an essential micronutrient for humans. It has important biological activities such as antioxidant, immunoregulation, and anticancer. In the 1930s, it was discovered that some microorganisms, such as photosynthetic bacteria, microalgae and fungi, can reduce Se(Ⅳ) or Se(Ⅵ) to elemental selenium (Se^0). Recently, Se nanoparticles are received additional attention due to their excellent biological activities and low toxicity.

In acidic solution, H_2SeO_3 can be reduced by excessive ascorbic acid, which is vitamin C (Vc), to Se nanoparticles. H_2SeO_3 is obtained by the reaction of SeO_2 and water. Here is the reaction:

H_2SeO_3 + 2 [structure: O, O, OH, $CHCH_2OH$, HO, OH] ⟶ Se^0 + 2 [structure: O, O, OH, $CHCH_2OH$, O, O] + $3H_2O$

After Se nanoparticles are formed, they will aggregate forming larger particles.

Surface decorations of Se nanoparticles control their size and improve their stability in aqueous solution.

For example, Se nanoparticles decorated with cysteine in aqueous solution can be stable at room temperature for seven days without aggregation. The diameters of the nanoparticles range from 100 nm to 150 nm. The average diameter of the Se nanoparticles decorated with glutathione (GSH) is 110nm±20 nm. Therefore, Se nanoparticles become more stable in aqueous solution by decorating the surface with cysteine and glutathione.

【Apparatus and Chemicals】

Ultrasonic cleaner, 970CRT fluorospectro photometer, BI9000AT laser scattering apparatus, TEANAI-10 transmission electron microscope (TEM).

Ascorbic acid, SeO_2，double distilled water.

【Experimental Procedure】

1. Preparation of stock solutions

Use ascorbic acid to prepare a 0.1mol·L^{-1} Vc stock solution, and SeO_2 to prepare a 4×10^{-4}

$mol \cdot L^{-1}$ Se(Ⅳ) stock solution.

2. Preparation of Se nanoparticles

In each of five 25mL of volumetric flasks, add 12.5mL double distilled water and an appropriate amount of Vc stock solution. Gently shake well. Then add 0.25mL, 0.5mL, 1.0mL, 2.0mL, 4.0mL of $4\times10^{-4} mol \cdot L^{-1}$ Se(Ⅳ) stock solution, respectively. Gently shake well. Finally add double distilled water to the 25mL mark. A series of sols of Se nanoparticles are obtained. Their Se concentrations are $4\mu mol \cdot L^{-1}$, $8\mu mol \cdot L^{-1}$, $16\mu mol \cdot L^{-1}$, $32\mu mol \cdot L^{-1}$ and $64\mu mol \cdot L^{-1}$, respectively.

3. Characterization of Se nanoparticles

Use TEM to show the morphology of the products. A laser scattering apparatus is used to determine the size and distribution of the nanoparticles. A fluorospectro photometer is used to obtain the resonance Rayleigh scattering spectra (synchronous scanning, $\Delta\lambda = \lambda_{ex} - \lambda_{em} = 0$).

【Questions and Discussion】

1. Discuss the effects of Vc on the size of the Se nanoparticles.

2. Which substances can be the decorators of Se nanoparticles?

实验 34　立德粉（锌钡白）的制备

【实验目的】

了解立德粉的制备原理和方法。

【实验提要】

立德粉又名锌钡白，是 ZnS 和 $BaSO_4$ 的等物质的量混合物，它大量用作涂料、白色颜料和橡胶制品的填料。

立德粉由硫化钡和硫酸锌反应而成：

$$BaS+ZnSO_4 = ZnS\downarrow(白色)+BaSO_4\downarrow(白色)$$

立德粉的工业制法如下：

1．焙烧闪锌矿（主要是 ZnS）制备 ZnO 粉

$$2ZnS+3O_2 \xrightarrow[焙烧]{900\sim1100℃} 2ZnO+2SO_2\uparrow$$

闪锌矿中含有 Fe、Mn、Cu、Ni、Cd 和 Pb 的硫化物，在焙烧过程中也转变为相应的氧化物。

2．由焙烧所得矿粉（含 ZnO 约 90%）与 H_2SO_4 作用制得硫酸锌溶液

$$ZnO+H_2SO_4 = ZnSO_4+H_2O$$

矿粉中含有的 Fe、Mn、Cu、Ni、Cd、Pb 的氧化物杂质与 H_2SO_4 反应生成相应的硫酸盐，除 $PbSO_4$ 外，均溶于水而混在 $ZnSO_4$ 溶液中。这些金属离子会与 BaS 反应生成有色的硫化物（NiS、FeS、CuS 都显黑色，MnS 显浅红色、CdS 显黄色），严重影响立德粉的质量，因此必须把它们从硫酸锌溶液中分离除去。

3．硫酸锌溶液的净化

应用氧化还原、盐类水解等方法除去金属离子杂质。

（1）粗氧化（除铁、锰杂质）　用高锰酸钾将 Fe^{2+} 氧化为 Fe^{3+}。

$$10FeSO_4+2KMnO_4+8H_2SO_4 = 5Fe_2(SO_4)_3+2MnSO_4\downarrow+K_2SO_4+8H_2O$$

在 pH 值 5.2 的条件下，$Fe_2(SO_4)_3$ 水解为 $Fe(OH)_3$ 沉淀：

$$Fe_2(SO_4)_3+6H_2O = 2Fe(OH)_3\downarrow+3H_2SO_4$$

与此同时，杂质 $MnSO_4$ 与 $KMnO_4$ 作用生成 MnO_2 沉淀：

$$3MnSO_4+2KMnO_4+2H_2O = K_2SO_4+5MnO_2\downarrow+2H_2SO_4$$

加热有利于水解反应的进行和破坏胶体的形成，过滤可以除去 $Fe(OH)_3$ 和 MnO_2 沉淀。

（2）置换　用 Zn 粉把溶液中的 Cd^{2+}、Ni^{2+}、Cu^{2+} 置换出相应的金属而除去。

$$CdSO_4+Zn = Cd\downarrow+ZnSO_4$$

$$NiSO_4+Zn = Ni\downarrow+ZnSO_4$$

$$CuSO_4+Zn = Cu\downarrow+ZnSO_4$$

加热有利于置换反应的进行，Cu^{2+}、Cd^{2+} 易被置换出来，而 Ni^{2+} 要加热到 90℃置换才较完全。

（3）精氧化（进一步除铁、锰杂质）　用漂白粉可将溶液中残留的微量 Fe^{2+}、Mn^{2+} 进一

步氧化除去。

$$2MnSO_4+Ca(ClO)_2+2H_2O \xlongequal{} CaCl_2+2MnO_2\downarrow+2H_2SO_4$$

$$4FeSO_4+Ca(ClO)_2+2H_2SO_4 \xlongequal{} CaCl_2+2Fe_2(SO_4)_3+2H_2O$$

$$Fe_2(SO_4)_3+6H_2O \rightleftharpoons 2Fe(OH)_3\downarrow+3H_2SO_4$$

4．BaS 溶液的制备

首先将重晶石（$BaSO_4$）与煤粉（C）混合，在高温下焙烧而得 BaS 熔块：

$$BaSO_4+4C \xlongequal{焙烧} BaS+4CO\uparrow$$

然后用热水浸泡 BaS 熔块即得 BaS 溶液。熔块中的杂质（SiO_2、CaO、$CaSiO_3$、Fe_2O_3、金属硫化物等）不溶于水，过滤即得 BaS 溶液。

BaS 溶液易吸收空气中的 CO_2，也易被空气氧化生成多硫化钡，因此应尽快用来合成立德粉。

5．立德粉的合成

将精制的 $ZnSO_4$、BaS 溶液按比例混合。

【仪器和药品】

（除特别注明外，试剂浓度单位为 $mol\cdot L^{-1}$）矿粉（或粗 ZnO 粉），ZnO 粉（纯），锌粉（工业用），漂白粉，BaS，H_2SO_4(2)，$NH_3\cdot H_2O$(2)，石灰水(饱和)，KI(0.1)，KSCN(0.1)，$KMnO_4$(0.01)，丁二酮肟试剂(质量分数为 1%)。

【实验内容】

1．硫酸锌溶液的制备

（1）由矿粉制硫酸锌溶液　称取由闪锌矿焙烧所得的矿粉（或市售的粗 ZnO）8g 放在烧杯内，在搅拌下慢慢加入 $2mol\cdot L^{-1}$ H_2SO_4 50~60mL，不断搅拌 20~30min，并用 ZnO 调节溶液的 pH 值为 5~6，过滤，滤液供下面用。

（2）硫酸锌溶液的净化

① 粗氧化　将上面的滤液在加热搅拌下慢慢加入 $KMnO_4$ 溶液至呈微红色（检验溶液的 pH 值有无变化，若 pH 值降低，可用饱和石灰水调节 pH＝5~6）过滤，滤液供下面用。

② 置换　将上面的滤液加热至 80~90℃，在不断搅拌下加入 0.5g 锌粉。检查溶液中的 Cd^{2+}、Ni^{2+}是否已除尽。Ni^{2+}的检验方法：可取数滴清液，加 $2mol\cdot L^{-1}$ 氨水 2 滴，再加丁二酮肟 2~3 滴，如出现红色沉淀，表示含有 Ni^{2+}，如不出现红色，表示 Ni^{2+}、Cd^{2+}均已除尽（Cd^{2+}比 Ni^{2+}易被 Zn 置换出来除去）。如未除尽，可再加少量锌粉并加热搅拌直到除尽为止，过滤，滤液供下面用。

（3）精氧化　将上面的滤液加热至 70~80℃，滴加适量的漂白粉溶液，直至取出数滴清液加入 KI 和 H_2SO_4 数滴呈现黄色为止，设法检验溶液中的 Fe^{3+}是否已被除尽。如 Fe^{3+}被除尽，表示已制得纯净的 $ZnSO_4$ 溶液，过滤所得溶液可供合成立德粉用。

2．硫化钡溶液的制备

称取 17g 硫化钡，加入约 50mL 热水，在不断搅拌下加热 15~20min，抽滤，所得硫化钡溶液供下面合成立德粉用。

3．立德粉（锌钡白）的合成

在 200mL 烧杯中，先加入 BaS 溶液约 20mL，然后加入约等体积的 $ZnSO_4$ 溶液，检验溶液的 pH 值，交替加入 BaS 和 $ZnSO_4$ 溶液，以调节溶液的 pH 值始终维持在 8~9 之间，将所得的锌钡白进行抽滤、吸干、称量、回收。

【问题与讨论】

1．制备 $ZnSO_4$ 溶液的过程中，为什么要控制溶液的 pH 值为 5~6？

2．在锌钡白合成的过程中，为什么溶液要保持碱性？如何控制 pH 值在 8~9 之间？

Experiment 34 Preparation of Lithopone

【Objectives】

To learn the method of preparation of lithopone.

【Introduction】

Lithopone is composed of a mixture of barium sulfate and zinc sulfide. It is widely used in pigments and coatings.

Lithopone is a co-precipitate of barium sulfate and zinc sulfide:

$$BaS+ZnSO_4 = ZnS\downarrow (white)+BaSO_4\downarrow (white)$$

Lithopone can be prepared by the following steps:

1．Preparation of ZnO powder by calcination of sphalerite (mainly ZnS)

$$2ZnS+3O_2 \xrightarrow{900\sim1100℃} 2ZnO+2SO_2\uparrow$$

Sphalerite contains Fe, Mn, Cu, Ni, Cd and Pb, which are also converted to the corresponding oxides during calcination.

2．Preparation of $ZnSO_4$ by using the above ZnO powder (about 90% ZnO) and H_2SO_4

$$ZnO+H_2SO_4 = ZnSO_4+H_2O$$

The oxide impurities of Fe, Mn, Cu, Ni, Cd and Pb contained in the ore powder react with H_2SO_4 to form the corresponding sulfate. All but $PbSO_4$ are dissolved in water and mixed in the $ZnSO_4$ solution. These metal ions react with BaS to form colored sulfides (NiS, FeS and CuS are black. MnS is light red. CdS is yellow), which seriously affects the quality of lithopone. Therefore, they must be separated from the zinc sulfate solution.

3．Purification of zinc sulfate solution

(1) Crude oxidation (removal of iron and manganese impurities) Fe^{2+} is oxidized to Fe^{3+} with potassium permanganate.

$$10FeSO_4+2KMnO_4+8H_2SO_4 = 5Fe_2(SO_4)_3+2MnSO_4\downarrow+K_2SO_4+8H_2O$$

$Fe_2(SO_4)_3$ is hydrolyzed to $Fe(OH)_3$ precipitate at the pH of 5.2:

$$Fe_2(SO_4)_3+6H_2O = 2Fe(OH)_3\downarrow+3H_2SO_4$$

At the same time, $MnSO_4$ reacts with $KMnO_4$ to form MnO_2 precipitate:

$$3MnSO_4+2KMnO_4+2H_2O = K_2SO_4+5MnO_2\downarrow+2H_2SO_4$$

Heating facilitates the progress of the hydrolysis and destroys the formation of colloids. $Fe(OH)_3$ and MnO_2 can be removed by filtration.

(2) Replacement Cd^{2+}、Ni^{2+}、Cu^{2+} can be removed by replacing them with Zn powder.

$$CdSO_4+Zn = Cd\downarrow+ZnSO_4$$

$$NiSO_4+Zn = Ni\downarrow +ZnSO_4$$

$$CuSO_4+Zn = Cu\downarrow +ZnSO_4$$

Heating facilitates the progress of the replacement. Cu^{2+} and Cd^{2+} can be easily replaced, but Ni^{2+} need to be heated to at least 90℃.

（3）Fine oxidation (further removal of iron and manganese impurities)　The trace amount of Fe^{2+} and Mn^{2+} remaining in the solution can be further oxidized and removed by using a bleaching powder.

$$2MnSO_4+Ca(ClO)_2+2H_2O = CaCl_2+2MnO_2\downarrow +2H_2SO_4$$

$$4FeSO_4+Ca(ClO)_2+2H_2SO_4 = CaCl_2+2Fe_2(SO_4)_3+2H_2O$$

$$Fe_2(SO_4)_3+6H_2O \rightleftharpoons 2Fe(OH)_3\downarrow +3H_2SO_4$$

4．Preparation of BaS solution

Baryte is mixed with coal in a suitable furnace to obtain BaS.

$$BaSO_4+4C = BaS+4CO\uparrow$$

The BaS is then soaked in hot water to obtain the BaS solution. The impurities (SiO_2, CaO, $CaSiO_3$, Fe_2O_3, metal sulfide, etc.) are insoluble in water, and can be removed by filtration.

The BaS solution easily absorbs CO_2 in the air and is also easily oxidized by air to form polysulfide. Therefore, it should be used as soon as possible.

5．Preparation of Lithopone

Mixing refined $ZnSO_4$ and BaS solutions in proportion will yield the lithopone.

【Apparatus and Chemicals】

Coarse ZnO powder, pure ZnO powder, zinc powder, bleaching powder, BaS, $2mol\cdot L^{-1}$ H_2SO_4, $2mol\cdot L^{-1}$ $NH_3\cdot H_2O$, saturated limewater, $0.1mol\cdot L^{-1}$KI, $0.1mol\cdot L^{-1}$ KSCN, $0.01mol\cdot L^{-1}$ $KMnO_4$, dimethylglyoxime (1%, wt/wt).

【Experimental Procedure】

1．Preparation of $ZnSO_4$ solution

（1）Weigh 8g of coarse ZnO powder in a beaker. Slowly add 50~60mL of $2mol\cdot L^{-1}$ H_2SO_4 while stirring. Stir for 20 to 30 minutes, and use ZnO to adjust the pH to 5~6. Vacuum filter the mixture.

（2）Purification of $ZnSO_4$ solution

① Crude oxidation　heat and stir the above filtrate, and slowly add $KMnO_4$ solution till pink. Test the pH of the solution, and adjust the pH to 5~6 with saturated limewater if necessary. Vacuum filter the mixture.

② Replacement　heat the above filtrate to 80~90℃. Add 0.5g of zinc power while stirring. Check if the Cd^{2+} and Ni^{2+} in the solution have been removed. Test method of Ni^{2+}: use a few drops of test solution by adding two drops of $2mol\cdot L^{-1}$ ammonia water and 2~3 drops of dimethylglyoxime. If red precipitate appears, it confirms the presence of Ni^{2+}. If no red color appears, it confirms that Ni^{2+} and Cd^{2+} have been removed. Cd^{2+} is more easily removed by Zn than Ni^{2+}. If necessary, a small amount of zinc powder may be added until Ni^{2+} and Cd^{2+} are removed. Vacuum filter the mixture, and the filtrate is used for the next step.

（3）Fine Oxidation　the above filtrate is heated to 70-80℃, and an appropriate amount of the bleaching powder solution is added dropwise. Use a few drops of the supernatant to react with KI and H_2SO_4. If a yellow color appears, it means that the bleaching powder solution is appropriate. Check whether the Fe^{3+} in the solution has been removed. If Fe^{3+} is removed, vacuum filter the mixture to obtain pure $ZnSO_4$ solution.

2．Preparation of BaS

Weigh 17g of BaS, and add about 50mL of hot water. Heat and stir for 15 to 20 minutes. Vacuum filter the mixture to obtain BaS solution.

3．Preparation of lithopone

In a 200mL beaker, first add about 20mL of BaS solution, then add about an equal volume of $ZnSO_4$ solution. Check the pH of the solution. Alternately add BaS and $ZnSO_4$ solution to maintain the pH of the solution between 8 and 9. Lithopone can be obtained after filtration.

【Questions and Discussion】

1. In the step of $ZnSO_4$ solution preparation, why do we adjust the pH of the solution to 5~6?

2. In the step of lithopone preparation, why should the solution be alkaline? How do we adjust the pH to 8~9?

附　录

附录 1　化学试剂的规格

关于化学试剂规格的划分，各国不一致。我国常用试剂等级的划分参阅下表。

国家标准	优（质）级纯（保证试剂）GR	分析纯 AR	化学纯 CP	实验试剂 LR
等　级	一级品(Ⅰ)	二级品(Ⅱ)	三级品(Ⅲ)	四级品(Ⅳ)
标　志	绿色标签	红色标签	蓝色标签	
用　途	精密的分析工作和科研工作	一般的分析工作和科研工作	厂矿的日常控制分析和教学实验	实验中的辅助试剂及制备原料

除上述 4 个等级外，还根据特殊需要而定出相应的纯度规格，如供光谱分析用的光谱纯，供核实验及其分析用的核纯等。

对于不同的试剂，各种规格要求的标准不同。但总的来说，优质纯试剂杂质含量最低，实验试剂杂质含量较高。应根据实际工作的需要，选用适当等级的试剂，既满足工作要求，又符合节约原则。

附录 2　市售酸碱浓度

溶液（化学式）	溶质摩尔质量 $M/g \cdot mol^{-1}$	溶质质量分数 $w/\%$	溶液密度 $\rho/g \cdot cm^{-3}$	溶液浓度 $c/mol \cdot L^{-1}$
冰醋酸(CH_3COOH)	60	99.5	1.05	17
浓盐酸(HCl)	36.5	36	1.18	12
浓硝酸(HNO_3)	63	70	1.42	16
发烟硝酸(HNO_3)	63	95	1.50	23
浓硫酸(H_2SO_4)	98	98	1.84	18
发烟硫酸(H_2SO_4)	98	(20% SO_3)	1.92	—
浓磷酸(H_3PO_4)	98	85	1.75	15
浓氨水(NH_3)	17	28	0.88	15

注：引自[英]J. G. 斯塔克. 化学数据手册. 北京：石油工业出版社，1980。

附录 3　水的饱和蒸气压

在密闭体系中，纯水与其水蒸气达到平衡后，水蒸气的压力称为饱和蒸气压。它随温度而变，如下表所列。

温度/°C	压力/kPa	温度/°C	压力/kPa	温度/°C	压力/kPa	温度/°C	压力/kPa
1	0.6567	26	3.3609	51	12.96	76	40.18
2	0.7058	27	3.5648	52	13.611	77	41.88
3	0.7579	28	3.7795	53	14.292	78	43.64
4	0.8134	29	4.0053	54	15.000	79	45.46
5	0.8723	30	4.2428	55	15.737	80	47.34
6	0.9350	31	4.4922	56	16.505	81	49.29
7	1.002	32	4.7546	57	17.308	82	51.31
8	1.072	33	5.0300	58	18.142	83	53.41
9	1.148	34	5.3192	59	19.011	84	55.57
10	1.228	35	5.6228	60	19.915	85	57.81
11	1.312	36	5.9411	61	20.855	86	60.11
12	1.4023	37	6.2751	62	21.834	87	62.49
13	1.4973	38	6.6249	63	22.848	88	64.94
14	1.5981	39	6.9916	64	23.906	89	67.47
15	1.7049	40	7.3758	65	25.003	90	70.094
16	1.8177	41	7.778	66	26.143	91	72.799
17	1.9371	42	8.199	67	27.325	92	75.591
18	2.0634	43	8.639	68	28.553	93	78.472
19	2.1967	44	9.100	69	29.825	94	81.445
20	2.3378	45	9.583	70	31.16	95	84.512
21	2.4864	46	10.09	71	32.52	96	87.674
22	2.6433	47	10.61	72	33.94	97	90.934
23	2.8088	48	11.16	73	35.42	98	94.293
24	2.9847	49	11.73	74	36.96	99	97.756
25	3.1671	50	12.33	75	38.54	100	101.33

注：本表原始数据引自 John A Dean. Lange's Handbook of Chemistry. 12th edition, 1979. 并按 1mmHg＝0.133 322 kPa 换算所得。

附录 4　常用的一些酸碱指示剂

名　称	变色 pH 范围	颜 色 变 化	溶液配制方法
麝香蓝（百里酚蓝）（第一次变色）	1.2～2.8	红～黄	0.1g 溶于 100mL 20%乙醇中
β-二硝基苯酚	2.4～4.0	无色～黄	0.1%水溶液
甲 基 橙	3.0～4.4	红～橙黄	0.1%水溶液
刚 果 红	3.0～5.2	蓝紫～红	0.1%水溶液
茜素红 S（第一次变色）	3.7～5.2	黄～紫	1%水溶液
甲 基 红	4.4～6.2	红～黄	0.1g 溶于 100mL 60%乙醇中
溴 酚 红	5.0～6.6	黄～红	0.1g 溶于 100mL 20%乙醇中
中 性 红	6.8～8.0	红～亮黄	0.1g 溶于 100mL 20%乙醇中
橘 黄 Ⅱ	7.6～8.9	棕～橙	0.1%水溶液

续表

名　　称	变色 pH 范围	颜 色 变 化	溶液配制方法
麝香蓝（百里酚蓝）（第二次变色）	8.0～9.6	黄～蓝	溶于 100mL 20%乙醇中
酚　　酞	8.2～10.0	无色～紫红	0.1g 溶于 100mL 60%乙醇中
茜素红 S（第二次变色）	10.0～12.0	紫～淡黄	0.1%水溶液
石　　蕊	4.0～6.4	红～蓝	2%乙醇溶液

注：*x*%水溶液是指质量分数；*x*%乙醇是指体积分数。

附录 5　常见沉淀物的 pH

（1）金属氢氧化物沉淀的 pH（包括形成氢氧配离子的大概数值）

氢氧化物	开始沉淀时的 pH 初浓度		完全沉淀时的 pH（残留离子浓度 ＜10^{-5}mol·L^{-1}）	沉淀开始溶解时的 pH	沉淀完全溶解时的 pH
	1mol·L^{-1}	0.01mol·L^{-1}			
$Sn(OH)_4$	0	0.5	1.0	13	15
$TiO(OH)_2$	0	0.5	2.0		
$Sn(OH)_2$	0.9	2.1	4.7	10	13.5
$ZrO(OH)_2$	1.3	2.3	3.8		
HgO	1.3	2.4	5.0	11.5	
$Fe(OH)_3$	1.5	2.3	4.1	14	
$Al(OH)_3$	3.3	4.0	5.2	7.8	10.8
$Cr(OH)_3$	4.0	4.9	6.8	12	15
$Be(OH)_2$	5.2	6.2	8.8		
$Zn(OH)_2$	5.4	6.4	8.0	10.5	12～13
Ag_2O	6.2	8.2	11.2	12.7	
$Fe(OH)_2$	6.5	7.5	9.7	13.5	
$Co(OH)_2$	6.6	7.6	9.2	14.1	
$Ni(OH)_2$	6.7	7.7	9.5		
$Cd(OH)_2$	7.2	8.2	9.7		
$Mn(OH)_2$	7.8	8.8	10.4	14	
$Mg(OH)_2$	9.4	10.4	12.4		
$Pb(OH)_2$		7.2	8.7	10	13
$Ce(OH)_2$		0.8	1.2		
$Th(OH)_4$		0.5			
$Tl(OH)_3$		～0.6	～1.6		
H_2MoO_4				～8	～9
稀土		6.8～8.5	～9.5		
H_2UO_4		3.6	5.1		

（2）沉淀金属硫化物的 pH

pH	被 H_2S 所沉淀的金属
1	铜组：Cu、Ag、Hg、Pb、Bi、Cd 砷组：As、Au、Pt、Sb、Se、Mo
2～3	Zn、Ti
5～6	Co、Ni
＞7	Mn、Fe

（3）溶液中硫化物沉淀时盐酸的最高浓度

硫化物	Ag_2S	HgS	CuS	Sb_2S_3	Bi_2S_3	SnS_2	CdS	PbS
盐酸浓度 /$mol \cdot L^{-1}$	12	7.5	7.0	3.7	2.5	2.3	0.7	0.35

硫化物	SnS	ZnS	CoS	NiS	FeS	MnS
盐酸浓度 /$mol \cdot L^{-1}$	0.3	0.02	0.001	0.001	0.0001	0.00008

附录 6　常见无机化合物在水中的溶解度

单位：$g \cdot (100g\ H_2O)^{-1}$

与饱和溶液平衡的固相物质	0°C	20°C	40°C	60°C	80°C	100°C
AgCl		1.5×10^{-4}				2×10^{-2}
$AgNO_3$	122	222	376	525	669	952
Ag_2SO_4	0.573	0.796	0.979	1.15	1.30	1.41
$Al_2(SO_4)_3 \cdot 18H_2O$	31.2	36.4	46.1	59.2	73.1	89.0
$BaCl_2 \cdot 2H_2O$	31.6	35.7	40.7	46.4	52.4	58.8
$Ba(OH)_2 \cdot 8H_2O$	1.67	3.89	8.22	20.94	101.4	
Br_2	4.22	3.20				
$CaCl_2 \cdot 6H_2O$	59.5	74.5				
$Ca(OH)_2$	0.185	0.165	0.141	0.116	0.094	0.077
$CaSO_4 \cdot 2H_2O$	0.1759		0.2097			0.1619
$CdCl_2 \cdot H_2O$		134.5	135.3	136.5	140.5	147.0
$CoCl_2 \cdot 6H_2O$	41.6	50.4				
$CuCl_2 \cdot 2H_2O$	70.7	77.0	83.8	91.2	99.2	107.9
$CuSO_4 \cdot 5H_2O$	14.3	20.7	28.5	40.0	55	75.4
$FeCl_2 \cdot 4H_2O$			77.3	88.7	100.0	
$FeCl_3$	74.4	91.9			525.8	535.7
$FeSO_4 \cdot 7H_2O$	15.65	26.5	40.2			
$FeSO_4 \cdot H_2O$				50.9(70°C)	43.6	37.3(90°C)
HCl	82.3		63.3	56.1		
$HgCl_2$	3.6	6.6	10.2	16.2	30.0	61.3
I_2		2.9×10^{-2}	5.6×10^{-2}			
$K_2SO_4 \cdot Al_2(SO_4)_3 \cdot 24H_2O$	3.0	5.9	11.7	24.8	71.0	
KBr	53.5	65.5	75.5	85.5	95.0	104.0
KCl	27.6	34.0	40.0	45.5	51.1	56.7
$KClO_3$	3.3	7.4	14.0	24.5	38.5	57.0
K_2CrO_4	58.2	61.7	65.2	68.8	72.1	75.6
$K_2Cr_2O_7$	5.0	12.0	26.0	43.0	61.0	80.0
$K_3[Fe(CN)_6]$	31.0	42.9	60.0	66.0		
$K_4[Fe(CN)_6]$	14.9	28.9	42.7	55.9	68.6	77.8
$KHC_4H_4O_6$	0.32	0.53	1.32	2.5	4.6	7.0

续表

与饱和溶液平衡的固相物质	0°C	20°C	40°C	60°C	80°C	100°C
K_2SO_4	7.4	11.1	14.8	18.2	21.4	24.1
$K_2C_2O_4$	25.5	36.4	43.8	53.2	63.6	75.3
KI	127.5	144	160	176	192	208
$KMnO_4$	2.83	6.4	12.56	22.2		
KNO_3	13.3	31.6	63.9	110.0	169	246
$KOH\cdot 2H_2O$	97	112				
$MgCl_2\cdot 6H_2O$	52.8	54.5	57.5	61.0	66.0	73.0
$MnCl_2\cdot 4H_2O$	63.4	73.9	88.6			
NH_4Cl	29.4	37.2	45.8	55.2	65.6	77.3
NH_4NO_3	118.3	192.0	297.0	421.0	580.0	871.0
NH_4SCN	119.8	170.0				
$(NH_4)_2SO_4$	70.6	75.4	81.0	88.0	95.3	103.3
$Na_2B_4O_7\cdot 10H_2O$	1.3	2.7		20.3		
$NaC_2H_3O_2$	119.0	123.5	129.5	139.5	153	170
$Na_2CO_3\cdot H_2O$			48.5	46.4	45.8	45.5
NaCl	35.7	36.0	36.6	37.3	38.4	39.8
$NaHCO_3$	6.9	9.6	12.7	16.4		
$NaOH\cdot H_2O$		109	129	174		
Na_2SO_3	14.4	26.3	37.2	32.6	29.4	27.9(90°C)
Na_2SO_4	4.9	19.5	48.8	45.3	43.7	42.5
$Na_2S\cdot 9H_2O$		18.8	28.5			
$Na_2S_2O_3$	52.5	70.0	102.6	206.6	246	266
$NiSO_4\cdot 7H_2O$	27.22					
$PbCl_2$	0.6728	0.99	1.45	1.98	2.62	3.34
$Pb(NO_3)_2$	38.8	56.5	75	95	115	138.8

注：本表数据录自 John A Dean. Lange's Handbook of Chemistry. 11th edition，10-8，1973。

附录 7　普通有机溶剂性质

溶剂	分子式	沸点/°C	密度(20°C)/$g\cdot cm^{-3}$
氯仿	$CHCl_3$	61.2	1.489
甲醇	CH_3OH	64.7	0.92
丙酮	CH_3COCH_3	56.7	0.791
四氯化碳	CCl_4	76.8	1.594
乙醇	C_2H_5OH	78.4	0.789
乙醚	$C_2H_5OC_2H_5$	34.6	0.708
苯	C_6H_6	80.1	0.879
二硫化碳	CS_2	46.3	1.263

注：本表数据录自 John A Dean. Lange's Handbook of Chemistry. 11th edition，7-56，1973。

附录 8 常用配合物的稳定常数表[1]

金属离子(M)与配位体(L)形或配合物的累积稳定常数的对数值($\lg\beta$)列于下面表中(20～25°C)。

$$M+L \rightleftharpoons ML,\ \beta_1=\frac{[ML]}{[M][L]}$$

$$M+2L \rightleftharpoons ML_2,\ \beta_2=\frac{[ML_2]}{[M][L]^2}$$

$$M+nL \rightleftharpoons ML_n,\ \beta_n=\frac{[ML_n]}{[M][L]^n}$$

配合物的逐级稳定常数以 K_n 表示，它与累积稳定常数β_n 的关系为$\beta_1=K_1$；$\beta_2=K_1K_2$；$\beta_n=K_1K_2K_3\cdots K_n$。

亚硫酸盐(SO_3^{2-})

金属离子	$\lg\beta_1$	$\lg\beta_2$	$\lg\beta_3$	$\lg\beta_4$	$\lg\beta_5$	$\lg\beta_6$
Ag^+	5.60	8.68	9.00			
Cu^+	7.85	8.70	9.36			
Hg^{2+}		24.07	24.96			

氨(NH_3)

金属离子	$\lg\beta_1$	$\lg\beta_2$	$\lg\beta_3$	$\lg\beta_4$	$\lg\beta_5$	$\lg\beta_6$
Ag^+	3.37	7.21				
Cd^{2+}	2.65	4.75	6.19	7.12		
Co^{2+}	2.11	3.74	4.79	5.55		
Co^{3+}	7.3	14.0	20.1	25.7	30.8	35.2
Cu^+	5.93	10.86				
Cu^{2+}	4.13	7.66	10.53	12.68		
Ni^{2+}	2.80	5.04	6.77	7.96		
Zn^{2+}	2.32	4.61	6.97	9.36		
Hg^{2+}				19.24		

氰化物(CN^-)

金属离子	$\lg\beta_1$	$\lg\beta_2$	$\lg\beta_3$	$\lg\beta_4$	$\lg\beta_5$	$\lg\beta_6$
Ag^+		21.1	21.8	20.7		
Cd^{2+}	5.18	9.6	13.92	17.11		
Co^{2+}						19.1
Co^{3+}						17.7
Cu^+		24.0	28.6	30.3		
Cu^{2+}				25		
Fe^{2+}						36.9
Fe^{3+}						43.9
Hg^{2+}	18.00	34.70	38.53	41.51		
Ni^{2+}				30.3		
Pb^{2+}				10.3		
Zn^{2+}		11.07	16.05	19.62		

[1] 摘自化学分析手册. 北京：科学出版社，1982。

硫氰酸盐(SCN^-)

金属离子	$\lg\beta_1$	$\lg\beta_2$	$\lg\beta_3$	$\lg\beta_4$	$\lg\beta_5$	$\lg\beta_6$
Ag^+	4.75	8.23	9.45	9.67		
Cu^+		12.11	9.90	10.09	9.59	9.27
Fe^{3+}	3.1	5.3	6.2			

氢氧化物(OH^-)

金属离子	$\lg\beta_1$	$\lg\beta_2$	$\lg\beta_3$	$\lg\beta_4$	$\lg\beta_5$	$\lg\beta_6$
Al^{3+}	9.27			33.03		
Bi^{3+}	12.7	15.8		35.2		
Cr^{3+}	10.1	17.8		29.9		
Cu^{2+}	7.0	13.68	17.00	18.5		
Pb^{2+}	7.82	10.85	14.58			61.0
Sb^{3+}		24.3	36.7	38.3		
Sn^{2+}	11.86	20.64	25.13			
Zn^{2+}	4.40	11.30	14.14	17.66		

氟化物(F^-)

金属离子	$\lg\beta_1$	$\lg\beta_2$	$\lg\beta_3$	$\lg\beta_4$	$\lg\beta_5$	$\lg\beta_6$
Al^{3+}	6.13	11.15	15.00	17.74	19.37	19.81
Cr^{3+}	4.36	7.70	10.18			
Fe^{3+}	5.17	9.09	12.00			
Sn^{2+}	6.26	8.76	9.25			
Sn^{4+}						~25

氯化物(Cl^-)

金属离子	$\lg\beta_1$	$\lg\beta_2$	$\lg\beta_3$	$\lg\beta_4$	$\lg\beta_5$	$\lg\beta_6$
Ag^+	3.04	5.04	5.04	5.30		
Bi^{3+}	2.43	4.43	5.78	6.21	6.69	
Cu^+		5.5	5.7			
Fe^{3+}	1.48	2.13	1.99	0.01		
Hg^{2+}	6.74	13.22	14.07	15.07		
Sb^{2+}	1.62	2.44	2.04	1.0		
Sn^{2+}	1.51	2.24	2.03	1.48		
Sn^{4+}						4

碘化物(I^-)

金属离子	$\lg\beta_1$	$\lg\beta_2$	$\lg\beta_3$	$\lg\beta_4$	$\lg\beta_5$	$\lg\beta_6$
Ag^+	6.58	11.74	13.68	13.10		
Bi^{3+}	3.63			14.95	16.80	18.80
Cd^{2+}	2.17	3.67	4.34	5.35	5.15	
Cu^+		8.85				
Hg^{2+}	12.87	23.82	27.60	29.83		
I_2	2.87(I_3^-)					
Pb^{2+}	2.00	3.15	3.92	4.47		
Sn^{2+}	0.70	1.13	2.13	2.30		2.59

草酸盐($C_2O_4^{2-}$)

金属离子	$\lg\beta_1$	$\lg\beta_2$	$\lg\beta_3$	$\lg\beta_4$	$\lg\beta_5$	$\lg\beta_6$
Ag^+	2.41					
Al^{3+}	7.26	13.0	16.3			
Cd^{2+}	4.00	5.77				
Co^{2+}	4.69	7.15				
Cr^{3+}	5.34	10.51	15.44			
Cu^{2+}	6.19	10.23				
Fe^{2+}	2.9	4.52	5.22			
Fe^{3+}	9.4	16.4	20.2			
Mg^{2+}	3.43	4.38				
Mn^{2+}	3.97	5.40				
Ni^{2+}	5.3	7.64				
Pb^{2+}		6.54				
Zn^{2+}	4.85	7.55				

乙二胺($C_2H_8N_2$)

金属离子	$\lg\beta_1$	$\lg\beta_2$	$\lg\beta_3$	金属离子	$\lg\beta_1$	$\lg\beta_2$	$\lg\beta_3$
Ag^+	4.7	7.7		Cu^{2+}	10.66	19.99	
Cd^{2+}	5.47	10.09	12.18	Fe^{2+}	4.34	7.65	9.70
Co^{2+}	5.93	10.66	13.96	Hg^{2+}	14.3	23.3	
Co^{3+}	18.7	34.9	48.7	Mn^{2+}	2.73	4.79	5.67
Cr^{3+}	16.5	<30.5		Ni^{2+}	7.52	13.84	18.33
Cu^+		11.4		Zn^{2+}	5.77	10.83	14.11

EDTA

金属离子	$\lg\beta_1$	$\lg\beta_2$	$\lg\beta_3$	金属离子	$\lg\beta_1$	$\lg\beta_2$	$\lg\beta_3$
Ag^+	7.31			Fe^{2+}	14.33		
Al^{3+}	16.11			Fe^{3+}	25.1		
Ba^{2+}	7.78			Hg^{2+}	21.8		
Bi^{3+}	27.9			Mg^{2+}	8.69		
Ca^{2+}	10.7			Mn^{2+}	14.04		
Cd^{2+}	16.46			Ni^{2+}	18.62		
Co^{2+}	16.31			Pb^{2+}	18.04		
Co^{3+}	36			Sn^{2+}	22.1		
Cr^{3+}	24.3			Sr^{2+}	8.63		
Cu^{2+}	18.8			Zn^{2+}	16.5		

附录 9　半微量定性分析的基本操作

半微量定性分析取样量较少。一般做鉴定反应时，如不严格遵守操作规则，很容易损耗而无法检出。反之，若反应器皿不净、操作不慎或试剂不纯也会造成过度检出。

定性分析操作包括分离和鉴定两部分。主要操作如下。

1．用试剂沉淀

为了进行分离，常在试液中加入某种试剂使一些离子生成难溶盐而沉淀，另一些离子则留在溶液中。

实验中欲使沉淀完全，应做到：

（1）严格按照规定的沉淀条件（如溶液的酸碱性，试剂加入顺序，温度以及如何防止生成胶体等）。否则会使沉淀不完全或不产生沉淀。

（2）加入的沉淀剂应稍微过量，以减少沉淀的溶解度，使沉淀完全。但也要防止试剂加入过多，避免因盐效应和配位效应而使沉淀溶解。

（3）沉淀反应大多数在离心试管中进行。沉淀剂应逐滴加入，同时用玻璃棒充分搅拌。搅拌可以加速沉淀，使反应均匀，且可促使胶粒凝聚。

（4）沉淀是否完全的检查方法，是在离心分离后的清液中加一滴沉淀剂，若溶液没有变浑浊，则表示沉淀完全。反之，则表示未完全沉淀，需继续加入沉淀剂至沉淀完全。

2．离心分离

参看本书第 2 章中无机化学实验基本操作中有关离心分离法的内容。

3．沉淀洗涤

洗涤的目的是为了充分除去渗入沉淀中的母液。洗涤沉淀必须注意：

（1）洗涤前，沉淀上的母液应吸净。

（2）洗涤液中应含有少量沉淀剂或与沉淀不发生反应的电解质（如 NH_4NO_3，NH_4Cl）等，以防沉淀溶解和生成胶体。

（3）洗涤次数不宜太多，每次洗涤液用量视沉淀多少而定，一般为 0.2～0.3mL。

（4）洗涤时，必须用细玻璃棒充分搅拌，壁上附着的沉淀要用玻璃棒刮下与沉淀一起洗涤。

4．沉淀溶解

沉淀中可能含有一种或几种离子的难溶盐，为了检验沉淀中的离子或进行进一步的分离，常要将沉淀全部或部分溶解。

溶解沉淀时必须注意：

（1）应逐滴加入试剂，并不断搅拌。

（2）沉淀溶解比较缓慢时，不急于加入过多的试剂，而应搅拌一段时间，或在水浴中适当加热。

（3）有些沉淀放置时间过久，会变得比刚生成的沉淀更难溶解，所以离心分离后的沉淀应随即溶解。

溶液从一个容器转入另一个容器，不可直接倾倒。因为溶液较少，需用吸管转移。

5．沉淀移取

移取方法有两种：一种是直接移取沉淀；另一种是在沉淀上加少量水后搅动，使沉淀悬浮，再用吸管吸取部分悬浮液，移入另一试管中。若检验反应不能有过多的水分，在移取沉淀后需再行离心分离。

6．溶液蒸发

为鉴定含量较少的离子，在鉴定前应将溶液浓缩。溶液浓缩一般在小烧杯中（可在小坩埚中）进行。烧杯放在石棉网中央，手持酒精灯，以小火在下面来回移动使溶液缓慢蒸发，不致溅出。

若需蒸发至干时，应在蒸发近干时停止加热，让残渣借助余热自行蒸干，以避免固体溅出，也可防止物质分解。

7．点滴反应

点滴反应可以在瓷点滴板上进行，也可在纸上进行。两种方法都很灵敏、方便。

在白色瓷板上进行有色的检验反应，颜色清楚，容易观察。瓷板上进行点滴反应需注意：

（1）反应前点滴板应洗干净并用吸管吸去孔穴中残留的水，必要时应用滤纸吸干。

（2）试液与试剂取量以1滴为宜，不可多取以免溢出。

（3）为使反应均匀，可用玻璃棒搅动。

在纸上进行点滴反应，所需试液（或试剂）的量较少，并可以提高反应的灵敏度和选择性。

纸上点滴反应的操作方法如下：

① 取定性滤纸一小块（约2cm×2cm）作为反应纸。

② 将吸有试液的毛细滴管直立并使其尖端与滤纸接触，借纸的毛细作用使试液自行流下而成斑点，斑点不宜过大（直径约2～3mm）。

③ 再用另一毛细滴管取试剂，用同法在第一斑点的中心加试剂，观察反应的结果。

纸上点滴反应的产物必须都是带色的（沉淀或可溶物）。同时，注意不能用滴管滴加溶液。

8．结晶反应的显微鉴定

各种晶体都有特征的晶形，用显微镜观察反应生成的晶体形状，可很快地作出某种离子是否存在的结论。

影响晶体生长的因素很多，这些因素不仅会影响结晶速度及晶体大小，有时还会改变结晶的形状。所以要得到一定形状的晶体，要有合适的结晶条件。一般来讲，从较稀的溶液中得到的晶体较大，晶形较好；从较浓的溶液中结晶，则晶体较细，晶形不易完整。

结晶反应的显微鉴定操作方法如下：

在干燥洁净的显微镜载玻片上，相距2cm左右各滴试液与试剂1滴，然后用细玻璃棒沟通，使试剂与试液发生缓慢的反应，结果在中间先生成晶体。观察晶形时，应将过多的溶液用滤纸吸去。

如溶液需浓缩后才能结晶，则必须使溶液在载玻片上受热蒸发。操作方法是：先滴1滴试液于载玻片的中央，然后用试管夹夹住载玻片的一端，在石棉网上方来回移动使其受热，缓慢蒸发至干，冷却后在残渣上加1滴试剂，过一段时间就会生成晶体。

观察晶形需用显微镜。显微镜的使用方法如下：

① 选择放大倍数合适的目镜及物镜（放大总倍数为目镜和物镜放大倍数之乘积）。

② 在载物台上放好载玻片，载玻片背面应擦干，以免沾污载物台。载玻片应夹好以防滑动。

③ 调好反光镜，使目镜内照明良好。

④ 调节物镜最低至离载玻片5cm左右，然后用左眼看目镜并缓慢升高镜筒，直到呈现清晰的物像为止。若镜筒升至最高点仍未看到物像，应重新将镜筒调至离载玻片5cm后再重新调节，绝不可在观察时下降镜筒，以防物镜触及载玻片。

⑤ 目镜及物镜若被沾污，应当用镜头纸擦，不能用一般的纸或布擦。

⑥ 显微镜不用时应放入箱内，物镜放在专用盒中。

9．检验气体的方法

一般用两个合在一起的表面皿，下面的表面皿上加有反应物，上面表面皿的内壁贴有试纸，用试纸检验生成气体的性质。这种检验气体的方法称为“气室法”。

10．焰色反应

某些物质能使火焰呈现特征的焰色，如钾盐呈紫色，钠盐呈黄色等。因此可以借助焰色的不同来鉴定某些离子，这种反应称为焰色反应。操作方法：

① 在小试管中放少量浓盐酸，然后将铂丝蘸取盐酸在煤气灯（或酒精喷灯）的氧化焰中心灼烧几秒钟。如此反复多次直到火焰不再呈特殊颜色为止。

② 将试液滴在点滴板上，然后用铂丝蘸取试液放在氧化焰灼烧，观察焰色。每种离子的颜色所持续的时间各不相同，一般的焰色可持续几十秒钟，但也有些离子的焰色几秒钟内即消失。观察时要特别注意。

③ 铂丝不能放在还原焰中，因为在还原焰中易生成碳化铂，使铂变脆。

④ 试验完毕用盐酸按①所述的方法把铂丝处理干净。

11．空白试验与对照试验

在鉴定某离子时，若试剂不纯或含有被鉴定的离子，则会导致结论错误。为避免这种错误，可在实验中另取一份蒸馏水代替试液，在与试液相同条件下，以相同方法对同一种离子进行鉴定。这种对比实验方法称为空白试验。它可以检查出本项实验所用的试剂和蒸馏水是否含有被检出的离子，判断试剂是否符合实验的要求。

用含有某种离子的纯盐溶液代替试液，与试液在相同条件下进行鉴定，两相对比，检查鉴定结果是否可靠，称为对照试验。通过对照试验，可更清楚地确证试液中被鉴定离子是否存在。

附录 10　常见阳离子的分析

阳离子的数目较多，干扰情况较为复杂，一般采用系统分析法。现应用最为广泛的是硫化氢系统分析法。

1．硫化氢系统的分组方案

硫化氢系统分组主要根据各阳离子硫化物溶解度的差异，其次是氯化物和硫酸盐溶解度的不同，以 HCl、H_2S、$(NH_4)_2S$ 和 $(NH_4)_2CO_3$ 为组试剂，将 26 种常见的阳离子分为五组。附表 10-1 和附图 10-1 分别为阳离子的硫化氢系统分组方案和硫化氢系统分析步骤示意图。

附表 10-1　阳离子的硫化氢系统分组方案

<table>
<tr><td rowspan="4">分组的根据</td><td colspan="4">硫化物不溶于水</td><td colspan="2">硫化物溶于水</td></tr>
<tr><td colspan="3">硫化物不溶于稀酸</td><td rowspan="3">硫化物溶于稀酸</td><td rowspan="3">碳酸盐不溶于水</td><td rowspan="3">碳酸盐溶于水</td></tr>
<tr><td rowspan="2">氯化物不溶于水</td><td colspan="2">氯化物溶于水</td></tr>
<tr><td>硫化物不溶于 Na_2S</td><td>硫化物溶于 Na_2S</td></tr>
<tr><td>组内的离子</td><td>Ag^+
Hg_2^{2+}
Pb^{2+}①</td><td>Pb^{2+}
Bi^{3+}
Cu^{2+}
Cd^{2+}</td><td>Hg^{2+}
As (Ⅲ、Ⅴ)
Sb (Ⅲ、Ⅴ)
Sn (Ⅱ、Ⅳ)</td><td>Al^{3+} Mn^{2+}
Fe^{3+} Zn^{2+}
Cd^{2+} Co^{2+}
Fe^{2+} Ni^{2+}</td><td>Ba^{2+}
Sr^{2+}
Ca^{2+}</td><td>Mg^{2+}
K^+
Na^+
NH_4^+</td></tr>
</table>

续表

组的名称	Ⅰ组 银组 盐酸组	Ⅱ A组	Ⅱ B组	Ⅲ组 铁组 硫化物组	Ⅳ组 钙组 硫酸铵组	Ⅴ组 钠组
		Ⅱ组 铜锡组 硫化氢组				
组试剂	HCl	$0.3mol\cdot L^{-1}$ HCl H_2S		NH_3-NH_4Cl $(NH_4)_2S$	NH_3-NH_4Cl $(NH_4)_2CO_3$	—

① Pb^{2+}浓度大时部分沉淀。

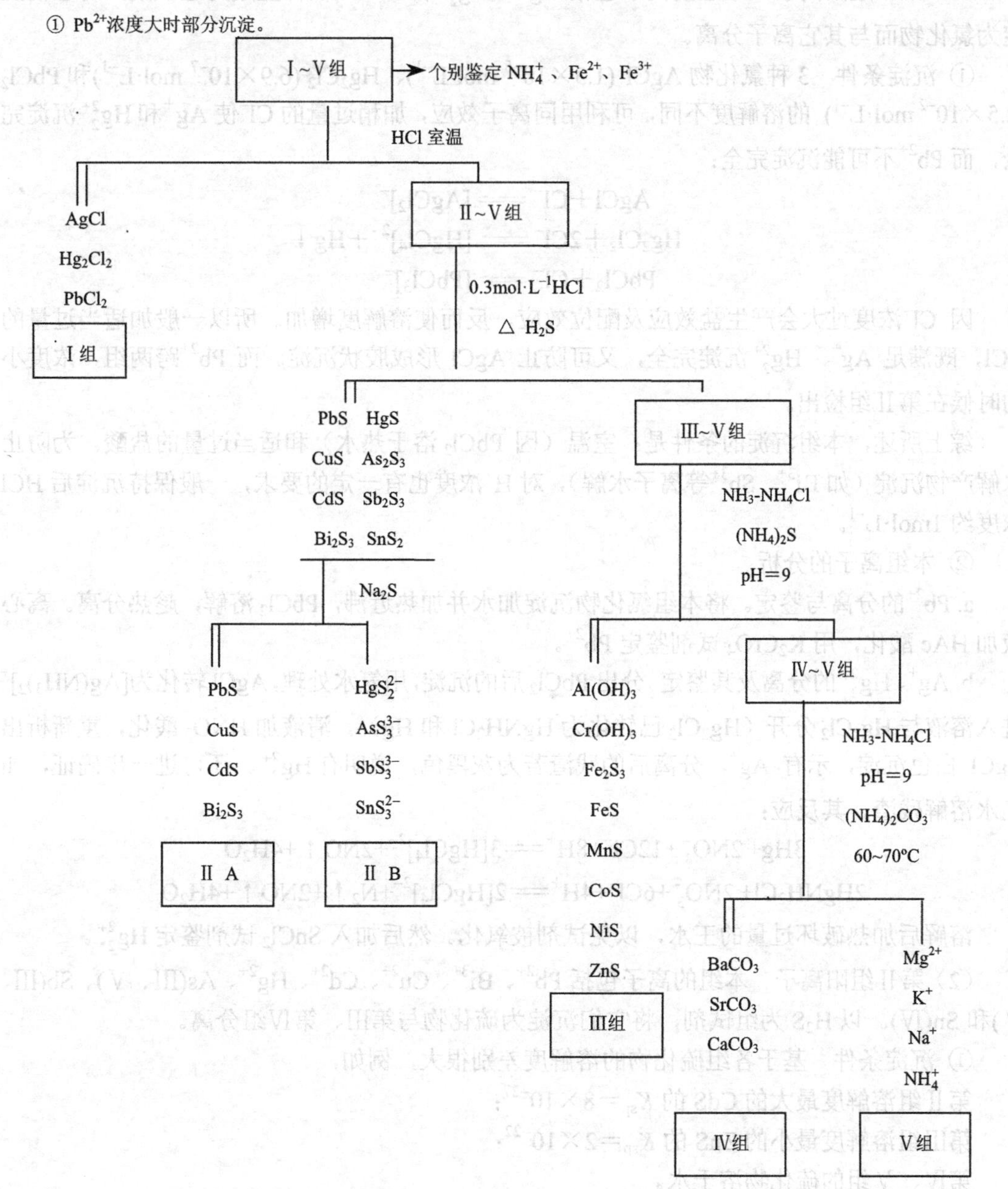

附图 10-1 硫化氢系统分析步骤示意图

由于 H_2S 气体有毒、恶臭，故目前一般采用硫代乙酰胺（TAA）作 H_2S 的代用品。TAA

的水解速度随温度的升高而加快，所以一般在沸水浴中进行。

2．硫化氢系统的分组条件及各组离子的分析

要使组与组分离完全，即被沉淀的离子完全沉淀（指被沉淀离子在溶液中剩余的浓度小于 10^{-5} mol·L^{-1}），而其它离子完全留在溶液中，需根据被沉淀离子和共存离子的性质考虑分离的条件，并运用离子平衡（酸碱、沉淀、氧化还原，配位等平衡）的规律以控制分离的条件。

（1）第Ⅰ组阳离子　本组的离子包括 Ag^+、Hg_2^{2+} 和 Pb^{2+}。以稀盐酸为组试剂，将它们沉淀为氯化物而与其它离子分离。

① 沉淀条件　3 种氯化物 AgCl (1.3×10^{-5} mol·L^{-1})、Hg_2Cl_2 (6.9×10^{-7} mol·L^{-1})和 $PbCl_2$ (2.5×10^{-2} mol·L^{-1}) 的溶解度不同，可利用同离子效应，加稍过量的 Cl^-使 Ag^+和 Hg_2^{2+} 沉淀完全，而 Pb^{2+}不可能沉淀完全：

$$AgCl + Cl^- \rightleftharpoons [AgCl_2]^-$$
$$Hg_2Cl_2 + 2Cl^- \rightleftharpoons [HgCl_4]^{2-} + Hg\downarrow$$
$$PbCl_2 + Cl^- \rightleftharpoons [PbCl_3]^-$$

因 Cl^-浓度过大会产生盐效应及配位效应，反而使溶解度增加。所以一般加适当过量的 HCl，既满足 Ag^+、Hg_2^{2+} 沉淀完全，又可防止 AgCl 形成胶状沉淀。而 Pb^{2+}跨两组，浓度小的时候在第Ⅱ组检出。

综上所述，本组沉淀的条件是：室温（因 $PbCl_2$ 溶于热水）和适当过量的盐酸。为防止水解产物沉淀（如 Bi^{3+}、Sb^{3+}等离子水解），对 H^+浓度也有一定的要求，一般保持沉淀后 HCl 浓度约 1mol·L^{-1}。

② 本组离子的分析

a. Pb^{2+}的分离与鉴定。将本组氯化物沉淀加水并加热近沸，$PbCl_2$ 溶解，趁热分离。离心液加 HAc 酸化，用 K_2CrO_4 试剂鉴定 Pb^{2+}。

b. Ag^+、Hg_2^{2+} 的分离及其鉴定。分出 $PbCl_2$ 后的沉淀，用氨水处理，AgCl 转化为$[Ag(NH_3)_2]^+$进入溶液与 Hg_2Cl_2 分开（Hg_2Cl_2 已转化为 $HgNH_2Cl$ 和 Hg）。清液加 HNO_3 酸化，重新析出 AgCl 白色沉淀，示有 Ag^+。分离后的残渣若为灰黑色，说明有 Hg_2^{2+}。还可进一步确证，加王水溶解残渣，其反应：

$$3Hg + 2NO_3^- + 12Cl^- + 8H^+ = 3[HgCl_4]^{2-} + 2NO\uparrow + 4H_2O$$
$$2HgNH_2Cl + 2NO_3^- + 6Cl^- + 4H^+ = 2[HgCl_4]^{2-} + N_2\uparrow + 2NO\uparrow + 4H_2O$$

溶解后加热破坏过量的王水，以免试剂被氧化。然后加入 $SnCl_2$ 试剂鉴定 Hg_2^{2+}。

（2）第Ⅱ组阳离子　本组的离子包括 Pb^{2+}、Bi^{3+}、Cu^{2+}、Cd^{2+}、Hg^{2+}、As(Ⅲ、Ⅴ)、Sb(Ⅲ、Ⅴ)和 Sn(Ⅳ)。以 H_2S 为组试剂，将它们沉淀为硫化物与第Ⅲ、第Ⅳ组分离。

① 沉淀条件　基于各组硫化物的溶解度差别很大，例如，

第Ⅱ组溶解度最大的 CdS 的 K_{sp}＝8×10^{-27}；

第Ⅲ组溶解度最小的 ZnS 的 K_{sp}＝2×10^{-22}；

第Ⅳ、Ⅴ组的硫化物溶于水。

所以，关键是第Ⅱ组与第Ⅲ组的分离。根据分步沉淀的原理，控制 S^{2-}的浓度使 Cd^{2+}沉淀完全，而 Zn^{2+}不被沉淀，即达到分离的目的。

因 H_2S 是二元弱酸，总的解离常数表达式是：

$$\frac{[H^+]^2[S^{2-}]}{[H_2S]}=K_{a_1}K_{a_2}=9.2\times10^{-22}$$

$$[S^{2-}]=\frac{[H_2S]}{[H^+]^2}\times9.2\times10^{-22}$$

室温下，H_2S 饱和溶液的浓度约为 $0.1mol\cdot L^{-1}$。

故

$$[S^{2-}]=\frac{0.1}{[H^+]^2}\times9.2\times10^{-22}$$

由上可知，S^{2-}浓度与 H^+浓度的平方成反比，因此调节溶液的酸度，则可控制 S^{2-}的浓度，从而使第Ⅱ、Ⅲ组的离子分离。

通过实验和计算证明，H^+浓度为 $0.3mol\cdot L^{-1}$ 最为适宜。因 H^+浓度大时，第Ⅱ组离子尤其是 Cd^{2+}沉淀不完全。H^+浓度小时，第Ⅲ组 Zn^{2+}会产生沉淀。

为防止生成硫化物的胶体，可将溶液加热，使硫化物在热溶液中生成，但加热会使 H_2S 浓度降低。为保证第Ⅱ组离子沉淀完全，需等溶液冷却后，再通 H_2S。

综上所述，本组沉淀条件是：调节溶液中 HCl 的浓度为 $0.3mol\cdot L^{-1}$（若用 TAA 作沉淀剂，溶液的酸度调至 $0.2mol\cdot L^{-1}$ 最适宜），加热通 H_2S，冷却后将溶液冲稀一倍，降低溶液的酸度（因沉淀过程中，H^+浓度增大），再通 H_2S，使第Ⅱ组离子硫化物沉淀完全。

② 本组离子的分析　本组包括的离子较多，为了分析方便，根据硫化物的酸碱性不同，进一步将它们分为两个小组。其中，Pb^{2+}、Bi^{3+}、Cu^{2+}和 Cd^{2+}的硫化物属于碱性硫化物，不溶于 Na_2S、$(NH_4)_2S$ 等碱性溶剂，这些离子称为铜组（Ⅱ A 组）。Hg^{2+}、As(Ⅲ)、Sb(Ⅲ)和 Sn(Ⅳ)的硫化物呈两性偏酸，能溶于 Na_2S 试剂生成硫代酸盐，称为锡组（Ⅱ B 组）。

在本组硫化物中加 Na_2S 试剂，其反应：

$$HgS+S^{2-}=\!=\!=[HgS_2]^{2-}$$

$$As_2S_3+3S^{2-}=\!=\!=2AsS_3^{3-}$$

$$SnS_2+S^{2-}=\!=\!=[SnS_3]^{2-}$$

$$Sn_2S_3+3S^{2-}=\!=\!=2[SbS_3]^{3-}$$

溶液为锡组，残渣为铜组。

③ 铜组（Ⅱ A 组）分析　将本组硫化物溶于 HNO_3，除去 S 之后，加甘油溶液（1∶1）和过量的浓碱（质量分数为 40%的 NaOH 溶液），这时只有 $Cd(OH)_2$ 沉淀，而 Pb^{2+}、Cu^{2+}和 Bi^{3+}与甘油-碱生成可溶性的配合物留在溶液中，将 $Cd(OH)_2$ 沉淀以稀的甘油-碱洗净，直接在该沉淀中加入 TAA 溶液，加热，若有黄色沉淀析出，示有 Cd^{2+}。

在分离 $Cd(OH)_2$ 后的溶液中鉴定 Cu^{2+}、Pb^{2+}和 Bi^{3+}。

取少量溶液，加 HAc 酸化，用 $K_4[Fe(CN)_6]$试剂鉴定 Cu^{2+}。

取少量溶液，用稀 HNO_3 中和，然后加 HAc 酸化，用 K_2CrO_4 试剂鉴定 Pb^{2+}。

取少量溶液，用 $Na_2[Sn(OH)_4]$试剂鉴定 Bi^{3+}。

铜组分析示意图如附图 10-2 所示。

④ 锡组（Ⅱ B 组）分析　在分离铜组后的溶液中滴加浓 HAc 至呈酸性，则硫代酸盐分解重新析出硫化物沉淀：

$$HgS_2^{2-}+2HAc=\!=\!=HgS\downarrow+H_2S\uparrow+2Ac^-$$

$$2AsS_3^{3-}+6HAc=\!=\!=As_2S_3\downarrow+3H_2S\uparrow+6Ac^-$$

$$2SbS_3^{3-}+6HAc=\!=\!=Sb_2S_3\downarrow+3H_2S\uparrow+6Ac^-$$

$$SnS_3^{2-}+2HAc=\!=\!=SnS_2\downarrow+H_2S\uparrow+2Ac^-$$

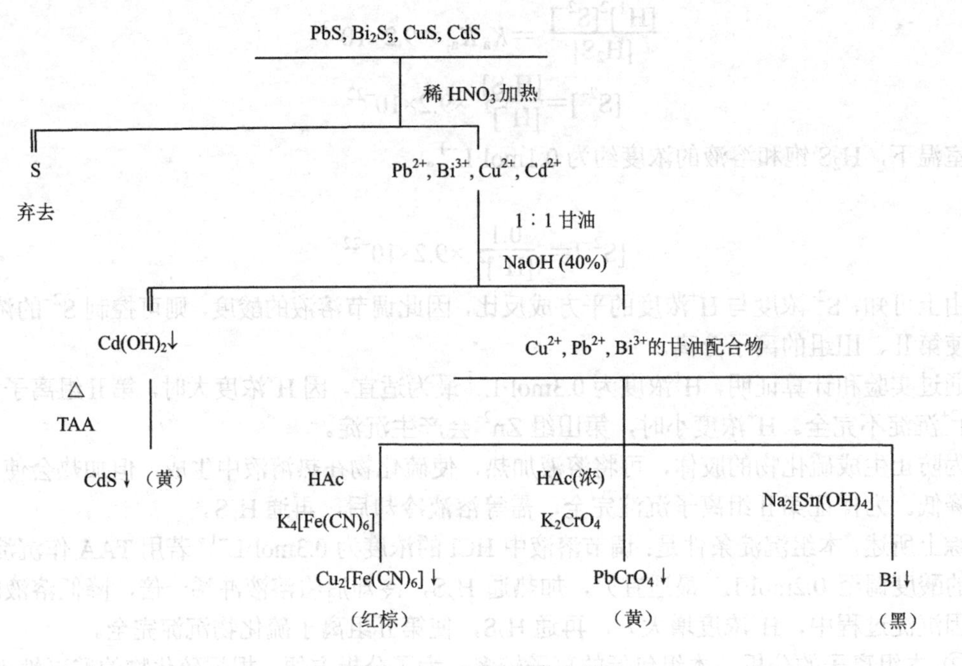

附图 10-2　铜组分析示意图

a. Hg^{2+}、As(Ⅲ)与 Sb(Ⅲ)、Sn(Ⅳ)的分离：在上述析出的硫化物沉淀上加浓盐酸并加热，则 Sb_2S_3 和 SnS_2 形成氯配离子而溶解；HgS 和 As_2S_3 不溶。

$$Sb_2S_3 + 6H^+ + 12Cl^- = 2[SbCl_6]^{3-} + 3H_2S\uparrow$$

$$SnS_2 + 4H^+ + 6Cl^- = [SnCl_6]^{2-} + 2H_2S\uparrow$$

b. Sn(Ⅳ)和 Sb(Ⅲ)的鉴定：取 a 中的少量溶液，加铁丝（或铅粒）使 Sn(Ⅳ)还原为 Sn(Ⅱ)，然后在溶液中加 $HgCl_2$ 试剂鉴定 Sn(Ⅱ)。

另取 a 中的少量溶液，将溶液稀释后通 H_2S，如 Sn(Ⅳ)不存在，则得到橘红色 Sb_2S_3 沉淀，示有 Sb(Ⅲ)。

如有 Sn(Ⅳ)存在，为了避免生成 SnS_2，需把 HCl 浓度调至约 2.2mol·L^{-1}，加一些草酸后，再通 H_2S。因为$[Sn(C_2O_4)_3]^{2-}$比$[Sb(C_2O_4)_3]^{3-}$稳定得多，在通 H_2S 时只有$[Sb(C_2O_4)_3]^{3-}$转化为 Sb_2S_3 沉淀。但如果 HCl 浓度太小，也会有一些 SnS_2 混在 Sb_2S_3 中，使沉淀呈褐色。

c. Hg^{2+}与 As(Ⅲ)的分离及其鉴定：HgS 与 As_2S_3 的酸碱性不同，其中 As_2S_3 的酸性较强，可溶于氨水及$(NH_4)_2CO_3$。

$$As_2S_3 + 6NH_3 + 3H_2O = AsS_3^{3-} + AsO_3^{3-} + 6NH_4^+$$

$$As_2S_3 + 3CO_3^{2-} = AsS_3^{3-} + AsO_3^{3-} + 3CO_2\uparrow$$

HgS 不溶，使 Hg^{2+}与 As(Ⅲ)分开。

取用$(NH_4)_2CO_3$ 处理后的溶液，加稀 HCl 酸化，出现黄色沉淀，示有 As(Ⅲ)。

HgS 用王水处理，然后加热破坏过量的王水，再以 $SnCl_2$ 鉴定 Hg^{2+}。

锡组分析示意图如附图 10-3 所示。

（3）第Ⅲ组阳离子　本组离子包括 Al^{3+}、Mn^{2+}、Fe^{3+}、Zn^{2+}、Cd^{2+}、Co^{2+}、Fe^{2+}、Ni^{2+}，在 0.3mol·L^{-1}HCl 溶液中，这些离子不被 H_2S 沉淀。但在 NH_3-NH_4Cl 存在下，以$(NH_4)_2S$ 为组试剂，Fe^{2+}、Mn^{2+}、Co^{2+}、Ni^{2+}、Zn^{2+}、Fe^{3+}形成硫化物沉淀，Al^{3+}、Cr^{3+}形成氢氧化物沉淀，与第Ⅳ、Ⅴ组离子分离。

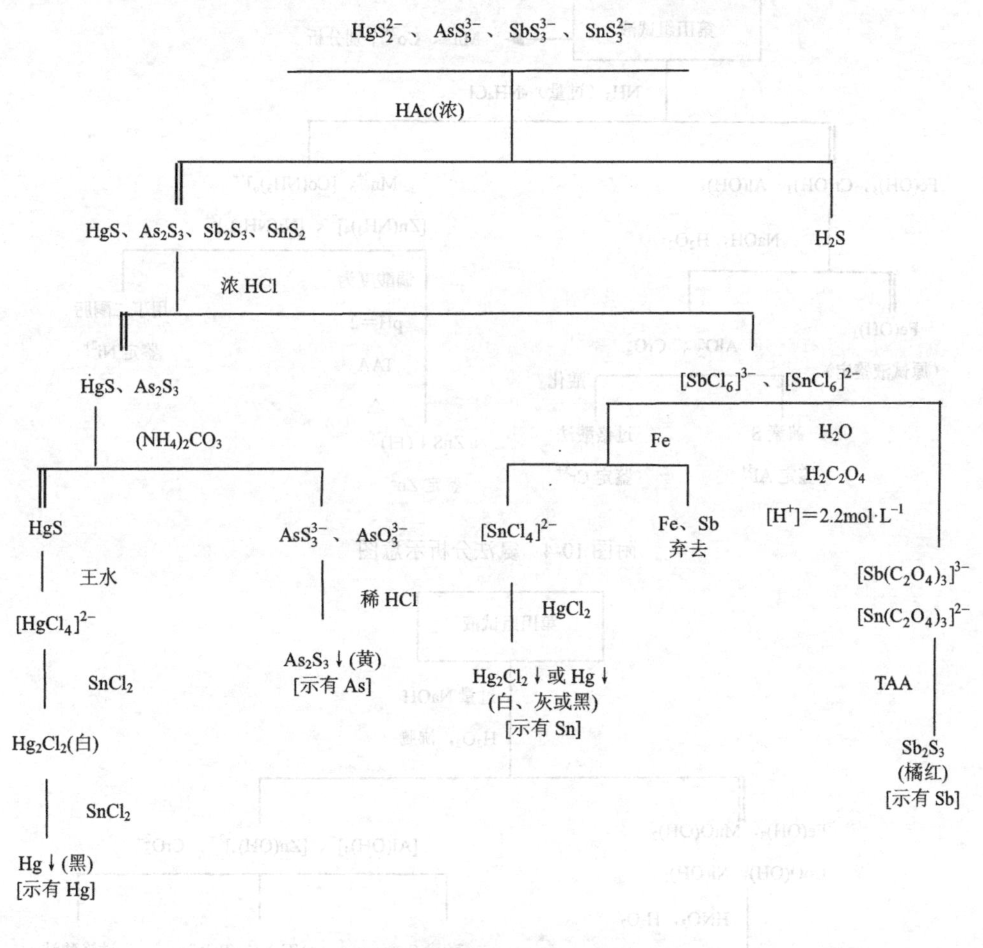

附图 10-3　锡组分析示意图

① 沉淀条件　沉淀本组离子时，溶液的酸度一般保持在 pH＝8～9，因 $Al(OH)_3$ 和 $Cr(OH)_3$ 为两性物质。pH＜5，Al^{3+}沉淀不完全，pH＜7，Cr^{3+}沉淀不完全，pH＞12，$Cr(OH)_3$ 和 $Al(OH)_3$ 开始溶解，而且 pH＞10，第Ⅴ组 Mg^{2+}也会沉淀，因此采用 NH_3-NH_4Cl 缓冲体系，保持 pH 基本不变。

为了防止硫化物形成胶体，除了在溶液中加电解质 NH_4Cl 外，还要将溶液加热。

综上所述，本组沉淀条件是：在 NH_3-NH_4Cl 存在下，调节 pH＝9，并加热，然后加入$(NH_4)_2S$ 或 TAA，使本组离子沉淀完全。

② 本组离子的分析　本组离子除 Al^{3+}、Zn^{2+}外都有颜色，离子的颜色是鉴定本组离子的重要依据之一。但当离子的含量很少或者不同的颜色发生互补现象时，溶液几乎无色。故溶液无色也不能完全排除有色离子的存在。

本组离子在鉴定前常需分成小组。分组方案多种多样，常用的方法有“氨法”和“碱-过氧化物法”。附图 10-4 和附图 10-5 这两种分析方法的示意图。

这两种方法都不够理想。氨法中 Mn^{2+}易被空气氧化为 $MnO(OH)_2$ 而留在沉淀中，部分 $Co(OH)_2$ 发生共沉淀，造成 Mn^{2+}和 Co^{2+}分离不完全，即一部分随 $Fe(OH)_3$、$Al(OH)_3$、$Cr(OH)_3$ 共沉淀下来，一部分留在溶液中。按碱-过氧化物法分离时，部分 $Zn(OH)_2$ 被带下来共沉淀，因而在检出时易丢 Zn^{2+}。

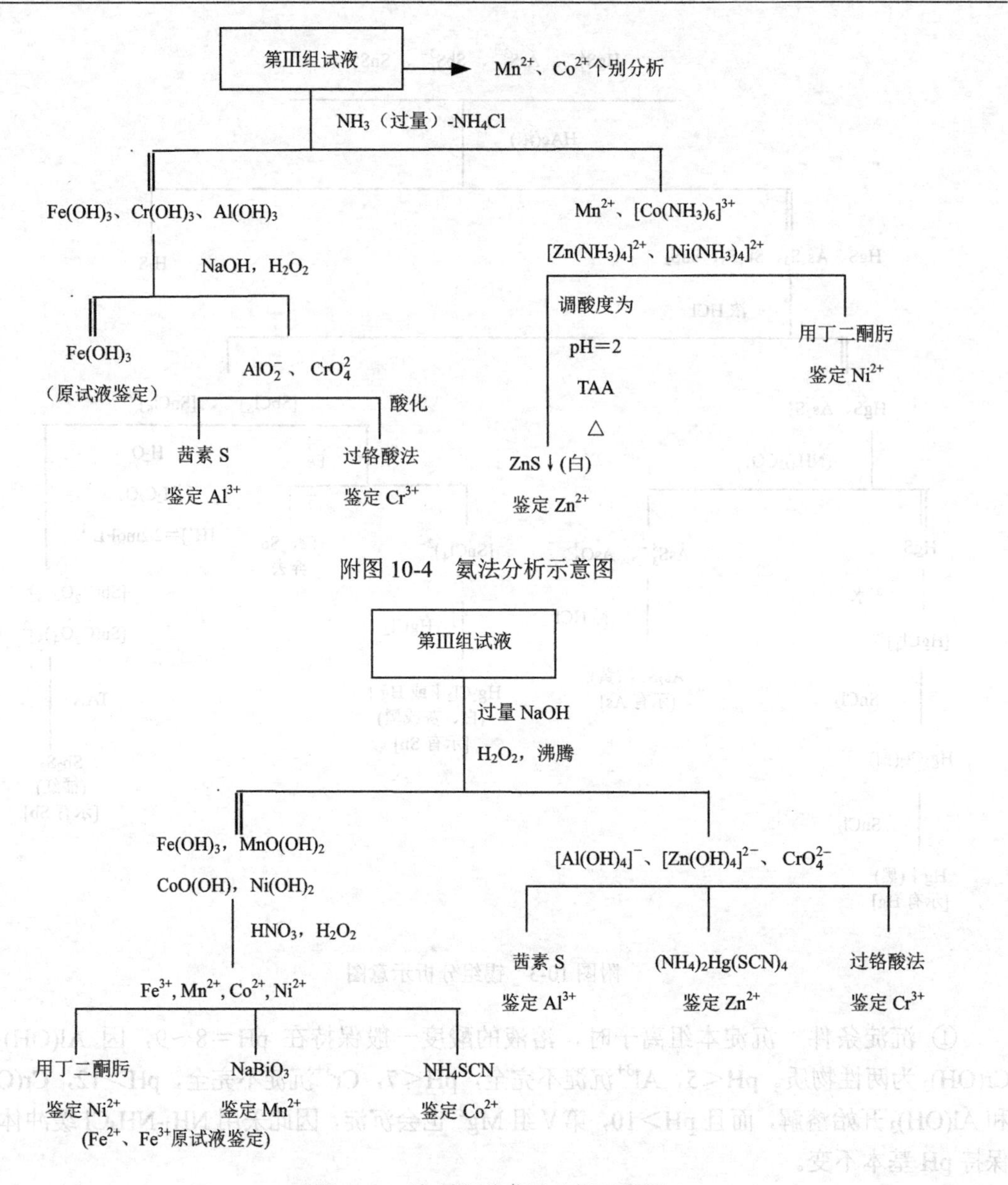

附图 10-4　氨法分析示意图

附图 10-5　碱-过氧化物法分析示意图

若在氨基乙酸存在下，加固体尿素，使 Fe^{3+}、Al^{3+}、Cr^{3+}形成均相的氢氧化物沉淀，而 Zn^{2+}、Co^{2+}、Ni^{2+}、Mn^{2+}则与氨基乙酸生成配合物留在溶液中，就克服了氨法和碱-过氧化物法分离的缺点。

（4）第Ⅳ、Ⅴ组阳离子。

① 第Ⅳ、Ⅴ组阳离子混合液的处理　按硫化氢系统分组法，第Ⅳ组离子包括 Ca^{2+}、Ba^{2+}、Sr^{2+}，第Ⅴ组离子包括 Mg^{2+}、K^+、Na^+、NH_4^+。第Ⅲ组离子分离后的溶液中，由于积累了大量的铵盐，若长期放置，溶液中的$(NH_4)_2S$会被空气中的氧部分氧化为$(NH_4)_2SO_4$，使 Ba^{2+}生成 $BaSO_4$ 沉淀而从溶液中丢失。因此在第Ⅲ组离子沉淀分离后，应立即用 HAc 酸化、煮沸除去 H_2S，并将析出的 S 分离除去，再将溶液蒸发至干并灼烧除去铵盐，冷却后将残渣用稀 HCl 溶解，以备分析第Ⅳ、Ⅴ组离子。

② 第Ⅳ、Ⅴ组离子的分离　此两组离子的氯化物和硫化物均易溶于水。但在 NH_3-NH_4Cl

存在的条件下，可用$(NH_4)_2CO_3$试剂使第Ⅳ组离子形成碳酸盐沉淀而与第Ⅴ组离子分离。

由于$(NH_4)_2CO_3$是弱酸弱碱盐，在水溶液中强烈水解：

$$NH_4^+ + CO_3^{2-} + H_2O = NH_3 \cdot H_2O + HCO_3^-$$

而使Ba^{2+}、Ca^{2+}、Sr^{2+}沉淀不完全，所以需加入NH_3抑制水解。但$NH_3 \cdot H_2O$的加入，将使Mg^{2+}部分生成碱式碳酸镁。为了避免这种现象的产生，应加入适量的NH_4Cl。试验与计算证明，当NH_3和NH_4Cl的浓度比约等于1、pH=9时，即可使Ba^{2+}、Ca^{2+}、Sr^{2+}沉淀完全，而Mg^{2+}不被沉淀。

另外，$(NH_4)_2CO_3$常含有相当量的NH_2COONH_4（氨基甲酸铵），加热至60～70℃，可促进氨基甲酸铵变为碳酸铵：

$$NH_2COONH_4 + H_2O \xlongequal{\triangle} (NH_4)_2CO_3 \text{（吸热反应）}$$

加热还可破坏溶液中的过饱和现象，加速沉淀的生成，并使碳酸盐的无定形沉淀转为晶体沉淀。

但温度过高，碳酸铵会分解，已生成的碳酸盐沉淀与大量铵盐一起煮沸也会部分溶解，使沉淀不完全。

综上所述，分离第Ⅳ、Ⅴ组离子的条件是：于存在NH_3-NH_4Cl (pH=9)的试液中，加入$(NH_4)_2CO_3$试剂，并加热至60～70℃，则第Ⅳ组离子形成碳酸盐沉淀而与第Ⅴ组离子分离。

③ 第Ⅳ、Ⅴ组离子的鉴定　用HAc溶解碳酸盐沉淀，则第Ⅳ组离子转入溶液中。由于它们在鉴定中互相干扰，故采用系统分析法。而第Ⅴ组则采用个别分析法。附图10-6和附图10-7分别为第Ⅳ、Ⅴ组阳离子分析示意图。

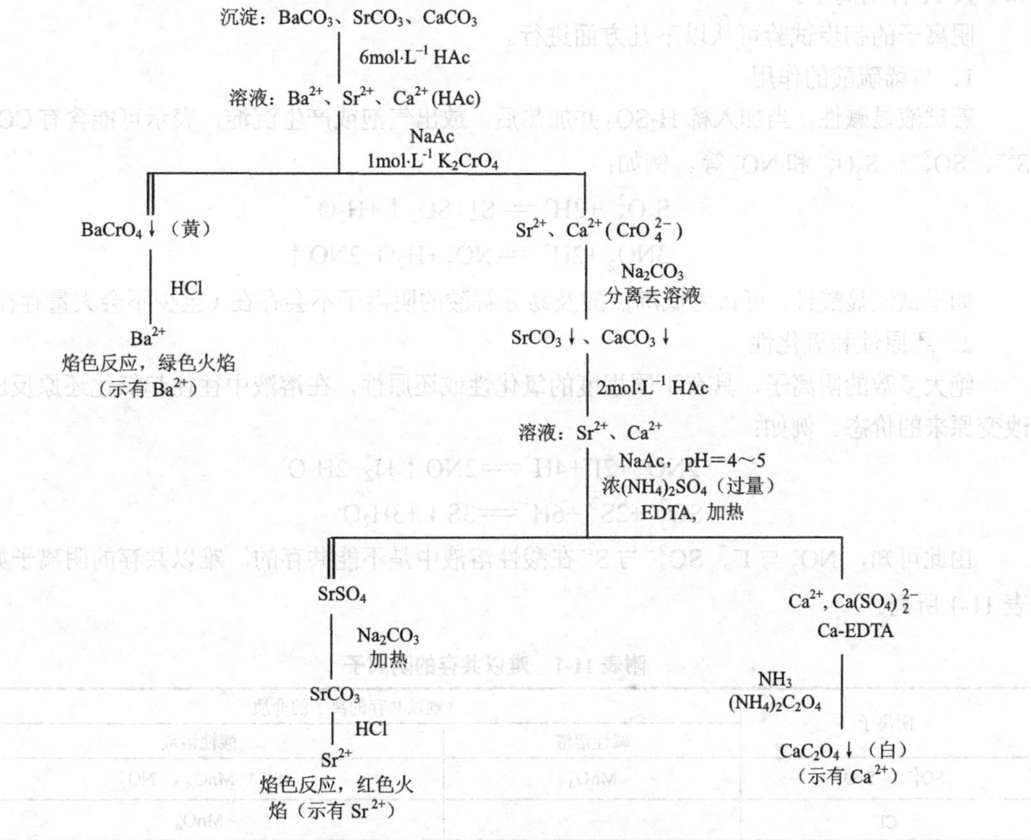

附图10-6　第Ⅳ组阳离子分析示意图

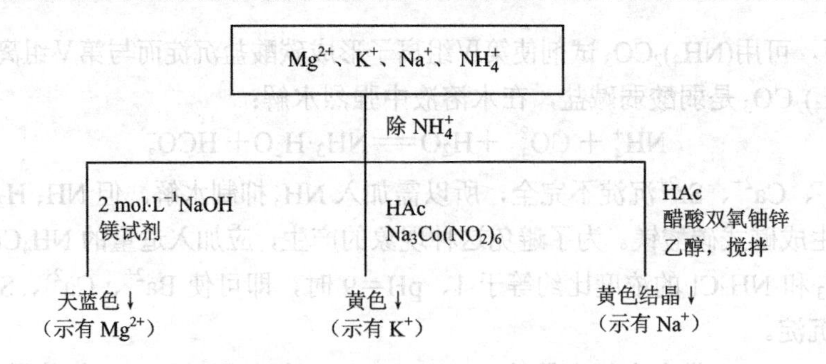

附图 10-7 第Ⅴ组阳离子分析示意图

由于在系统分析过程中，多次加入氨水和铵盐，故应取原溶液用气室法鉴定 NH_4^+。另外，NH_4^+ 干扰 K^+的鉴定，降低 Mg^{2+}的检出灵敏度，故在检出 Mg^{2+}、Na^+、K^+前，应先除去 NH_4^+。

附录 11 常见阴离子的分析

阴离子分析，目前还没有理想的系统分析方案，在进行混合离子分析时，一般是利用阴离子的化学性质进行初步试验，确定离子存在的可能范围，然后进行必要的分离和个别离子的检出。

这里主要讨论常见的 CO_3^{2-}、S^{2-}、SO_3^{2-}、$S_2O_3^{2-}$、SO_4^{2-}、PO_4^{3-}、NO_2^-、NO_3^-、Cl^-、Br^- 和 I^-共 11 种阴离子。

阴离子的初步试验可从以下几方面进行。

1．与稀硫酸的作用

若试液显碱性，当加入稀 H_2SO_4 并加热后，放出气泡或产生沉淀，表示可能含有 CO_3^{2-}、S^{2-}、SO_3^{2-}、$S_2O_3^{2-}$ 和 NO_2^- 等。例如：

$$S_2O_3^{2-}+2H^+ = S\downarrow+SO_2\uparrow+H_2O$$

$$3NO_2^-+2H^+ = NO_3^-+H_2O+2NO\uparrow$$

如果试液显酸性，可认为低沸点酸及易分解酸的阴离子不会存在（至少不会大量存在）。

2．还原性和氧化性

绝大多数的阴离子，具有不同程度的氧化性或还原性，在溶液中往往起氧化还原反应，改变原来的价态。例如：

$$2NO_2^-+2I^-+4H^+ = 2NO\uparrow+I_2+2H_2O$$

$$SO_3^{2-}+2S^{2-}+6H^+ = 3S\downarrow+3H_2O$$

由此可知，NO_2^- 与 I^-、SO_3^{2-} 与 S^{2-}在酸性溶液中是不能共存的。难以共存的阴离子如附表 11-1 所列。

附表 11-1 难以共存的阴离子

阴离子	难以共存的离子的介质	
	碱性溶液	酸性溶液
SO_3^{2-}，$S_2O_3^{2-}$	MnO_4^-	S^{2-}，MnO_4^-，NO_2^-
Cl^-		MnO_4^-

续表

阴离子	难以共存的离子的介质	
	碱性溶液	酸性溶液
Br^-		MnO_4^-
I^-	MnO_4^-	MnO_4^-，NO_2^-（NO_3^-浓度大时）
S^{2-}	MnO_4^-	MnO_4^-，NO_2^-，NO_3^-，SO_3^{2-}，$S_2O_3^{2-}$
NO_3^-		S^{2-}，SO_3^{2-}（慢）①，$S_2O_3^{2-}$（慢）
NO_2^-	MnO_4^-（慢）	S^{2-}，I^-，SO_3^{2-}，$S_2O_3^{2-}$，MnO_4^-

① 慢指较短时间内可共存。

当NO_3^-浓度大时，在酸性条件下，亦使KI-淀粉试纸变蓝。

利用某些阴离子不能共存这一性质，可以简化分析过程。不能共存的离子一方被鉴定出来，另一方离子就不可能存在，也就没有必要进行鉴定了。

3．难溶盐的性质

可利用某些阴离子与某种试剂发生沉淀反应进行分组（该试剂称为组试剂），以确定离子存在的可能范围。其组试剂只是起着检验该组离子是否存在的作用（称反证试验）。目前一般多根据钡盐和银盐溶解度的不同，将阴离子分为3组，见附表11-2。

附表 11-2　阴离子的分组

组别	组 试 剂	组内阴离子	特 性
第一组	$BaCl_2$ （中性或弱碱性）	SO_3^{2-}，SO_4^{2-}，$S_2O_3^{2-}$ CO_3^{2-}，PO_4^{3-}(浓溶液中析出)	钡盐难溶于H_2O（除$BaSO_4$之外，其它钡盐溶于酸），银盐溶于HNO_3
第二组	$AgNO_3$ （稀冷HNO_3）	S^{2-}，Cl^-，Br^-，I^-	钡盐溶于H_2O，银盐难溶于H_2O和稀HNO_3（Ag_2S溶于热HNO_3）
第三组	无组试剂	NO_3^-，NO_2^-	钡盐和银盐都易溶于H_2O

$S_2O_3^{2-}$浓度大于$45mg \cdot mL^{-1}$时，才能析出钡盐沉淀，且BaS_2O_3还易形成过饱和溶液，所以加入$BaCl_2$之后，应以玻璃棒摩擦管壁，若仍不产生沉淀，可初步认为$S_2O_3^{2-}$含量不大或不存在。$S_2O_3^{2-}$也可能在第二组析出白色沉淀，而后变黄、变棕最后为黑色的Ag_2S。其反应式如下：

$$2Ag^+ + S_2O_3^{2-} = Ag_2S_2O_3\downarrow \xrightarrow[\text{白}\to\text{黄}\to\text{棕}\to\text{黑}]{H_2O} Ag_2S\downarrow + H_2SO_4$$

但如果$S_2O_3^{2-}$浓度大时，可能生成$[Ag(S_2O_3)_2]^{3-}$。此时只在第一组出现沉淀，第二组则不沉淀。

参考文献

[1] 王箴. 化工辞典. 第 4 版. 北京：化学工业出版社，2000.
[2] 曹锡章等. 无机化学. 北京：高等教育出版社，1984.
[3] J. 明切斯基等著. 无机痕量分析的分离和预富集方法. 陈永兆等译. 北京：地质出版社，1986.
[4] 林对昌等. 定性分析化学. 北京：北京师范大学出版社，1984.
[5] 韩葆云. 快速定性分析. 北京：高等教育出版社，1984.
[6] 罗兰·S·杨著. 无机分析中的分离方法. 张国雄译. 上海：上海科学技术文献出版社，1984.
[7] 日本化学会编. 无机化合物合成手册（第二卷）. 安家驹，陈之川译. 北京：化学工业出版社，1986.
[8] J A Beran. Laboratory Manual for Principles of General Chemistry. 6th ed. New York：John Wiley and Sons, 2000.
[9] 伍晓春，姚淑心. 无机化学实验（英汉双语教材）. 北京：科学出版社，2010.
[10] 冯清. 医学基础化学实验（双语版）. 武汉：华中科技大学出版社，2007.
[11] 刘伟生. 配位化学. 第 2 版. 北京：化学工业出版社，2019.
[12] 黄开勋，徐辉碧. 硒的化学、生物化学及其在生命科学中的应用. 第 2 版. 武汉：华中科技大学出版社，2009.